Schiff Bases - Recent Developments and Application Areas

Edited by Nuriye Tuna Subasi

Published in London, United Kingdom

Schiff Bases - Recent Developments and Application Areas
http://dx.doi.org/10.5772/intechopen.1008132
Edited by Nuriye Tuna Subasi

Contributors
Ahmed A. Elrashedy, Asmaa A. Magd El-Din, Claudia-Valentina Popa, Ghalia A. Gaber, Isha Murtaza, Mai Saud Abdullah Alsubaie, Maria Marinescu, Masakatsu Shibasaki, Mustaghees Ur Rehman, Nuriye Tuna Subasi, Plato A. Magriotis, Ruijuan Zhang, Sohail Anjum Shahzad, Syeda Aaliya Shehzadi, Toshifumi Takeuchi, Xinying Liu, Yali Yao, Zulfiqar Ali Khan

First published in London, United Kingdom, 2025 by IntechOpen
IntechOpen is the global imprint of INTECHOPEN LIMITED, registered in England and Wales, registration number: 11086078, 167-169 Great Portland Street, London, W1W 5PF, United Kingdom

For EU product safety concerns: IN TECH d.o.o., Prolaz Marije Krucifikse Kozulić 3, 51000 Rijeka, Croatia, info@intechopen.com or visit our website at intechopen.com.

British Library Cataloguing-in-Publication Data
A catalogue record for this book is available from the British Library

Schiff Bases - Recent Developments and Application Areas
Edited by Nuriye Tuna Subasi
p. cm.
Print ISBN 978-1-83635-397-3
Online ISBN 978-1-83635-396-6
eBook (PDF) ISBN 978-1-83635-398-0

Meet the editor

Dr. Nuriye Tuna Subasi received her Ph.D. in Organic Chemistry from Middle East Technical University (METU), Ankara, Turkey, in 2011 under the supervision of Prof. Ayhan S. Demir. She began her academic career in 2012 as an Assistant Professor in the Department of Food Engineering at Kırşehir Ahi Evran University, Turkey. Her research interests include synthetic organic chemistry and supramolecular chemistry. Dr. Subasi focuses on the design and synthesis of functional organic molecules with specific properties, developing fluorescent molecular probes for detecting biologically important species, such as metal ions, anions, and reactive biological species.

Contents

Preface

Schiff bases have remained at the forefront of coordination and organic chemistry for more than a century, continuously revealing new aspects of their structural, functional, and applicative diversity. Their fascinating ability to form stable complexes with transition metals, coupled with tunable electronic and steric properties, has made them indispensable compounds in modern chemical research. Over the past decades, Schiff bases have evolved from simple condensation products to multifunctional entities that bridge synthetic chemistry, materials science, and biological studies.

This edited volume, *Schiff Bases – Recent Developments and Application Areas*, brings together recent advances and multidisciplinary perspectives on the chemistry and applications of Schiff bases. It provides an updated and comprehensive overview, beginning with the fundamental understanding of their structural features, synthetic versatility, and emerging trends. The introductory chapter establishes the conceptual framework of Schiff bases, highlighting their wide-ranging relevance from traditional synthesis to advanced functional materials. The following chapter, *A Comprehensive Review on Schiff Bases: Development Process, Green Synthetic Strategies, and Versatile Applications*, expands on this foundation by summarizing eco-friendly synthesis approaches and discussing the broad applicability of these compounds across various scientific domains.

The second section, *Reactions and Theoretical Approaches of Schiff Bases*, focuses on the chemical reactivity and mechanistic understanding of Schiff base-related processes. Topics such as direct catalytic asymmetric aldol reactions, the use of TMS imines in organic synthesis, and computational modeling through density functional theory (DFT) provide valuable insights into both experimental and theoretical aspects of Schiff base chemistry.

The third section, *Biomedical and Therapeutic Applications*, explores the role of Schiff bases in medicinal chemistry. It includes chapters discussing molecular dynamics insights into 1,2,4-triazole-based Schiff base compounds as dual EGFR/Tubulin inhibitors, along with broader biomedical applications that highlight the therapeutic potential of these compounds against various diseases.

In the fourth section, *Sensors and Analytical Applications*, the focus shifts toward analytical chemistry, where Schiff bases demonstrate exceptional sensing capabilities. Chapters on chemosensors and carbazole-based Schiff bases illustrate their sensitivity and selectivity in detecting metal ions, paving the way for the design of advanced optical and electrochemical sensors.

Finally, *Corrosion Inhibition and Advanced Materials* emphasizes the industrial and materials-oriented applications of Schiff bases. This section presents their effectiveness in corrosion inhibition, surface coating technologies, and polymer-based functional materials, demonstrating how Schiff base chemistry continues to expand into applied and engineering sciences.

Together, these twelve chapters provide readers with a panoramic view of Schiff base chemistry, encompassing molecular design, mechanistic exploration, and practical applications. The diversity of topics and the expertise of contributing authors make this book a valuable resource for chemists, material scientists, and researchers in related fields seeking to deepen their understanding of this versatile class of compounds.

I would like to express my sincere gratitude to Tea Jelaca, Publishing Process Manager at IntechOpen, for her ongoing support, guidance, and professional dedication throughout the editorial process. I also extend my heartfelt appreciation to the entire IntechOpen team for their invaluable assistance and commitment to maintaining high publishing standards, which greatly contributed to the successful completion of this volume.

Nuriye Tuna Subasi
Department of Food Engineering,
Kırşehir Ahi Evran University,
Kırşehir, Turkey

Section 1

Fundamentals and General Perspectives

Chapter 1

Introductory Chapter: Structural Features, Versatility, and Emerging Trends in Schiff Bases

Nuriye Tuna Subasi

1. Introduction

For more than a century and a half, Schiff bases have attracted the attention of scientists, having been discovered in 1864 by the German-Italian chemist Hugo Schiff [1], thereby opening a new page and marking a turning point in organic chemistry. Since their discovery, Schiff bases have consistently received great interest from both academic and industrial communities, keeping researchers' curiosity alive. From organic synthesis to coordination chemistry, and from biochemical processes to pharmaceutical applications, these compounds have played a crucial role and have become indispensable building blocks for scientists. Throughout their historical development, Schiff bases have been the subject of extensive research not only in the field of fundamental organic chemistry but also due to their biological functions, pharmacological activities, and applications in materials science.

At present, Schiff bases can be synthesized through a wide range of approaches, from classical organic synthesis methods to green chemistry strategies [2], and can be examined in detail using spectroscopic and theoretical techniques. With their long history, these compounds today continue to serve as a powerful bridge not only connecting different branches of chemistry, but also linking diverse disciplines such as biology, materials science, and pharmaceutical research, owing to their remarkable versatility.

2. Structural versatility and applications of schiff bases

But where does the remarkable versatility of Schiff bases come from? At the core of this lies the imine (–C=N–) group in their structure. The imine bond is typically formed through the condensation reaction of carbonyl compounds with primary amines. This simple yet highly effective mechanism allows the synthesis of countless derivatives by employing different carbonyl compounds and amines. In addition, various functional groups can easily be introduced to carbonyl compounds and amines, further expanding the structural diversity and application potential of Schiff bases. The chemical flexibility of the imine group enables these compounds to interact with both electron-donating and electron-withdrawing groups, while also allowing them to readily form complexes with different metal ions. Moreover, this functional group can act compatibly with various enzymatic mechanisms in biological systems,

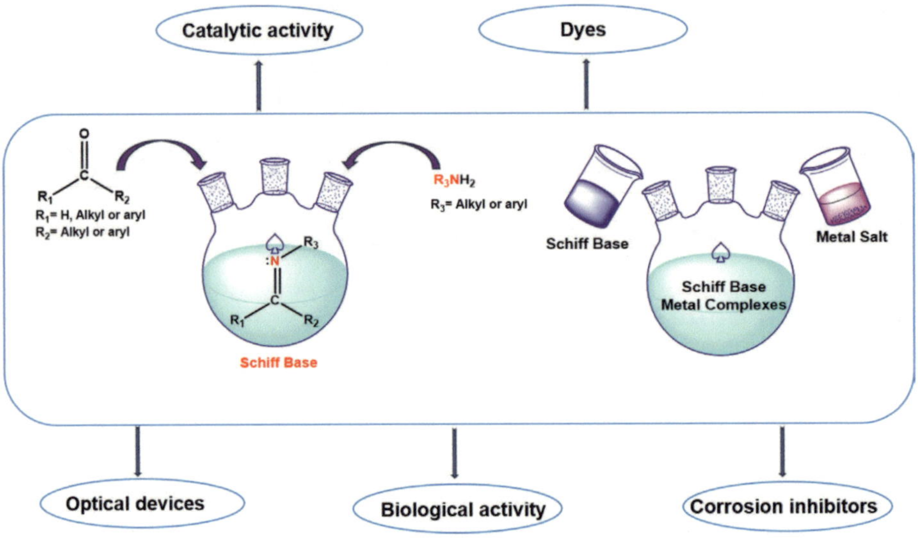

Figure 1.
Illustration of the synthesis of Schiff bases and their metal complexes, highlighting their structural diversity and main application areas.

thereby providing a basis for pharmacological activities. All these features make Schiff bases indispensable not only in classical organic synthesis and coordination chemistry but also in biomedical applications [3–6], catalysis [7], sensor technologies [8], and materials science (**Figure 1**) [9].

The primary aim of this book is to present a comprehensive overview of Schiff bases, covering a broad spectrum from their synthesis to their diverse applications. Today, Schiff bases play an important role not only in organic chemistry but also in various other disciplines such as biology, materials science, and medicine. Owing to their versatility, they offer researchers the opportunity both to reinterpret classical knowledge and to explore new frontiers. In this context, the book aspires to go beyond being a mere compilation of existing information by also highlighting contemporary developments such as green chemistry strategies, nanotechnological approaches, and sustainability-oriented methodologies. In doing so, the work aims to serve as an up-to-date and reliable reference for a wide audience, ranging from young researchers to experienced scientists engaged in Schiff base research.

The chapters included in this book address the versatile nature of Schiff bases from the different perspectives. They cover a wide range of topics, from recent biomedical applications (*Recent Applications of Schiff Bases in Biomedical Sciences; Molecular Dynamics Insights into Novel 1,2,4-Triazole-Based Schiff Base Compounds as Dual-EGFR/Tubulin Inhibitor for the Treatment of Cancer Patient; Density Functional Theory as a Tool for Assessing the Therapeutic Activity of New Compounds*) to corrosion and surface coating (*Schiff Bases – Synthesis, Characterization, and Applications in Corrosion Inhibition; Utilizing Schiff Bases for Surface Coating and Corrosion Prevention*), and from advanced sensing technologies (*Recent Advancements in Schiff Bases as Chemosensors; Carbazole-Based Schiff Bases: Structural Insights and Applications toward Metal Ion Detection*) to catalysis and organic synthesis methods (*Direct Catalytic Asymmetric Aldol Reactions of Glycine Schiff Bases to Access β-Hydroxy-α-Amino Esters; Recent Applications of TMS Imines in Organic Synthesis*). In addition, comprehensive review studies and polymer-based applications (*A Comprehensive*

Review on Schiff Bases: Development Process, Green Synthetic Strategies, and Versatile Applications; Polymer-Based Schiff Bases and Their Applications) strengthen the coherence of the book by providing readers with a consistent framework that encompasses synthesis, characterization, and applications. This diversity not only reflects the interdisciplinary significance of Schiff bases but also ensures that the book appeals to researchers from different fields.

3. Future perspectives and conclusion

From a future perspective, the research and application areas of Schiff bases are expected to diversify even further. The development of new functional derivatives holds great potential, particularly in fields such as nanotechnology, biomedical devices, drug design, and sensor technologies. Schiff base-based sensors, in particular, with their high selectivity and sensitivity in the detection of metal ions and biomolecules, are anticipated to find broader applications in areas such as environmental monitoring, health diagnostics, and industrial quality control. Furthermore, the design of new complexes that can enhance the efficiency of catalytic processes, along with the widespread adoption of environmentally friendly and sustainable synthetic methods, is expected to come to the forefront in the coming years. The versatility of Schiff bases makes them strategic compounds not only for fundamental sciences but also for industrial applications and technological innovations. In this context, it is evident that future studies on Schiff bases will have significant implications not only for the scientific community but also for daily life.

In conclusion, this book offers readers a comprehensive journey from the historical development of Schiff bases to their modern applications. The chapters have been prepared with an interdisciplinary perspective, encompassing not only synthetic methodologies but also a wide range of applications, from biomedical sciences and sensor technologies to corrosion prevention and catalysis. In this respect, the work is not merely a compilation of existing knowledge but also a contemporary resource that can inspire researchers for future studies. Therefore, this book can be regarded as both a roadmap and a reliable reference that sheds light on the future of Schiff base research.

Author details

Nuriye Tuna Subasi
Department of Food Engineering, Ahi Evran University, Kırşehir, Turkey

*Address all correspondence to: tunasubasi@gmail.com

References

[1] Schiff H. Mittheilungen aus dem Universität Laboratorium in Pisa: Eine neue Reihe organischer Basen. Justus Liebigs Annalen der Chemie. 1864;**131**:118-119

[2] Boulechfar C, Ferkous H, Delimi A, Djedouani A, Kahlouche A, Boublia A, et al. Schiff bases and their metal complexes: A review on the history, synthesis, and applications. Inorganic Chemistry Communications. 2023;**150**:110451. DOI: 10.1016/j.inoche.2023.110451

[3] Sun Y, Lu Y, Bian M, Yang Z, Ma X, Liu W. Pt (II) and Au (III) complexes containing Schiff-base ligands: a promising source for antitumor treatment. European Journal of Medicinal Chemistry. 2021;**211**:113098-113121

[4] Chohan ZH, Sumrra SH, Youssoufi MH, Hadda TB. Metal based biologically active compounds: Design, synthesis, and antibacterial/antifungal/cytotoxic properties of triazole-derived Schiff bases and their oxovanadium (IV) complexes. European Journal of Medicinal Chemistry. 2010;**45**:2739-2747

[5] Al-Amiery AA, Al-Majedy YK, Ibrahim HH, Al-Tamimi AA. Antioxidant, antimicrobial, and theoretical studies of the thiosemicarbazone derivative Schiff base 2-(2-imino-1-methylimidazolidin-4-ylidene) hydrazinecarbothioamide (IMHC). Organic and Medicinal Chemistry Letters. 2012;**2**:1-7

[6] Cheng LX, Tang JJ, Luo H, Jin XL, Dai F, Yang J, et al. Antioxidant and antiproliferative activities of hydroxyl-substituted Schiff bases. Bioorganic & Medicinal Chemistry Letters. 2010;**20**:2417-2420

[7] Che CM, Huang JS. Metal complexes of chiral binaphthyl Schiff-base ligands and their application instereoselective organic transformations. Coordination Chemistry Reviews. 2003;**242**(2):97

[8] Musikavanhu B, Liang Y, Xue Z, Feng L, Zhao L. Strategies for improving selectivity and sensitivity of Schiff Base fluorescent Chemosensors for toxic and heavy metals. Molecules. 2023;**28**:6960. DOI: 10.3390/molecules28196960

[9] Zhang J, Xu L, Wong W-Y. Energy materials based on metal Schiff base complexes. Coordination Chemistry Reviews. 2018;**355**:180-198. DOI: 10.1016/j.ccr.2017.08.007

Chapter 2

A Comprehensive Review on Schiff Bases: Development Process, Green Synthetic Strategies, and Versatile Applications

Nuriye Tuna Subasi

Abstract

Schiff bases are essential organic compounds that possess a wide range of synthetic and application potential due to their structural diversity and ability to be modified with various functional groups. In this chapter, the historical development of Schiff bases is first outlined, followed by an in-depth examination of environmentally friendly synthetic approaches—including microwave-assisted, mechanochemical, and sonochemical methods—developed in line with the principles of green chemistry. Furthermore, recent advancements in the diverse applications of Schiff bases are comprehensively reviewed, encompassing areas such as biomedicine, analytical chemistry, photophysical applications, and materials science. Given their sustainable synthesis strategies and expanding range of uses, the versatile and strategic importance of Schiff bases in contemporary scientific research is clearly emphasized.

Keywords: imines, metal complexes, green synthesis, catalysis, sensor applications, biological activity

1. Introduction

Schiff bases are a class of organic compounds with a wide range of applications, particularly in ligand design, biomedical research, and analytical chemistry. These compounds are characterized by the presence of a distinctive imine (–C=N–) functional group, which is formed through the condensation of primary amines with aldehydes or ketones (**Figure 1**).

Although numerous review articles have been published on Schiff bases and their metal complexes, most of them focus either on specific metal-ligand systems or isolated application areas. This chapter aims to present a more integrative perspective by highlighting green synthetic strategies alongside the versatile applications of Schiff bases in fields such as catalysis, corrosion inhibition, and biomedical science. The selection of topics is based on recent trends that emphasize sustainability, multifunctionality, and innovation in ligand design.

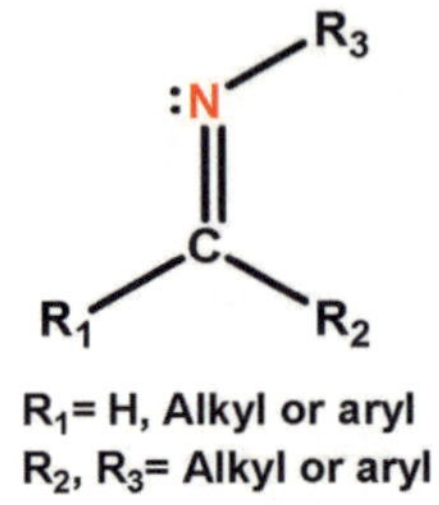

Figure 1.
General structural formula of Schiff base.

2. Historical background and development of Schiff bases

This class of compounds was first described in 1864 by the German-Italian chemist Hugo Schiff [1], who spent a significant portion of his career in Italy, contributing to organic chemistry. His pioneering study on this subject consisted of two main parts [2]. The first part focused on the structural properties and chemical behavior of quinoline and certain metal derivatives (Zn, Hg, Sb, and Bi), while the second part investigated the reactions between aniline and various aldehydes. As a result of these studies, Schiff recognized that the resulting compounds represented a new class of chemical structures, which he later named "Schiff bases" after himself (**Figure 2**).

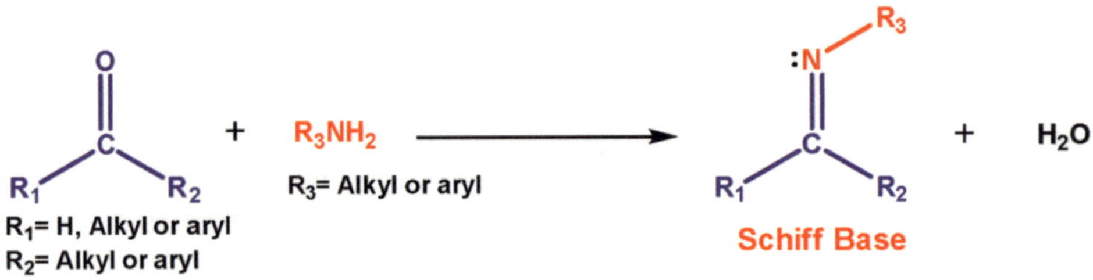

Figure 2.
General synthesis method of Schiff base.

Hugo Schiff's seminal contribution paved the way for novel and wide-ranging applications in the field of organic synthesis. In the early years, studies on Schiff bases primarily focused on structural characterization and understanding reaction mechanisms, with most investigations being limited to simple aromatic systems.

The history of Schiff base metal complexes dates back to 1889, with the first systematic study conducted by Alphonse Combes. In his work, Combes synthesized a copper(I) Schiff base complex through the reaction of ethylenediamine and acetylacetone, thus presenting the first reported example of a Schiff base metal complex in the literature [3]. This pivotal discovery laid the foundation for the growing interest in Schiff bases within coordination chemistry, primarily due to their remarkable ability to form stable complexes with metal ions.

A major milestone in the historical development of Schiff base ligands was the introduction of Salen-type ligand systems, first reported by Pfeiffer and colleagues in 1933 [4]. These structures are formed through the condensation of salicylaldehyde derivatives with diamine compounds, resulting in tetradentate ligands that feature two imine (–C=N) groups and two phenolic hydroxyl groups, as illustrated in **Figure 3** [5]. Due to their structural symmetry and high coordination capacity, Salen-type Schiff bases are widely utilized in the formation of stable and directional complexes with transition metals. Furthermore, their ability to stabilize metal ions in various oxidation states makes them highly valuable in the design of homogeneous catalysts [6]. Owing to these distinctive features, Salen derivatives are regarded as one of the most important structural platforms that Schiff base chemistry has contributed to coordination chemistry.

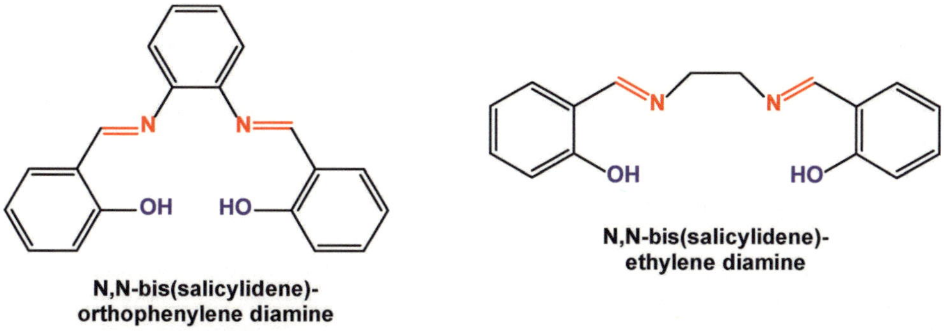

Figure 3.
Structures of Salen-Type Schiff bases.

By the mid-twentieth century, the first systematic observations regarding the biological activities of Schiff bases had begun to appear in the scientific literature. During this period, several Schiff base derivatives were reported to exhibit antibacterial and antifungal properties [7], laying the foundation for extensive investigations into their potential applications in biological systems. The imine (–C=N–) functional group present in Schiff bases allows for interactions with both organic molecules and metal ions, and it has been observed that their biological activity is significantly enhanced when these compounds form metal complexes [8]. Such interactions increase the binding affinity of the complexes to cellular structures, thereby amplifying their bioactivity. Subsequent studies demonstrated that Schiff base metal complexes possess diverse biological effects, including anticancer, antiviral, and enzyme inhibitory activities [9, 10]. Owing to these broad-spectrum properties, Schiff bases and their derivatives have emerged as prominent subjects of research in pharmaceutical chemistry, biotechnology, and medical diagnostics.

Today, Schiff bases are actively investigated and applied across a wide range of interdisciplinary fields. These compounds play significant roles in various domains, including:

- *Analytical chemistry*: These are colorimetric and fluorescent probes for metal ion detection, as well as ion-selective sensors.
- *Materials science*: In the development of nanocomposites, polymeric structures, and functionalized surfaces.
- *Photophysical applications*: In studies involving mechanisms such as aggregation-induced emission (AIE), photoinduced electron transfer (PET), and intramolecular charge transfer (ICT).
- *Biomedical fields*: In drug delivery systems, biosensors, and biointeractive platforms for tissue-related applications.

These versatile applications clearly demonstrate that Schiff bases have transcended their role as classical synthetic intermediates, attaining a strategic position at the intersection of organic and inorganic chemistry. Owing to their structural diversity and synthetic flexibility, Schiff bases continue to be a focus of intense research, holding great promise for even broader applications in the future.

3. Importance and general properties of Schiff bases

Owing to their structural flexibility and rich functional diversity, Schiff bases are highly versatile compounds that can be utilized across a wide spectrum of applications. The following section highlights the key features and significance of this important class of compounds:

3.1 Easy and rapid synthesizability

Schiff bases can generally be synthesized efficiently and in high yields through the condensation of primary amines with aldehydes or ketones, especially when optimized conditions and modern methodologies, such as solvent-free or microwave-assisted syntheses, are applied. Since water is usually the only byproduct, the reaction is considered environmentally friendly and highly compatible with the principles of green chemistry. Furthermore, many Schiff base syntheses can be carried out without the need for catalysts or under mildly acidic or basic conditions, offering the advantages of low energy consumption, operational simplicity, and cost-effectiveness.

In addition to yield and reaction time, the purity of Schiff bases obtained through different synthetic methods is a crucial parameter. Generally, modern techniques such as microwave-assisted synthesis tend to produce compounds with higher purity and fewer by-products compared to conventional reflux methods. This enhanced purity contributes to better performance in subsequent applications such as catalysis, sensing, and biological activity. However, detailed comparative studies on purity remain limited and warrant further investigation.

The synthesis of Schiff bases from aldehydes generally proceeds faster and yields are higher compared to those from ketones, owing to the higher electrophilicity and lower steric hindrance of aldehydes. In contrast, ketone-derived Schiff bases often require longer reaction times and more stringent conditions [5, 11].

Moreover, aromatic Schiff bases typically exhibit greater stability due to resonance stabilization of the imine bond, while aliphatic Schiff bases tend to be more reactive but less stable, which affects their application scope and synthesis protocols [5, 11].

3.2 Structural diversity and modifiability

One of the most prominent features of Schiff bases is their ability to be structurally tailored due to the vast variety of available amine and carbonyl components, allowing for the design of compounds with diverse electronic and steric properties. Aromatic, aliphatic, heterocyclic, or chiral building blocks can be used to generate a broad molecular library. The azomethine (–C═N–) group, containing an electron-rich nitrogen atom and an electron-deficient carbon atom, is inherently reactive toward both electrophilic and nucleophilic species. This characteristic makes Schiff bases ideal intermediates in the synthesis of target-specific molecules. In this context, Schiff bases are widely employed in asymmetric synthesis, aldol reactions, condensation processes [12–15], and as protecting groups. Particularly, Schiff bases derived from amino acid derivatives offer significant advantages in the selective synthesis of chiral compounds, highlighting their value as versatile tools in modern organic synthesis.

3.3 Coordination capacity

Schiff bases exhibit a high affinity for metal ions due to the nitrogen atom within their imine (–C=N–) functional group. This characteristic renders them effective ligands, particularly capable of forming stable and directional complexes with transition metals. Through their donor atoms-typically nitrogen, oxygen, or sulfur-Schiff bases can act as mono- or polydentate ligands, enabling the formation of metal complexes with various coordination geometries. This coordination flexibility has established Schiff bases as indispensable tools in fields such as catalyst design, metal complex synthesis, and the development of bioinorganic model systems. Notably, Schiff base metal complexes demonstrate high catalytic efficiency in applications including asymmetric catalysis, oxidation reactions, and polymerization processes [16–18].

3.4 Spectroscopic traceability and sensor applications

Schiff bases can be easily monitored and characterized by various spectroscopic techniques due to the presence of conjugated π-systems in their structures. These compounds are extensively analyzed using UV-Vis, IR, NMR, and especially fluorescence spectroscopy, often exhibiting pronounced bathochromic shifts and changes in fluorescence intensity upon interaction with metal ions. Such spectroscopic features make Schiff bases particularly attractive for the design of colorimetric and fluorescent sensors. Moreover, derivatives of Schiff bases that operate through photophysical mechanisms such as photoinduced electron transfer (PET) and intramolecular charge transfer (ICT) play an active role in the development of advanced optical sensing systems [19–21]. These compounds have demonstrated remarkable sensitivity and selectivity in the detection of heavy metal ions in environmental samples, the analysis of biological specimens, and in visual sensor technologies.

3.5 Biological activity potential

Numerous scientific studies have reported that various Schiff base derivatives with different structural features exhibit antimicrobial [22], antifungal [23, 24], anticancer [25–27], and antibacterial [28, 29] activities. Due to these biological properties, Schiff bases are considered promising candidates for pharmaceutical applications. Certain derivatives may play diverse biological roles such as enzyme inhibitors, DNA/RNA binders [30, 31], or metal ion carriers, owing to their ability to interact with specific cellular targets. Moreover, Schiff base metal complexes often exhibit stronger biological activity compared to their free ligand counterparts. This enhanced activity is attributed to factors such as the electronic density, coordination geometry, and lipophilicity of the metal ion, which improve the complex's interaction with biomolecules.

3.6 Photophysical and optoelectronic properties

Certain Schiff base derivatives, particularly those possessing extended conjugated π-systems, exhibit sensitivity to light. Due to this characteristic, such compounds have been utilized in a variety of optoelectronic applications [19], including fluorescent probes, organic light-emitting diodes (OLEDs), and light-responsive sensors.

Additionally, the structural flexibility and functional group diversity of Schiff bases enable them to serve as cross-linking agents or functional building blocks in the synthesis of metal–organic frameworks (MOFs) [32], covalent organic frameworks (COFs) [33], polymeric network structures, and nanomaterials. The incorporation of Schiff bases into these materials contributes directly to improvements in thermal stability, mechanical strength, and optoelectronic performance. Thus, Schiff bases are not limited to classical chemical applications but also exhibit significant potential in material science, nanotechnology, and the design of advanced functional systems.

4. Recent synthetic strategies for Schiff bases and their metal complexes

In recent years, environmentally friendly and energy-efficient synthetic techniques have gained significant importance in organic and coordination chemistry. Due to their straightforward condensation-based synthesis mechanisms, Schiff bases are highly compatible with various green synthetic strategies. In particular, modern approaches such as microwave-assisted synthesis, solvent-free grinding methods, and ultrasonic (sonication-assisted) techniques offer faster, more efficient, and more environmentally sustainable alternatives compared to conventional procedures. These methods not only reduce reaction times and energy consumption but also significantly minimize the use of toxic solvents and the formation of harmful by-products. As a result, the applicability of such green methodologies has become a focal point in recent studies for the synthesis of both Schiff base ligands and their transition metal complexes.

4.1 Microwave-assisted synthesis

The conventional synthetic protocol for Schiff bases typically involves the condensation of a suitable amine with an aldehyde or ketone in an organic solvent under reflux for several hours. However, this classic method often suffers from drawbacks such as long reaction times, lower product yields compared to modern techniques, and the use of toxic and environmentally hazardous solvents. In this context, novel synthetic approaches developed in accordance with the principles of Green Chemistry aim to make chemical processes more environmentally friendly. These strategies seek to reduce the number of synthetic steps, minimize waste generation, and achieve high yields in shorter reaction times. One such prominent method is microwave-assisted organic synthesis (MAOS), which has been increasingly applied in the preparation of Schiff bases (**Figure 4**) [34].

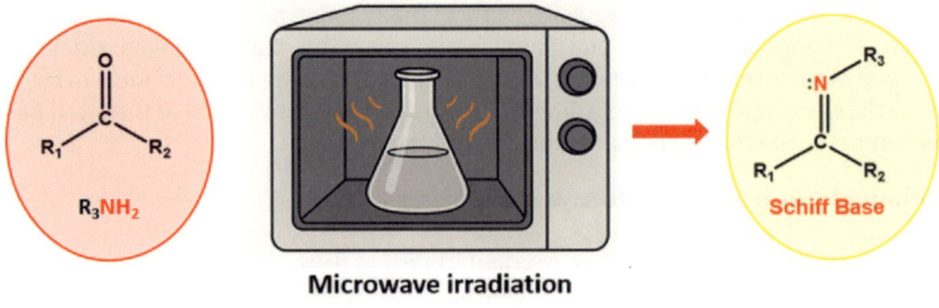

Figure 4.
Microwave-assisted synthesis of Schiff base.

Microwave-assisted heating offers several advantages over conventional techniques, making it a preferred modern approach for Schiff base synthesis. In this method, aldehydes, ketones, and amines can rapidly react to form a wide variety of Schiff base derivatives. Because the reaction mixture is directly and efficiently heated, reaction times are significantly reduced. The ability to carry out the reactions under controlled temperature conditions also minimizes the risk of thermal degradation, especially for temperature-sensitive compounds. Moreover, microwave irradiation ensures more uniform and efficient heating, which contributes to improved product yields. This rapid and effective heating mechanism reduces energy consumption, making the method not only energy-efficient but also cost-effective [35].

Furthermore, the use of microwave energy facilitates the employment of biodegradable and less toxic solvents (e.g., water and ethanol). Although acetonitrile is widely used in microwave-assisted synthesis, it is classified as a moderately toxic solvent according to chemical safety databases. Therefore, its use requires appropriate handling and safety precautions to enhance the environmental compatibility of the reaction conditions [36, 37]. Particularly when using weak proton-donating solvents, the reaction rate and selectivity can be significantly improved. These features enhance the method's applicability at both laboratory and industrial scales. All of these advantages render microwave-assisted synthesis an attractive approach in terms of both environmental sustainability and synthetic practicality.

To demonstrate the practical applicability of this method, a number of Schiff base derivatives have been synthesized under microwave conditions. In a 2024 study by Durairaj and colleagues [38], a novel Schiff base Cu(II) metal complex was synthesized and characterized *via* microwave-assisted condensation of phenylacetyl urea and salicylaldehyde in the presence of thiocyanate ions (**Figure 5**). The resulting compound involves a monodentate ligand coordinating to the copper ion through the imine nitrogen atom. This complex was further evaluated for its biological activities, revealing promising potential for biomedical applications.

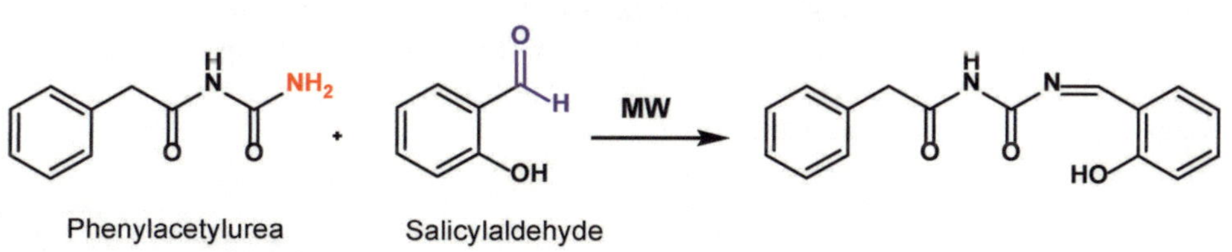

Figure 5.
Microwave-assisted synthesis of Schiff base Cu(II) complex.

4.2 Ultrasound-assisted synthesis

Ultrasound-assisted synthesis, also known as sonochemistry, has emerged as a modern technique that enables the environmentally friendly synthesis of Schiff bases [39, 40]. In this method, high-frequency ultrasonic sound waves serve as the energy source, generating microscopic cavitation bubbles within the reaction medium. These bubbles oscillate at high frequencies and collapse violently, creating localized micro-reactive zones with extremely high temperatures and pressures, known as "sono-chemical hot spots" (**Figure 6**). These extreme conditions enhance the activation of reactants, accelerate reaction rates, and facilitate the formation of target products in shorter times with higher selectivity and yields [5].

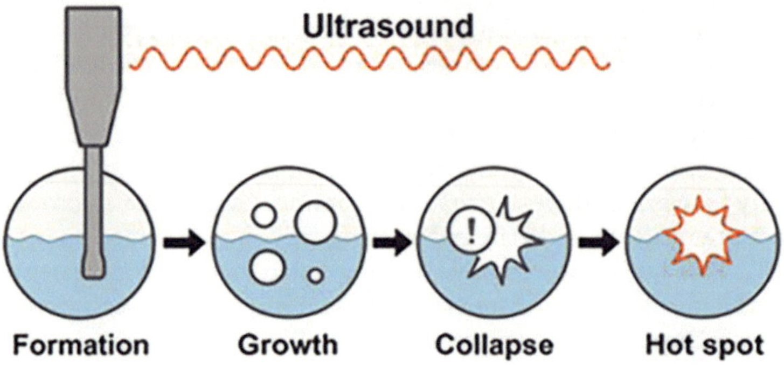

Figure 6.
Schematic representation of the ultrasound-assisted organic synthesis.

Compared to conventional heating and stirring techniques, sonochemical synthesis offers the advantage of proceeding at lower temperatures and shorter durations, thereby reducing the risk of thermal degradation. This makes the method particularly suitable for the synthesis of thermally sensitive compounds. In addition, its high energy efficiency enables significant product yields with minimal energy consumption. Sonochemistry is also cost-effective and applicable on both laboratory and industrial scales.

Ultrasound-assisted synthesis can be efficiently conducted not only in conventional organic solvents but also in aqueous media, which offer a greener reaction environment. Consequently, it aligns well with the principles of green chemistry by minimizing solvent usage, avoiding toxic reagents, and reducing waste production. Moreover, this technique can be readily integrated with purification methods such as column chromatography and improves the solubility of reactants, further enhancing reaction efficiency. With all these features, sonochemical methods offer a promising and innovative platform for the sustainable synthesis of Schiff bases [35].

In 2024, Assiri and colleagues synthesized a series of N-aryl (heteroaryl) Schiff bases of 3-(2-oxo-2H-chromen-3-yl)-1-(4-phenylthiazol-2-yl)-1H-pyrazole *via* ultrasound-assisted methods [41], with the aim of investigating their cytotoxic properties against cancer cells (**Figure 7**).

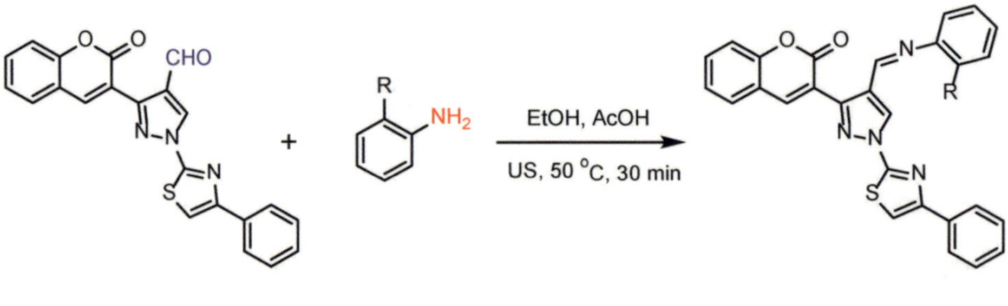

Figure 7.
Ultrasound-assisted synthesis of Schiff base.

4.3 Solvent-free grinding method (mechanochemical synthesis)

Another green synthetic approach for the preparation of Schiff bases is the grinding method, also known as mechanochemical synthesis—a solvent-free, environmentally friendly, and energy-efficient technique. In this method, a primary amine and an aldehyde or ketone are typically mixed and ground together in a mortar or mechanical devices such as a ball mill, without the use of solvents or catalysts (**Figure 8**). The mechanical energy generated during grinding promotes the formation of the imine (C=N) bond and often leads to high yields within short reaction times. Additionally, the grinding process increases the surface area of the reactants, which can further enhance the reaction rate and yield. This method eliminates the need for an external catalyst and can be carried out at room temperature under ambient pressure. Grinding can also be performed in the presence of green solvents such as water or ionic liquids using ball milling systems [5].

Figure 8.
Schematic representation of the solvent-free grinding method (mechanochemical synthesis) for the preparation of Schiff bases.

Volatile organic solvents constitute a major portion of chemical waste both in laboratory and industrial-scale synthesis. Mechanochemical activation offers an environmentally beneficial alternative by enabling solvent-free reaction conditions. The absence of solvent leads to high reactant concentrations, which can accelerate reaction rates and influence product selectivity. Therefore, this approach aligns well with the principles of green chemistry due to its elimination of harmful organic solvents, minimization of waste, and low energy consumption [42]. Moreover, reduced solvent use helps overcome solubility limitations and solvolysis-related side reactions, allowing for the use of various alternative starting materials—ultimately reducing overall synthesis costs [43–45]. Consequently, this method can also be applied to reactions involving reagents that are insoluble in conventional solvents [46]. Additionally, since the products are usually formed in the solid phase, the isolation process is straightforward and often does not require further purification. With its simplicity, low cost, and easily optimizable reaction conditions, the grinding method represents a highly attractive synthetic strategy. Various Schiff base derivatives, both simple and functionally substituted, have been successfully synthesized using this technique, making it particularly suitable for structures evaluated for biological activity and metal coordination ability [47–49].

In 2021, Boruah and colleagues synthesized a Schiff base (H_2L) by reacting salicylaldehyde with diethylenetriamine using the grinding method. Furthermore, a vanadium(V) complex of the Schiff base ($Na[VO_2L]$) was prepared *via* liquid-assisted grinding (LAG) (**Figure 9**). The synthesized Schiff base-vanadium complex was subsequently employed as a catalyst for the oxidation of organic sulfides and alcohols [50].

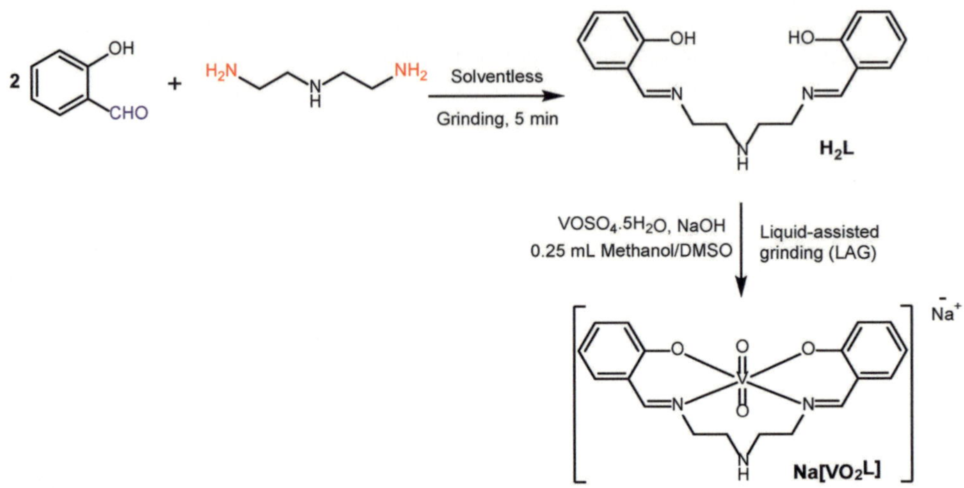

Figure 9.
Synthesis of a Schiff base ligand (H_2L) and its vanadium(V) complex ($Na[VO_2L]$) via *mechanochemical methods.*

4.4 Natural acid catalyst method

In recent years, researchers have increasingly focused on the use of natural acid catalysts in various organic reactions, as they offer high efficiency, low cost, non-toxicity, and minimal environmental impact [51, 52]. The synthesis of Schiff bases using natural acid catalysts has emerged as an environmentally friendly approach, gaining attention due to its alignment with the principles of green chemistry [53]. In this method, biodegradable, non-toxic, and inexpensive organic acids such as citric acid, tartaric acid, succinic acid, and ascorbic acid are employed as catalysts [54, 55]. These catalysts not only serve as proton sources but also facilitate effective interaction between reactants through hydrogen bonding, thereby enhancing the efficiency of the condensation reaction. This method is typically conducted under solvent-free conditions or in the presence of environmentally benign solvents such as water, offering a sustainable alternative to traditional mineral acids like HCl and H_2SO_4. The use of natural acids contributes to shorter reaction times, high yields, low energy consumption, and simplified product purification. Overall, Schiff base synthesis catalyzed by natural acids represents a promising and effective strategy for reducing the environmental footprint of chemical processes, with applicability at both laboratory and industrial scales.

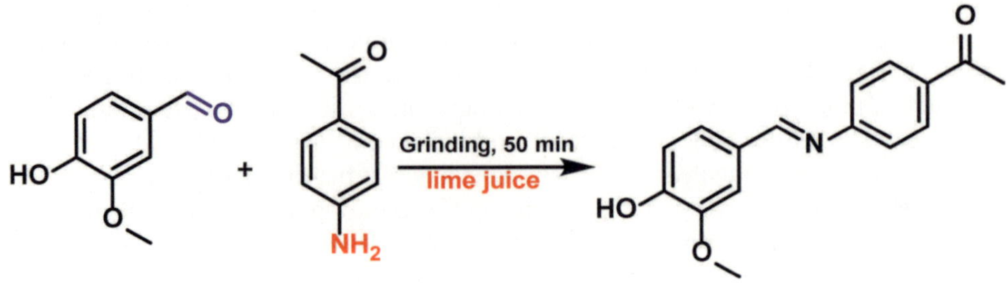

Figure 10.
Synthesis of a Schiff base via *natural acid catalysis using lemon juice as a catalyst.*

An example of Schiff base synthesis based on natural acid catalysis is the study conducted by L. Ma'rufah and colleagues (2021) [56]. In this work, the Schiff base was synthesized by grinding vanillin and p-aminoacetophenone in the presence of lemon juice (**Figure 10**).

5. Applications of Schiff bases and their metal complexes

Schiff bases and their transition metal complexes possess a wide range of applications across various scientific and technological fields due to their chemical diversity and functional group versatility (**Figure 11**). The ability of the imine (–C=N–) group to form strong coordination interactions with both organic molecules and metal ions brings these compounds particularly prominent in several disciplines as illustrated in the figure below including biomedical research, catalysis [57], sensor design [58], corrosion inhibition [59], dye technologies [60], and optoelectronic devices [61]. Metal complexes of Schiff bases have been reported to exhibit various biological activities such as anticancer [62, 63], antibacterial, antifungal [64, 65], anti-inflammatory [66, 67], antioxidant, and antiproliferative properties [68], many of which are enhanced through coordination with metal ions. Additionally, their ability to bind to DNA [69] offers significant potential for drug design and biosensor applications.

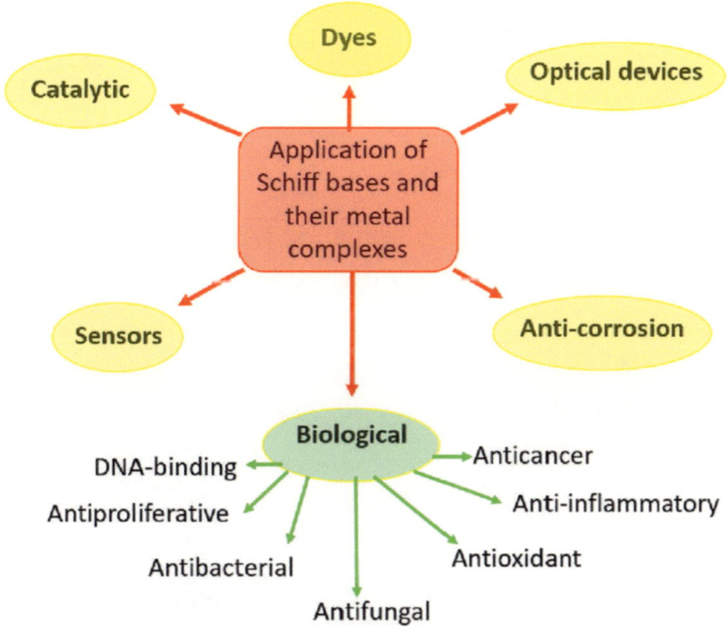

Figure 11.
Application of Schiff bases and their metal complexes.

5.1 Catalytic activity

Schiff base metal complexes are widely used as catalysts in various organic reactions due to their low cost, ease of synthesis, and high chemical and thermal stability. These complexes have been employed as catalysts in numerous homogeneous and heterogeneous reactions such as oxidation, polymerization, aldol condensation, Heck reaction, Henry reaction, carbonylation, epoxidation, and others. The catalytic activity of these complexes can vary depending on the metal ion, ligand structure, and the coordination geometry of the ligands. Owing to their high efficiency, good selectivity, mild reaction conditions, reusability, and operational simplicity, these catalysts are gaining increasing attention in both academic and industrial settings [70, 71].

The oxidation of alcohols to aldehydes or ketones is a key transformation in both laboratory-scale and industrial organic synthesis. For this reason, the use of Schiff base metal complexes as catalysts in oxidation reactions has been extensively studied. These complexes play crucial roles in oxygen transfer, electron transport, and the generation of active intermediates during the catalytic process. Typically, Schiff base complexes of iron, copper, and manganese are utilized for such oxidation reactions. However, in a study conducted by Kargar and colleagues in 2021 [72], dioxomolybdenum(VI) and oxovanadium(V) Schiff base complexes were synthesized and employed as catalysts for the oxidation of benzyl alcohol. The reaction was completed within 2 hours, yielding the corresponding product with 92% efficiency (**Figure 12**).

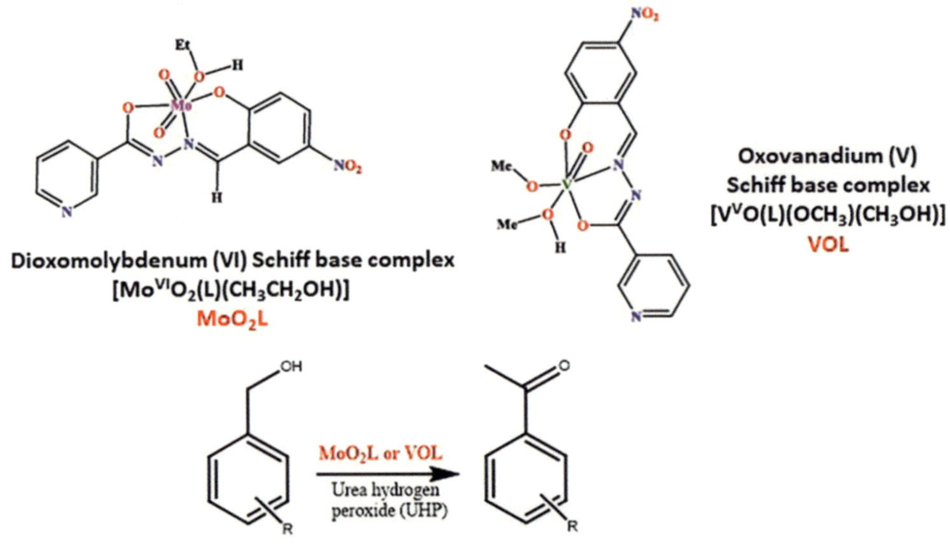

Figure 12.
Catalytic oxidation of benzylic alcohol catalyzed by MoO_2L or VOL.

Another example of reactions catalyzed by Schiff bases and their metal complexes includes aldol condensation and the Henry reaction. Aldol condensation constitutes a key class of reactions in organic chemistry due to its ability to form carbon-carbon bonds. When Schiff bases and their metal complexes are used as catalysts in such reactions, they have been reported to enhance the reaction rate, improve yields, and exhibit regio- and stereoselectivity, according to various studies.

In a 2021 study, Arora and colleagues synthesized two Schiff base ligands and their corresponding Cu(II) complexes ([Cu(D-valmet)(H_2O)]·H_2O and [Cu(L-valmet)(H_2O)]·H_2O) (**Figure 12**), and investigated their catalytic performance in the aldol reaction between cyclohexanone and benzaldehyde, both in the presence and absence of these catalysts (**Figures 13** and **14**) [73]. The term "valmet" refers to a series of chiral Schiff base ligands synthesized *via* the condensation of o-vanillin and L- or D-methionine, as described in Ref. [73].

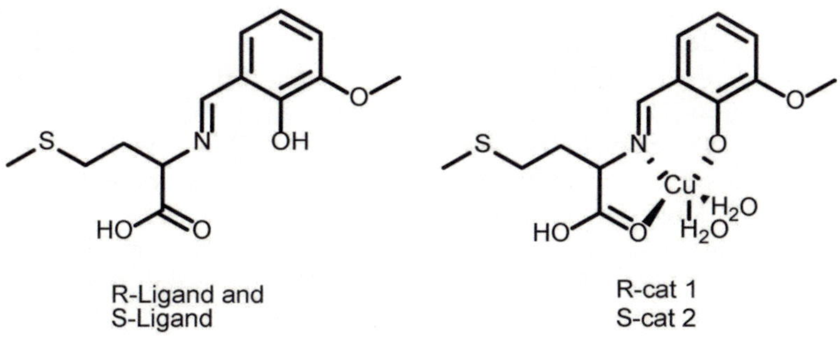

Figure 13.
Structure of Schiff base ligand and its Cu complex.

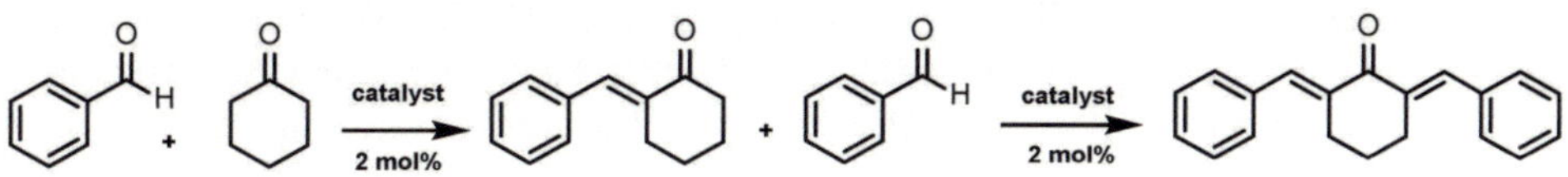

Figure 14.
Schiff base Cu complex catalyzed aldol reaction.

In the absence of Schiff bases and Cu(II) complexes, the reaction proceeded slowly at room temperature with low conversion, resulting in the formation of only a mono-condensation product. At elevated temperatures, although the overall conversion remained low, a predominant formation of the di-condensation product was observed. However, in the presence of Schiff base catalysts, the reaction rate significantly increased with high selectivity toward the di-condensation product. When Cu(II) complexes were used, the conversion rate further increased compared to the Schiff base alone, though the selectivity between mono- and di-condensation products was relatively lower [74].

Arora and co-workers further evaluated the catalytic performance of the synthesized compounds in the Henry reaction and observed promising catalytic efficiency (**Figure 15**).

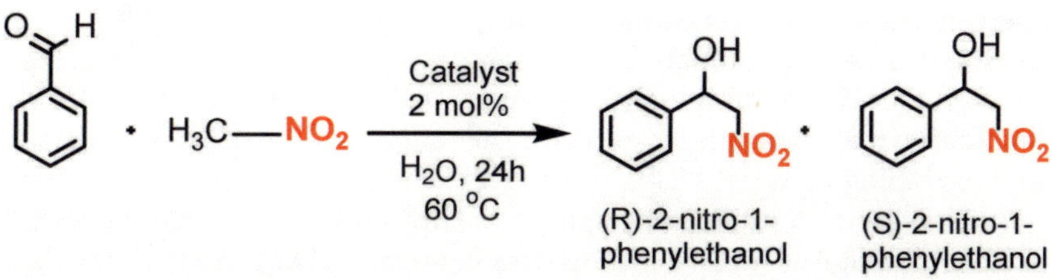

Figure 15.
Schiff base Cu complex catalyzed the Henry reaction.

5.2 Sensor applications

Schiff bases and their metal complexes exhibit high selectivity and sensitivity toward target ions, molecules, or gases due to their selective binding capabilities, redox-active nature, and ability to provide spectroscopic responses. These features have made them widely applicable in various sensor technologies, particularly in heavy metal ion detection [75], gas sensing [76, 77], optical sensor design [19], pH and solvent detection [78], and biosensor applications [79–81].

The strong coordination ability of the imine group (–C=N–) with metal ions makes Schiff bases suitable candidates for use in ion-selective electrodes, fluorescent sensors, and colorimetric detectors. Schiff base derivatives have been frequently employed in the selective detection of environmentally and biologically significant metal ions such as Cu^{2+}, Fe^{3+}, Zn^{2+}, Hg^{2+}, Al^{3+}, as well as trace amounts of heavy metals like Pb^{2+} [75, 82]. The interaction between the metal ion and the Schiff base ligand causes detectable changes in the optical or electrochemical properties of the complex—such as color, conductivity, or fluorescence—which can be utilized for quantitative analysis.

Moreover, Schiff bases are commonly used in optical sensors due to their color-changing behavior upon binding with specific analytes. These visible responses allow for real-time analysis without the need for sophisticated instrumentation, offering practical solutions for the detection of chemical pollutants, environmental toxins, and biological molecules. Schiff base complexes are also effective in detecting environmental gases like NO_2, CO, and NH_3, where gas coordination to the metal center leads to electronic changes that manifest as measurable variations in resistance or color.

In addition, the absorption spectra and fluorescence properties of some Schiff base complexes can respond to environmental changes, making them useful for monitoring pH or solvent composition. Finally, Schiff base metal complexes that can selectively interact with biomolecules such as DNA, glucose, or enzymes and generate measurable signals hold great promise in biosensor technology, particularly for medical diagnostics and environmental monitoring. Selected examples of Schiff base-based sensors and their detection capabilities are summarized in **Table 1**.

5.3 Antimicrobial and biological activities

Schiff bases, which act as key intermediates in the synthesis of natural products, exhibit significant biological activity due to the presence of the C=N group in their structures. The azomethine nitrogen in Schiff bases serves as the binding site for metal ions, allowing them to interact with various biomolecules such as amino acids and proteins. Schiff base metal complexes have demonstrated remarkable antimicrobial properties against bacteria, fungi, viruses, and various pathogens [83].

When Schiff bases are compared with their corresponding metal complexes, the latter are often found to exhibit higher antimicrobial activity. This can be attributed to several factors: metal complexes enhance the potential and efficacy of these compounds as biological agents capable of inhibiting microbial growth. By increasing lipophilicity, metal complexes can penetrate lipid membranes more effectively, leading to the blockage of metal-binding sites in microbial enzymes. Upon chelation, the polarity of the metal ion decreases, and the positive charge is partially shared with donor groups. This facilitates delocalization of π-electrons over the entire chelate ring,

Schiff base/complex	Target analyte	Sensor type	Detection method	Medium	Ref.
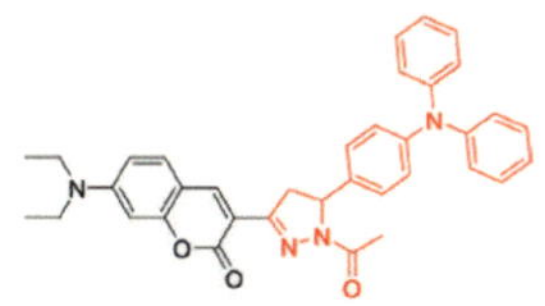	Fe^{3+}	Fluorescent sensor	Fluorescence spectroscopy	THF	[83]
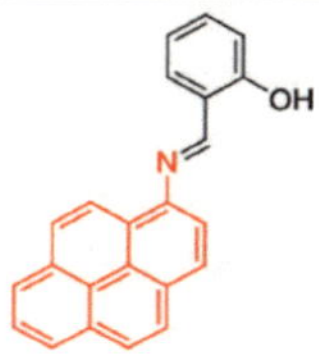	Cr^{3+}	Fluorescent sensor	Fluorescence spectroscopy	DMSO	[84]
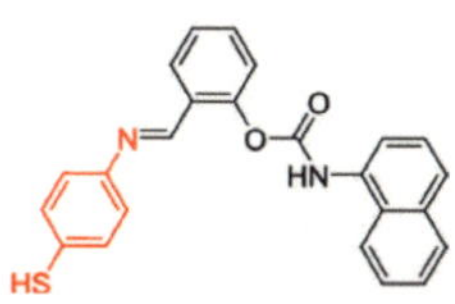	Cu^{2+}	Fluorescent sensor	Fluorescence spectroscopy	1:9 (DMSO: HEPES) buffer	[85]
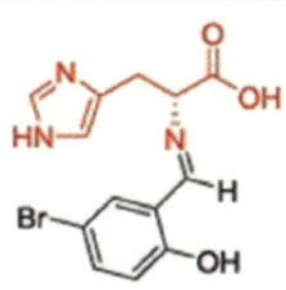	Pb^{2+}	Chemosensor	UV-Vis spectroscopy	Aqueous	[86]
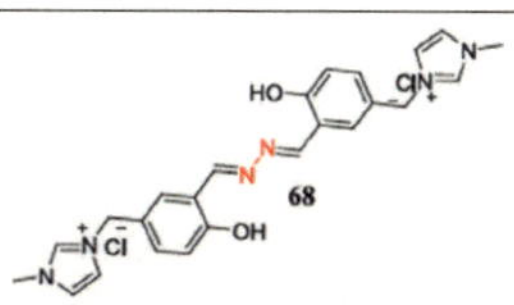	Cu^{2+} and Pd^{2+}	Colorimetric sensor	UV-Vis spectroscopy	Aqueous	[85]
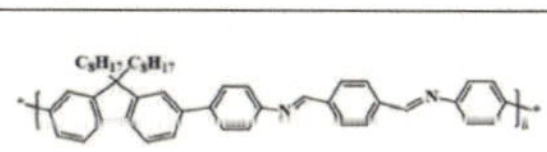	$NH_{3(g)}$	Sensor	UV-vis–NIR absorption spectra	—	[76]
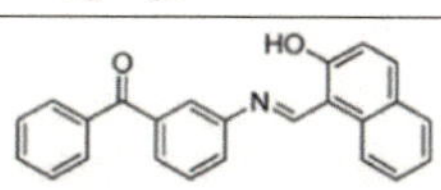	volatile organic compounds	Fluorescent sensor	Fluorescence spectroscopy	—	[77]
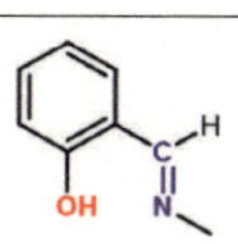	pH sensor	Colorimetric and fluorometric sensor	UV-Vis spectra and photoluminescence spectra	Britton-Robinson buffer	[78]

Table 1.
Selected examples of Schiff base-based sensors and their detection capabilities.

making the complex more capable of permeating lipid membranes. Additionally, metal complexes may disrupt the respiratory processes of microorganisms and inhibit protein synthesis, further suppressing microbial growth [84].

Moreover, Schiff base metal complexes exhibit a wide range of biological activities—such as antibacterial, antifungal, antiviral, antioxidant, anticancer, and antiulcer—depending on the coordinated metal ion. For instance, a copper(II) Schiff base complex has shown strong antioxidant and anticancer activities [85]. Similarly, Schiff base complexes of platinum(II), copper(II), chromium(III), palladium(II), and zinc(II) have demonstrated promising antitumor activity against various cancer cell lines [22, 86, 87]. These complexes exert their effects either by binding to DNA or enzymes involved in cancer cell proliferation, inducing apoptosis, or inhibiting tumor growth.

Another notable activity of Schiff base metal complexes is their antioxidant capability, which involves neutralizing free radicals and preventing oxidative damage in biological systems. This activity is particularly beneficial in combating oxidative stress-related conditions such as cardiovascular diseases, neurodegenerative disorders, and aging. Especially, copper(II) Schiff base complexes have exhibited potent antioxidant effects [88].

Figure 16 illustrates selected examples of Schiff bases and their metal complexes that exhibit notable biological activities, including antimicrobial, antioxidant, and anticancer properties [89–99].

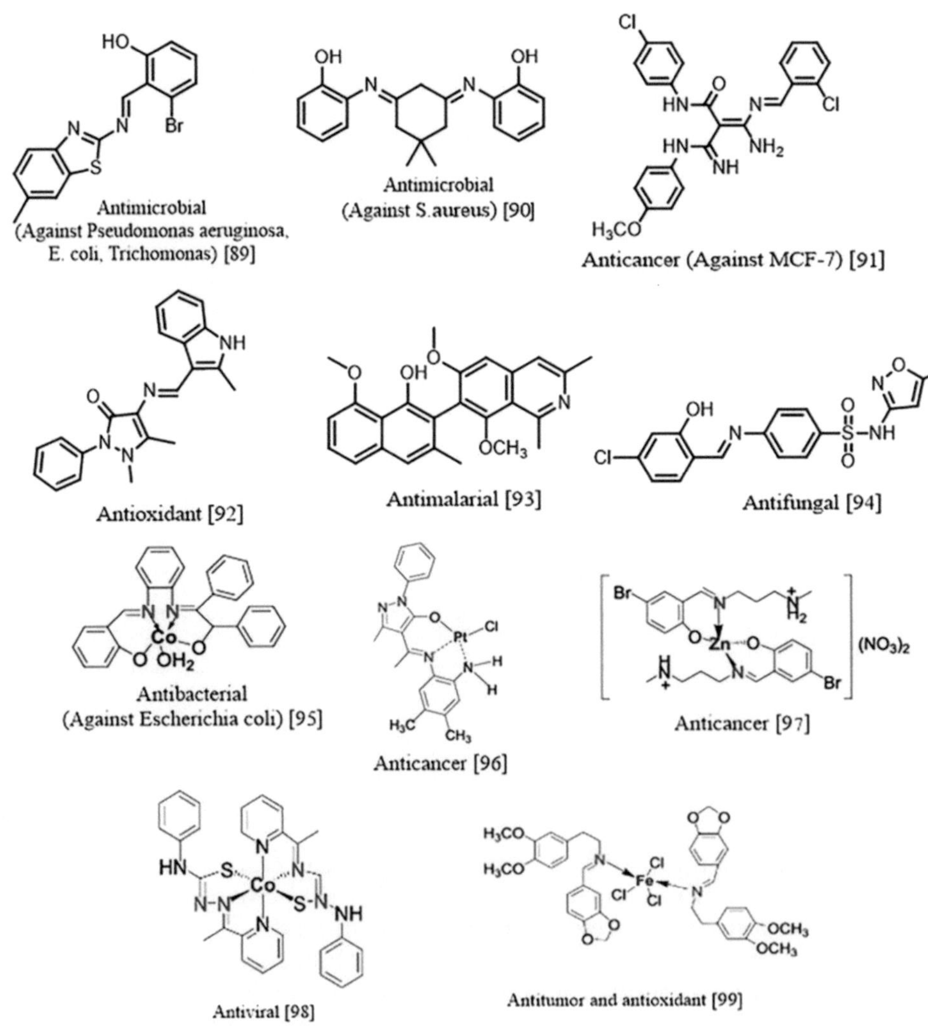

Figure 16.
Selected examples of biologically active Schiff bases and their metal complexes.

DOI: http://dx.doi.org/10.5772/intechopen.1012219

5.4 Schiff bases as corrosion inhibitors

Corrosion is defined as the degradation of metallic materials (metals and alloys) as a result of their chemical interaction with environmental components [1, 2]. Schiff bases represent a widely studied class of organic compounds with effective anticorrosive properties. Their inhibition mechanism is primarily based on the adsorption onto metal surfaces through the >C=N– (imine) functional group. The adsorption affinity and binding strength of Schiff bases on metallic surfaces are largely influenced by the electronic nature of the substituents present in their molecular structure. Electron-donating groups (e.g., –OH, $-NH_2$, –OR, $-NMe_2$, and –SH) generally enhance the corrosion inhibition efficiency of these compounds by increasing their electron density and surface reactivity. In contrast, electron-withdrawing groups (e.g., –COOH, $-NO_2$, and –CN) tend to reduce their inhibition performance [59].

A review of the literature reveals that Schiff bases have been predominantly applied as corrosion inhibitors for mild steel and carbon steel alloys, owing to their strong surface adsorption capabilities and structural tunability. **Figure 17** illustrates the corrosion inhibition potential of selected Schiff base compounds for mild steel and carbon steel in various electrolyte environments [100, 101].

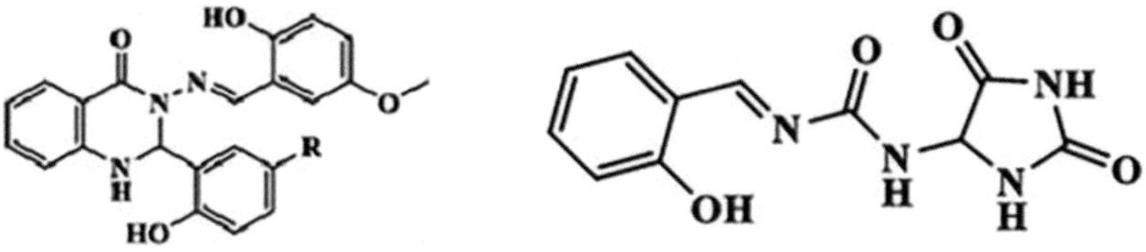

Figure 17.
Representative examples of Schiff base compounds used as corrosion inhibitors for mild and carbon steel in different electrolyte media.

6. Conclusion

In this chapter, the versatile nature of Schiff bases and their metal complexes has been comprehensively discussed, highlighting their synthetic strategies, structural diversity, and wide range of applications. Environmentally friendly and sustainable approaches, such as microwave-assisted, ultrasound-assisted, grinding, and natural acid-catalyzed syntheses, were emphasized as promising alternatives to conventional methods, aligning with green chemistry principles. Moreover, the extensive applicability of these compounds in catalysis, sensing, corrosion inhibition, and biological systems demonstrates their significance in modern chemical research. Particularly, their ability to coordinate metal ions through the azomethine group and form stable complexes has paved the way for innovations in pharmaceuticals, materials science, and environmental monitoring. The reported examples and figures throughout this chapter underscore the potential of Schiff base derivatives and their metal complexes as multifunctional agents in scientific and industrial domains. With ongoing advancements in ligand design and synthetic methodology, Schiff bases are expected to play an increasingly vital role in the development of novel functional materials.

Author details

Nuriye Tuna Subasi
Department of Food Engineering, Ahi Evran University, Kırşehir, Turkey

*Address all correspondence to: tunasubasi@gmail.com

References

[1] Schiff H. Mittheilungen aus dem Universit¨atslaboratorium in Pisa: Eine neue Reihe organischer Basen. Justus Liebigs Annalen der Chemie. 1864;**131**:118-119

[2] Fabbrizzi L. Beauty in chemistry: Making artistic molecules with Schiff bases. The Journal of Organic Chemistry. 2020;**85**(19):12212-12226. DOI: 10.1021/acs.joc.0c01420

[3] Combes A. Sur l'action des diamines sur les diac´ etones (on the action of diamines on diketons). Comptes rendus de l'Académie des Sciences. 1889;**108**:1252-1255

[4] Pfeiffer P, Breith E, Lübbe E, Tsumaki T. Tricyclische orthokondensierte Nebenvalenzringe. Justus Liebigs Annalen der Chemie. 1933;**503**:84-130. DOI: 10.1002/jlac.19335030106

[5] Boulechfar C, Ferkous H, Delimi A, Djedouani A, Kahlouche A, Boublia A, et al. Schiff bases and their metal complexes: A review on the history, synthesis, and applications. Inorganic Chemistry Communications. 2023;**150**:110451. DOI: 10.1016/j.inoche.2023.110451

[6] Asatkar AK, Tripathi M, Asatkar D. Salen and Related Ligands. London, UK, London: IntechOpen; 2020. DOI: 10.5772/intechopen.88593

[7] Pandeya SN, Sriram D, Nath G, Clereq ED. Synthesis and antimicrobial activity of Schiff and Mannich bases of isatin and its derivatives with pyrimidine. Il Farmaco. 1999;**54**:624-628

[8] Singh K, Barwa MS, Tyagi P. Synthesis, characterization and biological studies of Co(II), Ni(II), Cu(II) and Zn(II) complexes with bidentate Schiff bases. European Journal of Medicinal Chemistry. 2006;**41**:147-153. DOI: 10.1016/j.ejmech.2005.06.006

[9] da Silva CM, da Silva DL, Modolo LV, Alves RB, de Resende MA, Martins CVB, et al. Schiff bases: A short review of their antimicrobial activities. Journal of Advanced Research. 2011;**2**(1):1-8. DOI: 10.1016/j.jare.2010.05.004

[10] Kovala-Demertzi D, Demertzis MA. Biological activity of Schiff base metal complexes of transition elements. Journal of Inorganic Biochemistry. 2001;**86**:555-563. DOI: 10.1016/S0162-0134(01)00224-0

[11] Mir MA, Banik BK. Synthesis of Schiff base ligands under environment friendly conditions: A systematic review. Inorganic Chemistry Communications. 2025;**174**:113987

[12] Gupta KC, Sutar AK. Catalytic activities of Schiff base transition metal complexes. Coordination Chemistry Reviews. 2008;**252**(12-14):1420. DOI: 10.1016/j.ccr.2007.09.005

[13] Dalia SA, Afsan F, Hossain MS, Khan MN, Zakaria C, Zahan MKE, et al. A short review on chemistry of Schiff base metal complexes and their catalytic application. International Journal of Chemical Studies. 2018;**6**:2859

[14] Mitsunuma H, Matsunaga S. Dinuclear Ni2–Schiff base complex-catalyzed asymmetric 1,4-addition of β-keto esters to nitroethylene toward γ2,2-amino acid synthesi. Chemical Communications. 2011;**47**:469. DOI: 10.1039/C0CC02152K

[15] Maleev VI, North M, Larionov VA, Fedyanin IV, Savel'yeva TF,

Moscalenko MA, et al. Chiral octahedral complexes of cobalt (III) as “organic catalysts in disguise” for the asymmetric addition of a glycine Schiff Base ester to activated olefins. Advanced Synthesis and Catalysis. 2014;**356**:1803-1810. DOI: 10.1002/adsc.201400091

[16] Katsuki T. Unique asymmetric catalysis of cis-β metal complexes of salen and its related Schiff-base ligands. Chemical Society Reviews. 2004;**33**(7):437. DOI: 10.1039/B304133F

[17] Gupta KC, Sutar AK, Lin CC. Polymer-supported Schiff base complexes in oxidation reactions. Coordination Chemistry Reviews. 2009;**253**(13-14):1926

[18] Che CM, Huang JS. Metal complexes of chiral binaphthyl Schiff-base ligands and their application instereoselective organic transformations. Coordination Chemistry Reviews. 2003;**242**(2):97

[19] Berhanu AL, Mohiuddin GI, Malik AK, Aulakh JS, Kumar V, Kim KH. Review of the applications of Schiff bases as optical chemical sensors. TrAC Trends in Analytical Chemistry. 2019;**116**:74-91. DOI: 10.1016/j.trac.2019.04.025

[20] Dash PP, Patel DA, Mohanty P, Behura R, Behera S, Sahoo SK, et al. Advances on chromo-fluorogenic sensing of copper (II) with Schiff bases. Inorganica Chimica Acta. 2023;**556**:121635. DOI: 10.1016/j.ica.2023.121635

[21] Afrin A, Swamy PCA. Novel Schiff base derivatives for the detection of one-to-multi metal ions and tracking the live cell imaging. Coordination Chemistry Reviews. 2023;**494**:215327. DOI: 10.1016/j.ccr.2023.215327

[22] Shiju C, Arish D, Kumaresan S. Novel water soluble Schiff base metal complexes: Synthesis, characterization, antimicrobial-, DNA cleavage, and anticancer activity. Journal of Molecular Structure. 2020;**1221**:128770. DOI: 10.1016/j.molstruc.2020.128770

[23] Ejidike IP, Ajibade PA. Transition metal complexes of symmetrical and asymmetrical Schiff bases as antibacterial, antifungal, antioxidant, and anticancer agents: Progress and prospects. Reviews in Inorganic Chemistry. 2015;**35**:191-224

[24] Dharmaraj N, Viswanathamurthi P, Natarajan K. Ruthenium (II) complexes containing bidentate Schiff bases and their antifungal activity. Transition Metal Chemistry. 2001;**26**:105-109. DOI: 10.1023/A:1007132408648

[25] Arunadevi A, Raman N. Biological response of Schiff base metal complexes incorporating amino acids–a short review. Journal of Coordination Chemistry. 2020;**73**:2095-2116. DOI: 10.1080/00958972.2020.1824293

[26] Utreja D, Singh S, Kaur M. Schiff bases and their metal complexes as anti-cancer agents: A review. Current Bioactive Compounds. 2015;**11**:215-230

[27] Tadele KT, Tsega TW. Schiff bases and their metal complexes as potential anticancer candidates: A review of recent works. Anti-Cancer Agents in Medicinal Chemistry-Anti-Cancer Agents. 2019;**19**:1786-1795

[28] Ardakani AA, Kargar H, Feizi N, Tahir MN. Synthesis, characterization, crystal structures and antibacterial activities of some Schiff bases with N_2O_2 donor sets. Journal of the Iranian Chemical Society. 2018;**15**:1495-1504. DOI: 10.1007/s13738-018- 1347-6

[29] Shaker AM, Nassr LAE, Adam MSS, Mohamed I. Synthesis, characterization

and spectrophotometric studies of seven novel antibacterial hydrophilic iron (II) Schiff base amino acid complexes. Journal of the Korean Chemical Society. 2013;**57**:560-567

[30] Singh A, Gogoi HP, Barman P, Das A, Pandey P. Tetracoordinated ONNO donor purine-based Schiff base and its metal complexes: Synthesis, characterization, DNA binding, theoretical studies, and bioactivities. Applied Organometallic Chemistry. 2022;**36**:e6852

[31] Singh A, Maiti SK, Gogoi HP, Barman P. Purine-based Schiff base Co (II), Cu (II), and Zn (II) complexes: Synthesis, characterization, DFT calculations, DNA binding study, and molecular docking. Polyhedron. 2023;**230**:116244

[32] Kaur M, Kumar S, Yusuf M, Lee J, Malik AK, Ahmadi Y, et al. Schiff base-functionalized metal-organic frameworks as an efficient adsorbent for the decontamination of heavy metal ions in water. Environmental Research. 2023;**236**(2):116811. DOI: 10.1016/j.envres.2023.116811

[33] Segura JL, Mancheno MJ, Zamora F. Covalent organic frameworks based on Schiff-base chemistry: Synthesis, properties and potential applications. Chemical Society Reviews. 2016;**45**:5635-5671. DOI: 10.1039/C5CS00878F

[34] Bargujar S, Ratnani S, Jain R. Recent advances in microwave assisted synthesis of Schiff base metal complexes. Inorganic Chemistry Communications. 2024;**162**:112250. DOI: 10.1016/j.inoche.2024.112250

[35] Pathan IR, Patel MK. A comprehensive review on the synthesis and applications of Schiff base ligand and metal complexes: A comparative study of conventional heating, microwave heating, and sonochemical methods. Inorganic Chemistry Communications. 2023;**158**(1):111464. DOI: 10.1016/j.inoche.2023.111464

[36] Das S, Das VK, Saikia L, Thakur AJ. Environment-friendly and solvent-free synthesis of symmetrical bis-imines under microwave irradiation. Green Chemistry Letters and Reviews. 2012;**5**:457-474

[37] Gabano E, Ravera M. Microwave-assisted synthesis: Can transition metal complexes take advantage of this "green" method? Molecules. 2022;**27**:4249

[38] Durairaj P, Maruthavanan T, Manjunathan S, Subashini S, Rokhum SL, Baskarf G. Microwave assisted synthesis, characterization and bioactivity evaluation of a cobalt (II) complex with a novel Schiff base ligand derived from phenylacetyl urea and salicylaldehyde. Journal of Molecular Structure. 2024;**1295**:136650. DOI: 10.1016/j.molstruc.2023.136650

[39] Nikpassand M, Fekri LZ, Sharafi S. An efficient and green synthesis of novel azo Schiff base and its complex under ultrasound irradiation. Oriental Journal of Chemistry. 2013;**29**:1041-1046

[40] Abd EL-Rahman NM, Saleh TS, Mady MF. Ultrasound assisted synthesis of some new 1, 3, 4-thiadiazole and bi (1, 3, 4-thiadiazole) derivatives incorporating pyrazolone moiety. Ultrasonics Sonochemistry. 2009;**16**:70-74

[41] Assiri MA, Ali TE, Alsolimani AK, Shati AA, Alfaifi MY, Elbehairi SEI. Ultrasound-assisted synthesis of novel Schiff bases from 3-(2-oxo-2H-chromen-3-yl)-1-(4-phenylthiazol-2-yl)-1H-pyrazole-4-carboxaldehyde and their cytotoxicity, apoptosis, cell cycle, molecular docking,

and ADMET profiling. Synthetic Communications. 2024;**54**(11):881-908. DOI: 10.1080/00397911.2024.2347501

[42] Leonardi M, Villacampa M, Menendez JC. Multicomponent mechanochemical synthesis. Chemical Science. 2018;**9**:2042-2064. DOI: 10.1039/C7SC05370C

[43] Cincic D, Kaitner B. Schiff base derived from 2-hydroxy-1-naphthaldehyde and liquid-assisted mechanochemical synthesis of its isostructural Cu(II) and Co(II) complexes. CrystEngComm. 2011;**13**:4351-4357. DOI: 10.1039/C0CE00421A

[44] Friscic T, Childs SL, Rizvi SAA, Jones W. The role of solvent in mechanochemical and sonochemical cocrystal formation: A solubility-based approach for predicting cocrystallisation outcome. CrystEngComm. 2009;**11**:418-426. DOI: 10.1039/B815174A

[45] James SL, Adams CJ, Bolm C, Braga D, Collier P, Friscic T, et al. Mechanochemistry: Opportunities for new and cleaner synthesis. Chemical Society Reviews. 2012;**41**:413-447. DOI: 10.1039/C1CS15171A

[46] Głowniak S, Szczesniak B, Choma J, Jaroniec M. Mechanochemistry: Toward green synthesis of metal–organic frameworks. Materials Today. 2021;**46**:109-124. DOI: 10.1016/j.mattod.2021.01.008

[47] Neelam VD, Neelam AKN. Solvent-free mechanochemical synthesis of organic compounds: Review article. Journal of Pharmacy Research. 2023;**1**(2):007-014

[48] Ismaeel M, Parveen B, Dogar SS, Aftab K, Abbas K, Munawar KS. Comparing green and conventional methods for Schiff base synthesis and unveiling environmental stability applications: A review. Journal of Coordination Chemistry. 2024;**77**(9-10):921-959. DOI: 10.1080/00958972.2024.2362344

[49] Shaw TE, Mathivathanan L, Jurca T. One-pot, one-step precatalysts through mechanochemistry. Organometallics. 2019;**38**(21):4066-4070. DOI: 10.1021/acs.organomet.9b00575

[50] Boruah JJ, Bhatt ZS, Nathani CR, Bambhaniya VJ, Guha AK, Das SP. Green synthesis of a vanadium(V) Schiff base complex by grinding method: Study on its catalytic and anti-bacterial activity. Journal of Coordination Chemistry. 2021;**74**(12):2055-2068. DOI: 10.1080/00958972.2021.1942861

[51] Hamidinasab M, Ahadi N, Bodaghifard MA, Brahmachari G. Sustainable and bio-based catalysts for multicomponent organic synthesis: An overview. Polycyclic Aromatic Compounds. 2023;**43**(6):5172-5226. DOI: 10.1080/10406638.2022.2097278

[52] Gulati S, Singh R, Sindhu J, Sangwan S. Eco-friendly preparations of heterocycles using fruit juices as catalysts: A review. Organic Preparations and Procedures International. 2020;**52**:381-395. DOI: 10.1080/00304948.2020.1773158

[53] Nagar S, Raizada S, Tripathee N. A review on various green methods for synthesis of Schiff base ligands and their metal complexes. Results in Chemistry. 2023;**6**:101153. DOI: 10.1016/j.rechem.2023.101153

[54] Elemike EE, Dare EO, Samuel ID, Onwuka JC. 2-Imino-(3,4-dimethoxybenzyl) ethanesulfonic acid Schiff base anchored silver nanocomplex mediated by sugarcane juice and their antibacterial activities. Journal of

Applied Research and Technology. 2016;**14**(1):38-46. DOI: 10.1016/j.jart.2015.12.001

[55] Alikhani A, Foroughifar N, Pasdar H. Lemon juice as a natural catalyse for synthesis of Shiff's base: A green chemistry approach. International Journal of Advanced Engineering Research and Science (IJAERS). 2018;**5**(2):61-65

[56] Ma'rufah L, Hanapi A, Ningsih R, Fasya AG. Synthesis of Schiff base compounds from vanillin and p-aminoacetophenone using lime juice as a natural acid catalyst and their utilization as corrosion inhibitors. In: International Conference on Engineering, Technology and Social Science (ICONETOS 2020). Dordrecht, The Netherlands: Atlantis Press; 2021. pp. 297-301

[57] De S, Jain A, Barman P. Recent advances in the catalytic applications of chiral Schiff-base ligands and metal complexes in asymmetric organic transformations. Chemistry Select. 2022;**7**:e202104334. DOI: doi.org/10.1002/slct.202104334

[58] Musikavanhu B, Liang Y, Xue Z, Feng L, Zhao L. Strategies for improving selectivity and sensitivity of Schiff base fluorescent chemosensors for toxic and heavy metals. Molecules. 2023;**28**:6960. DOI: 10.3390/molecules28196960

[59] Verma C, Quraishi MA. Recent progresses in Schiff bases as aqueous phase corrosion inhibitors: Design and applications. Coordination Chemistry Reviews. 2021;**446**:214105. DOI: 10.1016/j.ccr.2021.214105

[60] Abuamer K, Maihub A, El-Ajaily M, Etorki A, Abou-Krisha M, Almagani M. The role of aromatic Schiff bases in the dyes techniques. International Journal of Organic Chemistry. 2014;**4**:7-15. DOI: 10.4236/ijoc.2014.41002

[61] Zhang J, Xu L, Wong W-Y. Energy materials based on metal Schiff base complexes. Coordination Chemistry Reviews. 2018;**355**:180-198. DOI: 10.1016/j.ccr.2017.08.007

[62] Sun Y, Lu Y, Bian M, Yang Z, Ma X, Liu W. Pt (II) and Au (III) complexes containing Schiff-base ligands: A promising source for antitumor treatment. European Journal of Medicinal Chemistry. 2021;**211**:113098-113121

[63] Hassan AS, Awad HM, Magd-El-Din AA, Hafez TS. Synthesis and in vitro antitumor evaluation of novel Schiff bases. Medicinal Chemistry Research. 2018;**27**:915-927

[64] Chohan ZH, Sumrra SH, Youssoufi MH, Hadda TB. Metal based biologically active compounds: Design, synthesis, and antibacterial/antifungal/cytotoxic properties of triazole-derived Schiff bases and their oxovanadium (IV) complexes. European Journal of Medicinal Chemistry. 2010;**45**:2739-2747

[65] Kumara R, Seemab K, Singhc DK, Jaind P, Manave N, Gautamf B, et al. Synthesis, antibacterial and antifungal activities of Schiff base rare earth metal complexes: A review of recent work. Journal of Coordination Chemistry. 2023;**76**(9-10):1065-1093. DOI: 10.1080/00958972.2023.2231608

[66] Al-Amiery AA, Al-Majedy YK, Ibrahim HH, Al-Tamimi AA. Antioxidant, antimicrobial, and theoretical studies of the thiosemicarbazone derivative Schiff base 2-(2-imino-1-methylimidazolidin-4-ylidene) hydrazinecarbothioamide (IMHC). Organic and Medicinal Chemistry Letters. 2012;**2**:1-7

[67] Sandhu Q-U-A, Pervaiz M, Majid A, Younas U, Saeed Z, Ashraf A, et al. Review: Schiff base metal complexes as anti-inflammatory agents. Journal of Coordination Chemistry. 2023;**76**(9-10):1094-1118. DOI: 10.1080/00958972.2023.2226794

[68] Cheng LX, Tang JJ, Luo H, Jin XL, Dai F, Yang J, et al. Antioxidant and antiproliferative activities of hydroxyl-substituted Schiff bases. Bioorganic & Medicinal Chemistry Letters. 2010;**20**:2417-2420

[69] Jayaseelan P, Prasad S, Vedanayaki S, Rajavel R. Synthesis, characterization, anti-microbial, DNA binding and cleavage studies of Schiff base metal complexes. Arabian Journal of Chemistry. 2016;**9**:668-677

[70] Juyal VK, Pathak A, Panwar M, Thakuri SC, Prakash O, Agrwal A, et al. Schiff base metal complexes as a versatile catalyst: A review. Journal of Organometallic Chemistry. 2023;**999**:122825. DOI: 10.1016/j.jorganchem.2023.122825

[71] Naghipour A, Fakhri A. Heterogeneous Fe_3O_4@chitosan-Schiff base Pd nanocatalyst: Fabrication, characterization and application as highly efficient and magnetically-recoverable catalyst for Suzuki–Miyaura and Heck–Mizoroki C–C coupling reactions. Catalysis Communications. 2016;**73**:39-45. DOI: 10.1016/j.catcom.2015.10.002

[72] Kargar H, Fallah-Mehrjardi M, Behjatmanesh-Ardakani R, Munawar KS, Ashfaq M, Tahir MN. Selective oxidation of benzyl alcohols to benzaldehydes catalyzed by dioxomolybdenum Schiff base complex: Synthesis, spectral characterization, crystal structure, theoretical and computational studies. Transition Metal Chemistry. 2021;**46**:437-455. DOI: 10.1007/s11243-021-00460-w

[73] Arora Z, Eftemie D-I, Spinciu A, Maxim C, Hanganu A-M, Tudorache M, et al. Valmet chiral Schiff-baseligands and their copper(II) complexes as organo, homogeneous and heterogeneous catalysts for Henry, cyanosilylation and aldol coupling reactions. ChemCatChem. 2021;**13**:4634-4644

[74] Mondal K, Mistri S. Schiff base based metal complexes: A review of their catalytic activity on Aldol and Henry reaction. Comments on Inorganic Chemistry. 2022;**43**(2):77-105. DOI: 10.1080/02603594.2022.2094919

[75] Taha A, Farooq N, Singh N, Hashmi AA. Recent developments in Schiff base centered optical and chemical sensors for metal ion recognition. Journal of Molecular Liquids. 2024;**401**:124678. DOI: 10.1016/j.molliq.2024.124678

[76] Du H, Maimaitiyiming X, Luo Y, Obolda A. A highly sensitive ammonia gas sensor based on non-covalent functionalized single-walled carbon nanotubes with Schiff base polyphenylene polymer. Sensors and Actuators B: Chemical. 2023;**394**:134426. DOI: 10.1016/j.snb.2023.134426

[77] Zhang M, Gao L, Zhao X, Duan Y, Liao Y, Han T. Fabricating a hydroxynaphthalene benzophenone Schiff base into a wearable fluorescent sensor for point-of-care sensing of volatile organic compounds. Dyes and Pigments. 2022;**205**:110561. DOI: 10.1016/j.dyepig.2022.110561

[78] Silva RC, Canisares FSM, Mutti AMG, Pires AM, Lima SAM. Small Schiff base molecules derived

from salicylaldehyde as colorimetric and fluorescent neutral-to-basic pH sensors. Dyes and Pigments. 2023;**213**:111191. DOI: 10.1016/j.dyepig.2023.111191

[79] Erbaş A, Dikim S, Arslan F, Bodur OC, Arslan S, Özdemir F, et al. Schiff bases from 4-aminoantipyrine: Investigation of their in silico, antimicrobial, and anticancer effects and their use in glucose biosensor design. Bioinorganic Chemistry and Applications. 2025;**2025**(18):2786064. DOI: 10.1155/bca/2786064

[80] Munir A, Bozal-Palabiyik B, Eren G, Shah A, Ali S, Uddin N, et al. Electrochemical biosensor design with multi-walled carbon nanotube to display DNA-Schiff base interaction. Electroanalysis. 2021;**33**:1761

[81] Vijayakumar S, Raja L, Venkatesan S, Lin M-C, Vediappen P. A highly selective Schiff base based chemodosimeter for the detection of perfluorooctanoic acid by optical biosensor. Journal of Fluorescence. 2024;**34**:787-794. DOI: 10.1007/s10895-023-03298-w

[82] Kumar A, Virender, Saini M, Mohan B, Shayoraj, Kamboj M. Colorimetric and fluorescent Schiff base sensors for trace detection of pollutants and biologically significant cations: A review (2010-2021). Microchemical Journal. 2022;**181**:107798. DOI: 10.1016/j.microc.2022.107798

[83] Alipour S, Ghamari P, Bagherzade AG. Comparative study of antioxidant and antimicrobial activity of berberine-derived Schiff bases, nitro-berberine and amino-berberine. Heliyon. 2023;**9**(12):e22783. DOI: 10.1016/j.heliyon.2023.e22783

[84] Kargar PG, Moodi Z, Bagherzade G, Nikoomanesh F. Synthesis, characterization, and antimicrobial activity of metal complexes derived from Schiff base of quercetin extracted from Origanum vulgare L. Materials Chemistry and Physics. 2024;**317**:129207. DOI: 10.1016/j.matchemphys.2024.129207

[85] Hajrezaie M, Paydar M, Moghadamtousi SZ, Hassandarvish P, Gwaram NS, Zahedifard M, et al. A Schiff base-derived copper (II) complex is a potent inducer of apoptosis in colon cancer cells by activating the intrinsic pathway. Scientific World Journal. 2014;**2014**:540463. DOI: 10.1155/2014/540463

[86] Bazsefidpar P, Zolghadri S, Nikpoor AR, Eftekhar E, Jahromi MZ. Anti- proliferative impact of three Schiff base platinum (II) complexes against human breast cancer cell line. Journal of Research in Pharmacy. 2022;**26**(6):1665-1675. DOI: 10.29228/jrp.257

[87] Abu-Dief AM, El-khatib RM, Aljohani FS, Alzahrani SO, Mahran A, Khalifa ME, et al. Synthesis and intensive characterization for novel Zn (II), Pd (II), Cr (III) and VO(II)-Schiff base complexes; DNA-interaction, DFT, drug- likeness and molecular docking studies. Journal of Molecular Structure. 2021;**1242**:130693. DOI: 10.1016/j.molstruc.2021.130693

[88] Brodowska K, Sykuła A, Garribba E, Łodyga-Chruscinska E, Sójka M. Naringenin Schiff base: Antioxidant activity, acid–base profile, and interactions with DNA. Transition Metal Chemistry. 2016;**41**(2):179-189. DOI: 10.1007/s11243-015-0010-7

[89] Suyambulingam JK, Karvembu R, Bhuvanesh NSP, Enoch IVMV, Selvakumar PM, Premnath D, et al. Synthesis, structure, biological/chemosensor evaluation and molecular

docking studies of aminobenzothiazole schiff bases. Journal of Adhesion Science and Technology. 2020;**34**(23):1-23. DOI: 10.1080/01694243.2020.1775032

[90] Ragi K, Kakkassery JT, Raphael VP, Johnson R, Vidhya TK. In-vitro antibacterial and in-silico docking studies of two Schiff bases on staphylococcus aureus and its target proteins. Future Journal of Pharmaceutical Sciences. 2021;7. Article no: 78. DOI: 10.1186/s43094-021-00225-3

[91] Mishra VR, Ghanavatkar CW, Mali SN, Chaudhari HK, Sekar N. Schiff base clubbed benzothiazole: Synthesis, potent antimicrobial and MCF-7 anticancer activity, DNA cleavage and computational studies. Journal of Biomolecular Structure and Dynamics. 2019;**38**(6):1772-1785. DOI: 10.1080/07391102.2019.1621213

[92] Kizilkaya H, Dag B, Aral T, Genc N, Erenle R. Synthesis, characterization and antioxidant activity of heterocyclic schiff bases. Journal of the Chinese Chemical Society. 2020;**67**(9):1696-1701. DOI: 10.1002/jccs.202000161

[93] Adebayo JO, Tijjani H, Adegunloye AP, Ishola AA, Balogun EA, Malomo SO. Enhancing the antimalarial activity of artesunate. Parasitology Research. 2020;**119**:2749-2764. DOI: 10.1007/s00436-020-06786-1

[94] Azam M, Wabaidur SM, Alam M, Khan Z, Alanazi IO, Al-Resayes SI, et al. Synthesis, characterization, cytotoxicity, and molecular docking studies of ampyrone-based transition metal complexes. Transition Metal Chemistry. 2021;**46**:65-71. DOI: 10.1007/s11243-020-00422-8

[95] Shukla SN, Gaur P, Raidas ML, Chaurasia B. Tailored synthesis of unsymmetrical tetradentate ONNO Schiff base complexes of Fe (III), Co (II) and Ni (II): Spectroscopic characterization, DFT optimization, oxygen-binding study, antibacterial and anticorrosion activity. Journal of Molecular Structure. 2020;**1202**:127362-127379

[96] Abolhassan MR, Divsalar A, Badalkhani-khamseh F, Kheiripour N, Eslami- Moghadam M, Mirzaei H. Protein binding and anticancer activity of two newly synthesized Schiff base platinum (II) complexes: A theoretical and experimental study. Journal of Molecular Structure. 2023;**1289**:135917-135931

[97] Jayendran M, Begum PS, Kurup MP. Structural, spectral and biological investigations on Cu (II) and Zn (II) complexes derived from NNO donor tridentate Schiff base: Crystal structure of a 1D Cu (II) coordination polymer. Journal of Molecular Structure. 2020;**1206**:127682-127724

[98] Martins DO, Souza RA, Freire MC, de Moraes Roso Mesquita NC, Santos IA, de Oliveira DM, et al. Insights into the role of the cobalt (III)-thiosemicarbazone complex as a potential inhibitor of the Chikungunya virus nsP4. Journal of Biological Inorganic Chemistry. 2023;**28**:101-115

[99] Naureen B, Miana GA, Shahid K, Asghar M, Tanveer S, Sarwar A. Iron (III) and zinc (II) monodentate Schiff base metal complexes: Synthesis, characterisation and biological activities. Journal of Molecular Structure. 2021;**1231**:129946-129958

[100] Khan G, Basirun WJ, Kazi SN, Ahmed P, Magaji L, Ahmed SM, et al. Electrochemical investigation on the corrosion inhibition of mild steel by Quinazoline Schiff base compounds in hydrochloric acid solution. Journal

of Colloid and Interface Science. 2017;**502**:134-145

[101] Gürten AA, Keles H, Bayol E, Kandemirli F. The effect of temperature and concentration on the inhibition of acid corrosion of carbon steel by newly synthesized Schiff base. Journal of Industrial and Engineering Chemistry. 2015;**27**:68-78

Section 2

Reactions and Theoretical Approaches of Schiff Bases

Chapter 3

Direct Catalytic Asymmetric Aldol Reactions of Glycine Schiff Bases to Access β-Hydroxy-α-Amino Esters

Toshifumi Takeuchi and Masakatsu Shibasaki

Abstract

β-Hydroxy-α-amino acids and their esters represent essential structural motifs found in a wide range of biologically active natural products and pharmaceutical agents. These compounds, characterized by vicinal aminoalcohol functionalities, function as versatile chiral synthons for the synthesis of complex molecules. Among the various synthetic methods available, the direct catalytic asymmetric aldol reaction of glycine Schiff bases with aldehydes has emerged as a highly efficient, one-step strategy for assembling the 1,2-aminoalcohol scaffold. This review summarizes recent developments in both organocatalytic and metal-based catalytic strategies for the synthesis of β-hydroxy-α-amino esters. Particular attention is given to the scope of these reactions, their diastereo- and enantioselectivities, and the mechanistic insights they offer. Current limitations are critically examined, and future prospects are proposed to expand substrate scope and simplify the synthetic process.

Keywords: β-Hydroxy-α-amino acid, β-Hydroxy-α-amino ester, aminoalcohol, direct catalytic asymmetric aldol reaction, organocatalyst, metal-based catalyst, glycine Schiff base

1. Introduction

Optically active β-hydroxy-α-amino acids and their esters constitute a valuable class of chiral building blocks widely present in bioactive natural products and pharmaceutical compounds [1–4]. These structural motifs are present in glycopeptide antibiotics such as vancomycin (**Figure 1**) [5–8]. Additional complex metabolites incorporating β-hydroxy-α-amino acid units include nucleoside antibiotics such as polyoxins [9, 10], and cyclic peptides like GE3 [11–13], hypeptin [14], and lysobactin (also known as katanosin B) [15, 16], which display notable antimicrobial activity and/or cytotoxic effects against human tumor cell lines. Beyond their biological significance, β-hydroxy-α-amino esters serve as versatile chiral building blocks [17–26] and synthons. Classical chemical transformations can convert these compounds into other valuable scaffolds; for instance,

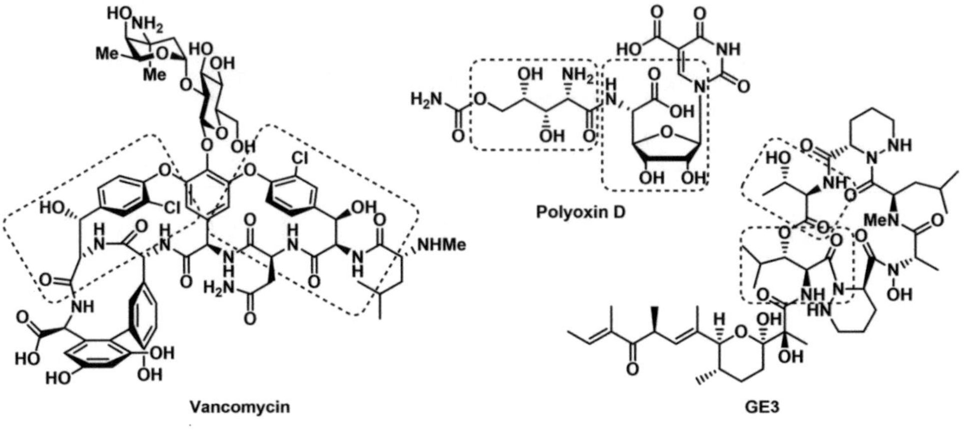

Figure 1.
Structures of vancomycin, GE3, and polyoxin D.

intramolecular cyclizations yield β-lactams [27], while dehydration or cyclization reactions can produce aziridine-2-carboxylates [28]. Consequently, the enantioselective synthesis of β-hydroxy-α-amino esters has garnered considerable attention in both synthetic and medicinal chemistry [1–4, 29–32].

Traditional approaches have often relied on stoichiometric chiral auxiliaries [33–37], which tend to be labor-intensive and generate considerable waste. In contrast, the direct aldol reaction between a glycine-derived donor and an aldehyde provides a highly atom-economical route for constructing β-hydroxy-α-amino ester frameworks in a single step, simultaneously creating the adjacent stereocenters [38–43]. Among these methods, the direct catalytic asymmetric aldol addition of glycine Schiff bases—such as benzophenone imines of glycine esters [44]—to aldehydes has emerged as particularly efficient. This one-step transformation simultaneously forms a new C–C bond and two stereogenic centers, effectively generating the 1,2-aminoalcohol framework. Numerous stereoselective strategies for the synthesis of β-hydroxy-α-amino esters have been developed. This review focuses on diastereo- and enantioselective approaches utilizing the direct aldol reaction of glycine Schiff bases with aldehydes.

2. Glycine Schiff bases: History and reactivity

The classical synthesis of α-amino acids can be traced back to Sørensen's method (1903), which involved the deprotonation and alkylation of diethyl acetamidomalonate, followed by hydrolysis and in situ decarboxylation to yield racemic amino acid derivatives (**Figure 2**). A significant advancement occurred in 1976 when Yamada introduced the alkylation of chiral glycine Schiff bases with auxiliaries derived from natural products such as α-pinene [33]. Based on this concept, Belokon later expanded the methodology by employing glycine imine equivalents coordinated to chiral copper or nickel complexes, thereby extending the scope and utility of enantioselective transformations [36, 37]. In 1976, Stork reported the alkylation and Michel addition of glycine Schiff base to yield racemic amino acids [45]. In 1978, O'Donnell reported the alkylation of glycine Schiff base or Schiff base of aminoacetonitrile using achiral phase-transfer catalyst, which also produced racemic amino acids [46, 47]. In 1989,

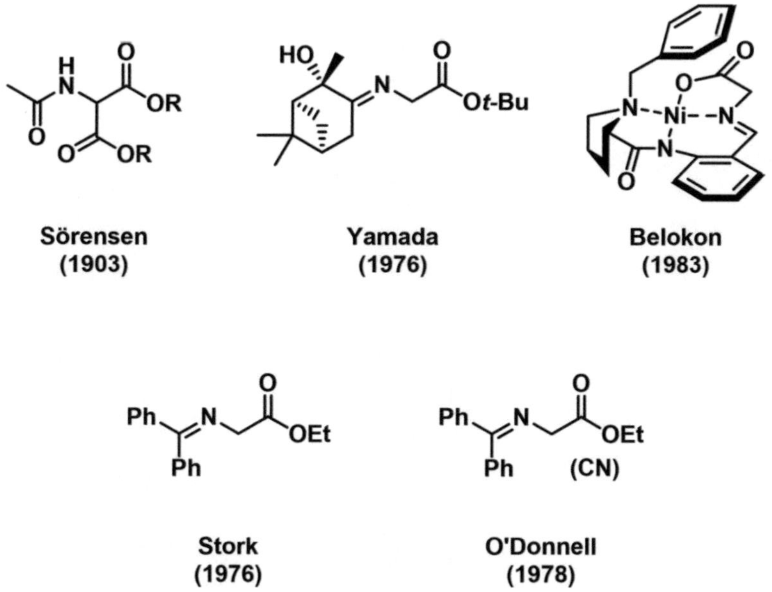

Figure 2.
Glycine equivalents as pronucleophiles.

O'Donnell refined this strategy to enable the enantioselective phase-transfer catalytic alkylation of glycine Schiff base, affording chiral α-amino acids [48].

The utilization of glycine Schiff bases as key substrates in asymmetric aldol reactions stems from their unique combination of stability, reactivity, and tunable selectivity. Schiff bases derived from glycine esters offer enhanced nucleophilicity due to the increased acidity of the α-proton, allowing facile enolization under mild conditions (pKa ≈ 18.7) [49]. Compared to traditional enolate precursors or preformed metal enolates, glycine Schiff bases provide improved control over stereochemistry, significantly reduced reaction complexity, and high atom economy. Additionally, the Schiff base functionality serves as a removable protecting group, enabling straightforward downstream processing to free β-hydroxy-α-amino acids [44]. Thus, their structural versatility and synthetic convenience position glycine Schiff bases as advantageous substrates in direct asymmetric aldol methodologies.

3. Direct asymmetric aldol synthesis strategies

3.1 Direct asymmetric aldol reactions

The aldol reaction, first discovered by Wurtz in 1872, remains one of the most important transformations in organic synthesis because of its ability to form C–C bonds efficiently and generate stereocenters [50–52]. The direct aldol reaction represents a foundational transformation in asymmetric synthesis, enabling the stereoselective synthesis of β-hydroxy carbonyl compounds, including β-hydroxy-α-amino esters, in a single step via the use of a catalytic amount of both base and chiral ligands (or these functional complexes). Traditionally, aldol products have been obtained using preformed enolates or enolate equivalents incorporating chiral auxiliaries;

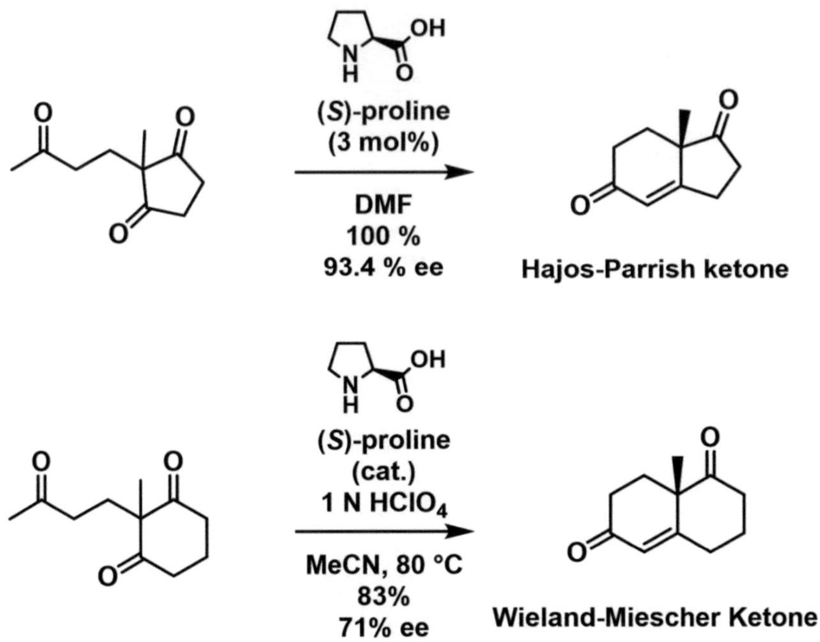

Figure 3.
Hajos-Parrish-Eder-Sauer-Wiechert reaction.

however, these approaches require stoichiometric quantities of reagents and involve multiple steps to remove the chiral auxiliaries. In contrast, direct asymmetric aldol reactions utilize catalytic methods to activate simple donors and acceptors, thereby eliminating the need for intermediate derivatization.

The Hajos-Parrish-Eder-Sauer-Wiechert reaction was first reported independently by Eder, Sauer, and Wiechert [38], and Hajos and Parrish [39]. The reaction utilized proline as a chiral small-molecule catalyst to induce chirality in the intramolecular aldol reactions of triketones (**Figure 3**). This pioneering work demonstrated that small organic molecules could efficiently catalyze asymmetric carbon–carbon bond-forming reactions via enamine intermediates, avoiding the need for stoichiometric chiral auxiliaries.

3.2 Lanthanum-lithium-BINOL (LLB) complexes catalyze direct asymmetric aldol reactions

In 1997, Shibasaki and co-workers reported one of the earliest examples of intermolecular direct catalytic asymmetric aldol reactions employing lanthanum-lithium-2,2'-dihydroxy-1,1'-binaphthyl (BINOL) (LLB, $Li_3[La(R\text{-BINOL})_3]$) complexes [40–42]. These catalysts effectively coordinate both donor and acceptor carbonyl compounds, thereby facilitating the enantioselective aldol reaction under mild conditions. Mechanistically, the LLB complex establishes a well-defined chiral environment around the activated substrates, enhancing facial selectivity and precisely controlling the formation of stereocenters. This strategy has been broadly utilized for the synthesis of β-hydroxy carbonyl compounds [53] and has also been extended to the preparation of β-hydroxy-α-amino esters (**Figure 4**) [54] (see Section 4.2.1).

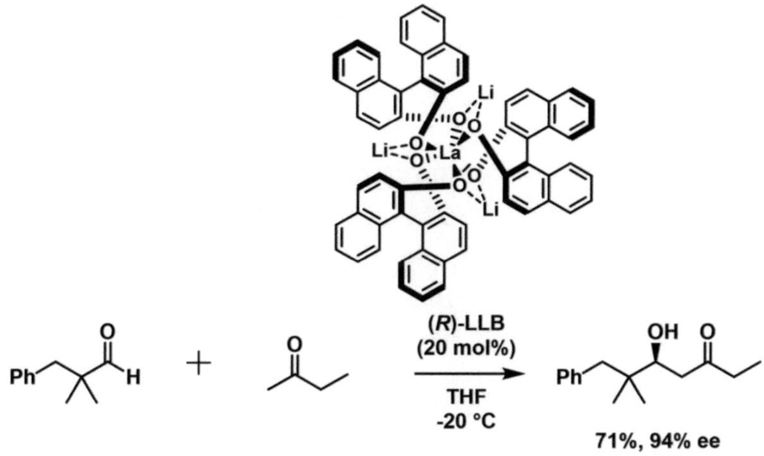

Figure 4.
Lanthanum-lithium-BINOL complexes catalyze direct asymmetric aldol reactions.

3.3 Proline-catalyzed direct asymmetric aldol reactions

Following insights from the Hajos-Parrish-Eder-Sauer-Wiechert reaction [28, 29], in 2000, List and Barbas reported that the intermolecular direct asymmetric aldol reaction is catalyzed by (*S*)-proline [43]. This reaction enables the condensation of unmodified ketones with aldehydes to form *syn*-β-hydroxy ketones with high enantio- and diastereoselectivity. This finding opened new avenues for the use of proline and its derivatives in direct asymmetric synthesis (**Figure 5**).

3.4 Effect of Schiff base structure on reactivity and selectivity

The structure and substitution patterns of glycine Schiff bases critically influence the efficiency and selectivity of direct aldol reactions. Electron-withdrawing substituents on the Schiff base aromatic ring generally enhance the acidity of the α-proton, facilitating enolization and improving reaction kinetics and stereoselectivity [44]. Kobayashi reported glycine Schiff base with 9-fluorene imine allowed a strong activation of the α-proton through stabilization of the 14 π electron system on 9-fluorenylidene group. This glycine Schiff base enabled direct aldol reaction with several aliphatic aldehydes to provide racemic *anti*-β-hydroxy-α-amino esters in good yields (74–91%) and diastereoselectivities (up to 95:5 dr (*anti*/*syn*)) using magnesium 5,6,7,8-tetrahydro-1-naphthyltosylamide as

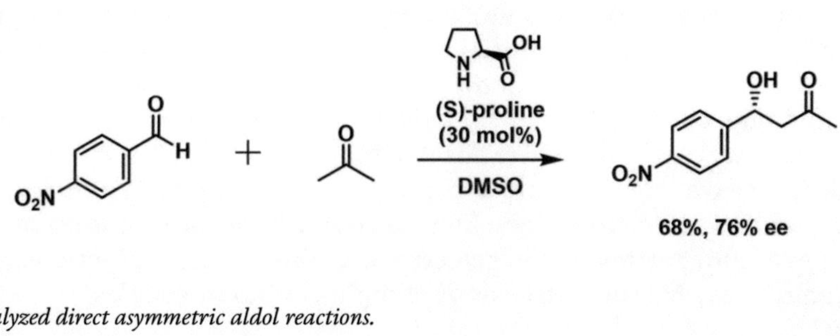

Figure 5.
Proline-catalyzed direct asymmetric aldol reactions.

a base (2 mol%) [55]. This intrinsic structural flexibility underscores the broader value of Schiff base chemistry, extending beyond aldol reactions to other catalytic asymmetric transformations.

4. Synthesis of β-hydroxy-α-amino esters through direct asymmetric aldol reactions of glycine Schiff bases

This section reviews the synthesis of β-hydroxy-α-amino esters from glycine Schiff bases through two major catalytic platforms employed in direct aldol reactions: organocatalytic and metal-based catalytic methods.

4.1 Organocatalytic direct aldol reactions

4.1.1 O'Donnell cinchoninium chloride phase-transfer catalyst

Gasparski and Miller marked a significant milestone in the field of asymmetric aldol reactions [56] utilizing the O'Donnell cinchoninium chloride phase-transfer catalyst [48, 57]. This catalyst facilitated the first direct asymmetric aldol reaction between the Schiff base of glycine *tert*-butyl ester and aldehydes such as hydrocinnamaldehyde or heptanal, employing 20 mol% of sodium hydroxide as the base. Although this approach successfully produced *anti*-β-hydroxy-α-amino esters, it exhibited several drawbacks, including low yields, limited diastereoselectivity, and poor enantiomeric excess. The reaction offered limited stereochemical control of 3:1 dr (*anti*/*syn*). Additionally, the enantiomeric excess was minimal, further decreasing the method's utility in asymmetric synthesis. The reaction efficiency was also compromised, with yields often less than 50% due to competing self-condensations and various side reactions. These limitations collectively restricted the practical application of this pioneering catalytic system, prompting further research to develop more efficient and stereoselective methods for asymmetric aldol reactions.

4.1.2 Cinchoninium salt catalysis

Castle's discovery of the *N*-2,3,4-trifluorobenzyl-substituted hydrocinchonidine derivative of the Park-Jew catalyst marked a significant advancement in cinchoninium salt catalysis (**Figure 6**) [58]. This catalyst, which uses 2.5 equivalents of a phosphazene base (*tert*-buthyliminotri(pyrrolidino)phosphorane (BTTP)), promoted aldol addition between the Schiff base of glycine *tert*-butyl ester and aliphatic aldehydes. The initial results were promising, with enantioselectivity reaching up to 83% ee for the *syn* aldol adduct, despite moderate diastereoselectivities (1,1 dr). Building upon this initial success, Castle et al. further refined their approach [59]. They developed a modified version of the Park-Jew catalyst that demonstrated improved performance. This improved catalyst achieved enhanced diastereoselectivity, with *syn*/*anti* ratios reaching up to 2.6:1. Furthermore, the enantioselectivity was significantly increased, attaining up to 91% ee for the *syn* product. These advancements in cinchoninium salt catalysis represent important progress toward more efficient and selective asymmetric aldol reactions, potentially paving the way for the synthesis of complex chiral molecules.

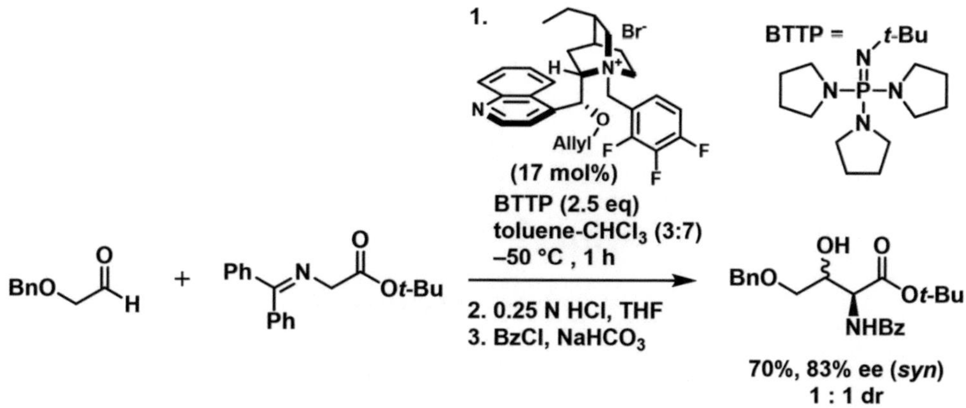

Figure 6.
Synthesis of β-hydroxy-α-amino esters using a cinchoninium salt catalyst.

4.1.3 N-Spiro binaphthyl-based ammonium catalysts

The development of chiral quaternary ammonium salts based on *N*-spiro binaphthyl scaffolds by Maruoka and Ooi marked a significant advancement in asymmetric aldol reactions (**Figure** 7) [60, 61]. These phase-transfer catalysts enabled highly efficient *anti*-selective aldol additions of glycine Schiff bases, delivering outstanding results in terms of yield, diastereoselectivity, and enantioselectivity. For instance, the reaction of hydrocinnamaldehyde with the Schiff base of glycine *tert*-butyl ester, catalyzed by a chiral quaternary ammonium salt (2 mol%) in the presence of 1% aqueous sodium hydroxide (15 mol%) in toluene at 0°C for 1.5 h, produced the *anti*-aldol adduct with remarkable selectivity (96:4 dr, 98% ee) and a yield of 82% following cleavage of the benzophenoneimine moiety with hydrochloric acid. This method performs well with aliphatic aldehydes, including both α-branched and linear substrates. However, its efficacy is decreased when applied to aromatic aldehydes such

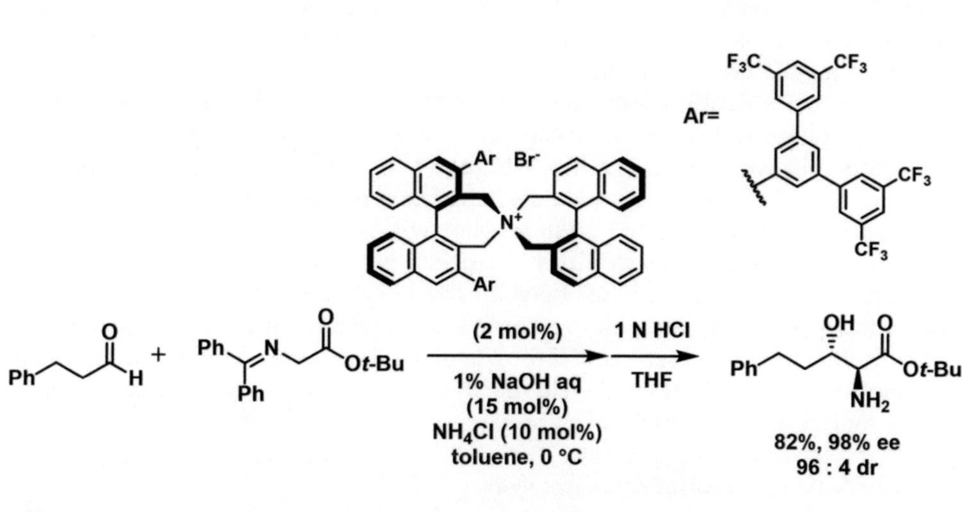

Figure 7.
Synthesis of anti-*β-hydroxy-α-amino esters using chiral quaternary ammonium salts.*

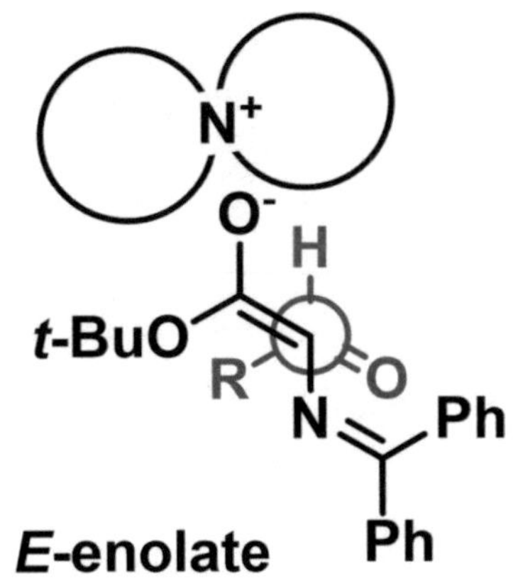

Figure 8.
Plausible transition state model of the direct aldol reaction catalyzed by chiral quaternary ammonium salts.

as benzaldehyde, where both yield and selectivity are notably reduced. Despite this limitation, the Maruoka-Ooi protocol remains a valuable and versatile approach for the synthesis of complex chiral molecules and pharmaceuticals.

The observed high *anti*-selectivity in this reaction can be attributed to the selective formation of the *E*-enolate, as depicted in **Figure 8**. The formation of the *anti*-aldol adduct from the *E*-enolate can be explained by significant steric repulsion caused by the chiral quaternary ammonium cation. This repulsion appears sufficient to overcome the gauche interactions between the aldehyde substituent (R) and both the 2-imino moiety and the *tert*-butoxy group, as illustrated in **Figure 8**. The proposed transition state model is consistent with the observed enhancement in diastereoselectivity. Based on the absolute configuration of the product, it is inferred that the *re*-face of the enolate is effectively shielded by the chiral ammonium cation, thereby restricting the approach of the aldehyde to the *si*-face. This facial selectivity leads to the preferential formation of the (2*S*,3*S*)-β-hydroxy-α-amino ester as the major product. The model offers a plausible explanation for the stereochemical outcome of the reaction and underscores the critical role of steric effects in controlling the selectivity of this asymmetric aldol reaction.

Maruoka subsequently advanced this methodology by developing a newly optimized chiral phase-transfer catalyst capable of selectively promoting the formation of *syn*-adducts in asymmetric synthesis (**Figure 9**) [62, 63]. Utilizing this refined catalyst system, the reaction between the Schiff base of glycine *tert*-butyl ester and various aldehydes afforded good diastereoselectivities (up to 85:15 dr (*syn*/*anti*)) and relatively high enantioselectivities (up to 94% ee). For example, the reaction of 4-pentenal with the Schiff base of glycine *tert*-butyl ester, catalyzed by 1 mol% of the chiral quaternary ammonium catalyst in the presence of 1% aqueous sodium hydroxide (1 equiv.) in toluene at 0°C for 2 h, yielded *syn*-aldol adducts with adequate selectivity (85:15 dr, 93% ee) and yield (68%) after the cleavage of the benzophenoneimine moiety using hydrochloric acid.

4.1.4 Bifunctional urea-based organocatalysts

Palomo's study highlights the strategic development of a bifunctional Brønsted base organocatalyst designed to simultaneously activate both the glycine imine

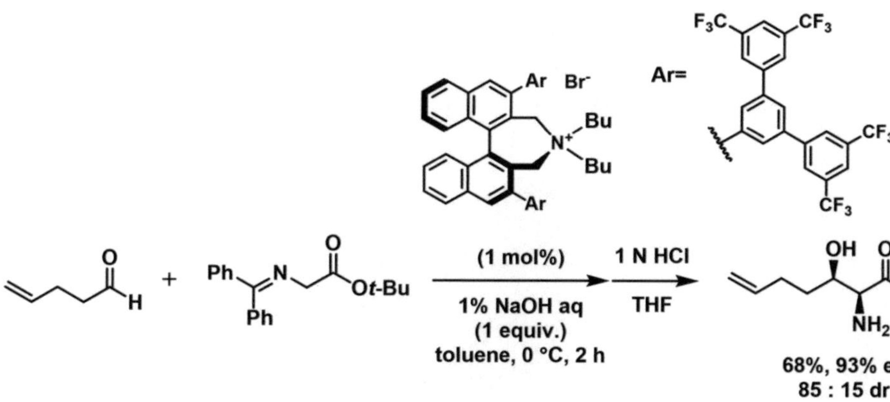

Figure 9.
Synthesis of syn*-β-hydroxy-α-amino esters using chiral quaternary ammonium salts.*

(through urea hydrogen bonding) and the aldehyde (via deprotonation from a quaternary ammonium) [64]. This dual activation mechanism effectively promotes the formation of *syn*-β-hydroxy-α-amino amides with high diastereo- and enantioselectivity. The catalytic system gives *syn*-β-hydroxy-α-amino amides in good yields and selectivities (63–82% yield, up to >98:2 dr (*syn*/*anti*), 85–94% ee (*syn*)) (**Figure 10**). A representative example involves the reaction between a glycine Schiff base *o*-nitroanilide and hydrocinnamaldehyde in the presence of 20 mol% of the organocatalyst in dichloromethane at 0°C for 64 h, yielding the *syn*-aldol adduct in 77% yield with exceptional diastereo- and enantioselectivity (>98:2 dr, 94% ee), followed by a one-pot reductive workup using $NaBH_3CN$ and AcOH. While the methodology has proven effective for linear aliphatic aldehydes, it has shown limited applicability for aromatic and α-branched aliphatic aldehydes because of its ability to undergo a retro-aldol reaction, elimination, and/or epimerization.

The described reaction mechanism involves a glycine Schiff base *o*-nitroanilide and an organocatalyst containing urea and piperidine moieties. The nitro group on the substrate plays a crucial role in activating the imine moiety through interaction with the amide proton, which in turn controls the enolate geometry, favoring the *E*-enolate configuration. This activation also increases the α-acidity of the glycine amide, increasing its susceptibility to deprotonation. The urea moiety of the

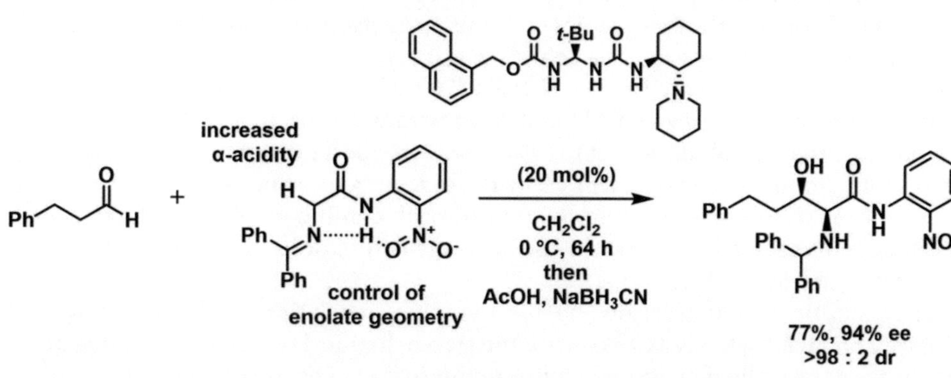

Figure 10.
Synthesis of β-hydroxy-α-amino esters promoted by a urea-based organocatalyst.

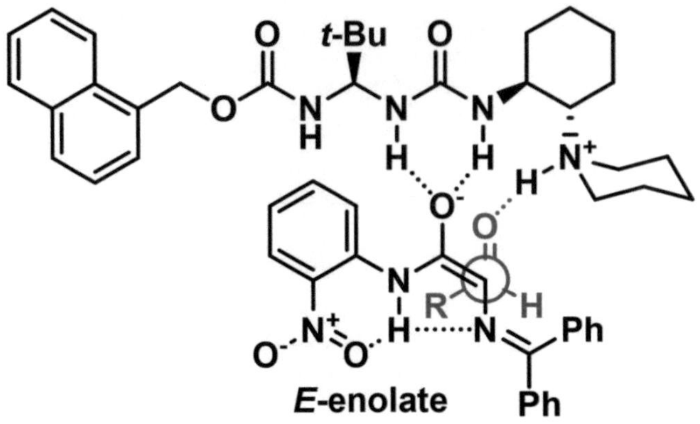

Figure 11.
Plausible model accounting for the syn*-selectivity.*

organocatalyst engages in hydrogen bonding with the carbonyl group of the glycine Schiff base amide, thereby enhancing its α-acidity. Concurrently, the piperidine amine of the organocatalyst deprotonates the α-position of the glycine amide, generating the corresponding *E*-enolate. The ammonium proton of the piperidine subsequently activates the aldehyde substrate through hydrogen bonding, guiding its approach from the *si*-face of the enolate. This coordinated sequence of interactions and activations facilitates the stereoselective formation of the (2*S*,3*R*)-β-hydroxy-α-amino amide as the major product, underscoring the precise stereochemical control achieved in this organocatalytic aldol reaction (**Figure 11**).

4.2 Metal-based catalytic systems

4.2.1 $Li_3[La(S\text{-}BINOL)_3]$, a heterobimetallic catalyst

The $Li_3[La(S\text{-}BINOL)_3]$, a heterobimetallic catalyst, when employed with glycine Schiff bases and aldehydes, afforded moderately *anti*-selective β-hydroxy-α-amino esters (**Figure 12**) [54]. In this system, the lanthanum(III) center functions as a hard Lewis acid, coordinating with the aldehyde electrophile to enhance its reactivity. Simultaneously, the lithium alkoxide acts as a Brønsted base, generating the glycine imine enolate. This dual activation strategy—combining hard Lewis acid and Brønsted base catalysis—enables efficient and selective transformations. For instance, the reaction between the Schiff base of glycine *tert*-butyl ester and pivalaldehyde, catalyzed by $Li_3[La(S\text{-}BINOL)_3]$ (20 mol%) in the presence of lithium hydroxide (18 mol%) and water (22 mol%) in tetrahydrofuran (THF) at −50°C for 23 h, yielded the *anti*-aldol adduct in 71% yield, with 86:14 dr (*anti*/*syn*) and 76% ee for the *anti*-product following removal of the benzophenone moiety using citric acid. The operational simplicity of this one-pot protocol, combined with its effectiveness, establishes this catalytic system as a significant benchmark in the development of direct asymmetric aldol reactions.

The catalytic system described operates via a non-chelation mechanism, relying primarily on facial selectivity to control the stereochemical outcome. The catalyst establishes a sterically demanding environment that shields the *re*-face of the aldehyde, thereby directing nucleophilic attack from the *si*-face. This selective shielding influences the approach of the enolate, resulting in the observed stereochemical

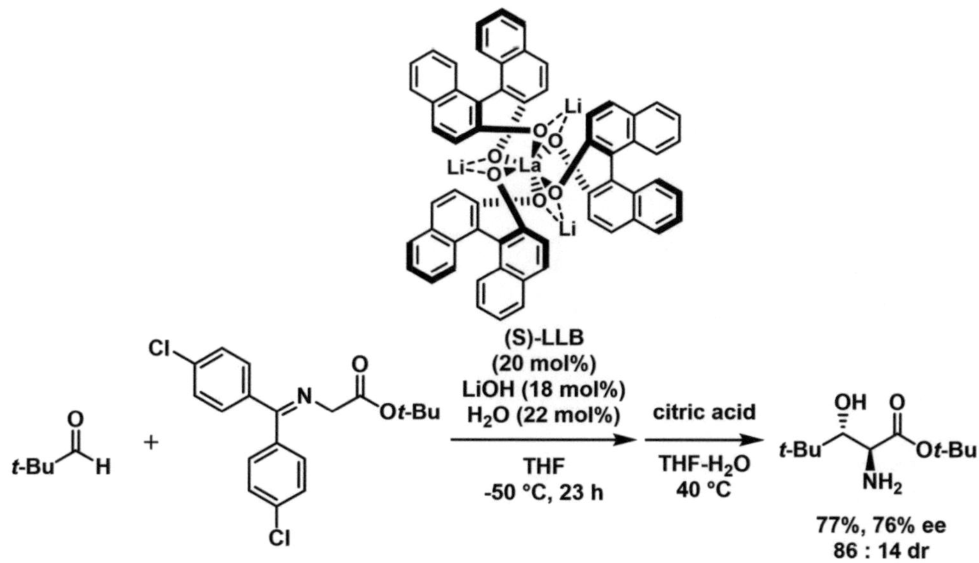

Figure 12.
Synthesis of β-hydroxy-α-amino esters catalyzed by $Li_3[La(S\text{-}BINOL)_3]$.

preference during product formation. A possible explanation for the moderate selectivity observed is the concurrent participation of both *E*- and *Z*-enolate geometries in the reaction. Although the proposed transition state model favors a *Z*-enolate, the involvement of an *E*-enolate cannot be definitively excluded. The coexistence of both enolate geometries could give rise to multiple transition states, potentially leading to the formation of diverse stereoisomeric products (**Figure 13**).

4.2.2 Zinc-ProPhenol catalysts

Trost's group made a significant breakthrough in *syn*-selective aldol addition reactions by developing a dinuclear zinc-ProPhenol catalytic system (**Figure 14**) [65, 66]. This catalyst, prepared from diethylzinc (Et_2Zn) and a ProPhenol ligand, effectively catalyzed the addition of an aldol to the Schiff base of glycine methyl or *tert*-butyl ester with various aldehydes at room temperature. The reaction produced *syn*-β-hydroxy-α-amino esters with high yields, diastereoselectivities, and enantioselectivities (70–96%, up to 19:1 dr (*syn*/*anti*), 71–95% ee (*syn*)). For example, the use of a Schiff base of glycine *tert*-butyl ester and 2,2-diethoxy-4-pentenal with a ProPhenol ligand (10 mol%), Et_2Zn (20 mol%), and 4A molecular sieves in toluene resulted in excellent outcomes (90% yield, 19:1 dr (*syn*/*anti*), 94% ee (syn)) after a reductive reaction with $NaBH_3CN$ and AcOH.

The zinc-ProPhenol catalytic system demonstrated a broad substrate scope, particularly with α-branched and α-quaternary aliphatic aldehydes. It consistently delivered high yields, excellent diastereoselectivities, and high enantiomeric excesses for these substrates. The system also proved effective with α-chiral aldehydes, producing single diastereomers in high yields. However, some limitations were noted with aromatic aldehydes such as benzaldehyde and simple linear aliphatic aldehydes like hydrocinnamaldehyde, where only moderate enantiomeric excesses were obtained. Despite these minor constraints, Trost's dinuclear zinc-ProPhenol catalyst represents a significant advancement in *syn*-selective aldol addition reactions, providing

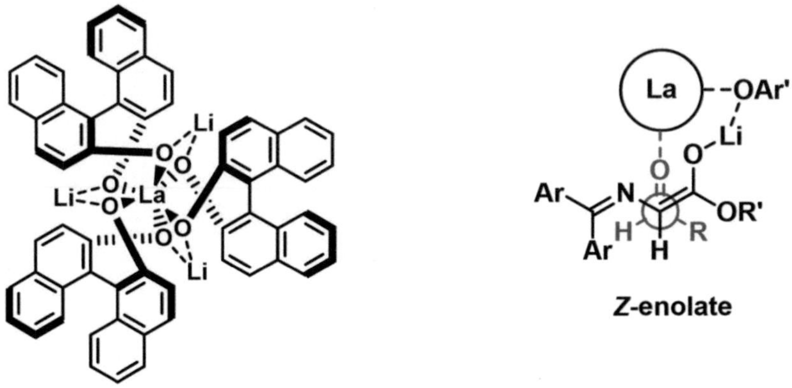

Figure 13.
Working model of the LLB-catalyzed direct aldol reaction.

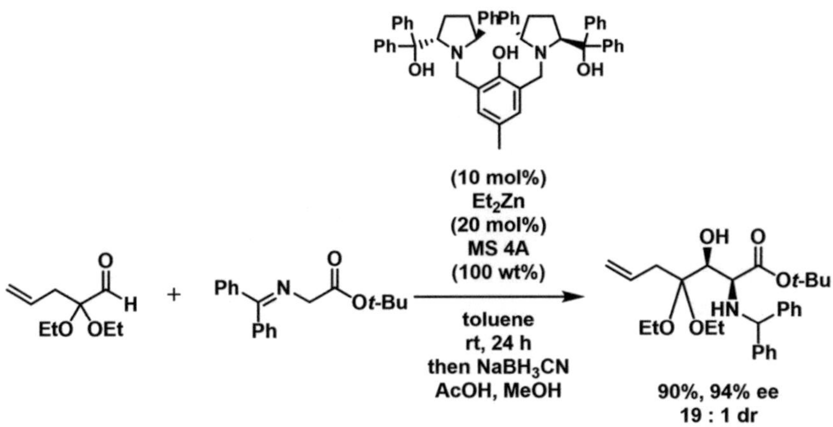

Figure 14.
Synthesis of β-hydroxy-α-amino esters using a zinc-ProPhenol catalyst.

a versatile and efficient method for the synthesis of *syn*-β-hydroxy-α-amino esters under mild reaction conditions.

The stereochemical outcome of the transformation can be rationalized by the intermediate depicted in **Figure 15**. The *Z*-enolate, derived from the glycine Schiff base methyl ester, adopts a conformation that permits chelation by both zinc centers in the dinuclear zinc complex. This bidentate coordination stabilizes the enolate and orients it in a defined geometry, which is crucial for achieving stereocontrol. The aldehyde is coordinated by the more Lewis acidic zinc center, positioning it in close proximity to the enolate. This spatial arrangement minimizes steric hindrance and electronic repulsion, thereby enhancing the selectivity of the reaction. The enolate preferentially attacks the *si*-face of the coordinated aldehyde, and this facial selectivity leads to the formation of the (2*S*,3*R*)-β-hydroxy-α-amino ester as the major product.

4.2.3 Copper-TolBINAP-2,2,5,7,8-pentamethyl-6-chromanol system

A copper-catalyzed direct aldol addition between aldehydes and glycine Schiff bases bearing methyl, allyl, or tert-butyl esters has demonstrated exceptional

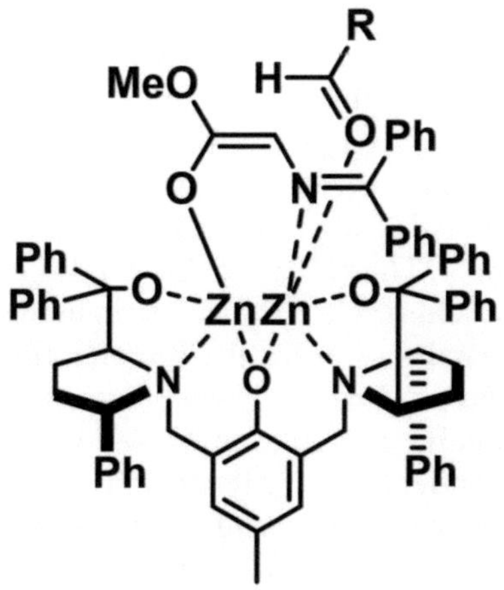

Figure 15.
Stereochemical rationalization of the zinc-ProPhenol-catalyzed aldol reaction.

selectivity (**Figure 16**) [67]. This catalytic system, composed of readily available components including (*R*)-TolBINAP, mesitylcopper, and 2,2,5,7,8-pentamethyl-6-chromanol, efficiently promotes the direct asymmetric aldol reaction, affording *syn*-β-hydroxy-α-amino esters in excellent yields, with outstanding diastereo- and enantioselectivities (85–95% yield, up to >99:1 dr (*syn*/*anti*), 93–99% ee (*syn*)). For example, the reaction between the Schiff base of glycine allyl ester and iso-butanal, catalyzed by mesitylcopper (10 mol%), (*R*)-TolBINAP (10 mol%), and 2,2,5,7,8-pentamethyl-6-chromanol (10 mol%) in ethyl acetate at −70°C for 1 h, afforded the *syn*-aldol adduct in 95% yield with >99:1 dr (*syn*/*anti*), 97% ee (*syn*) after one-pot reductive process using sodium cyanoborohydride ($NaBH_3CN$) and acetic acid (AcOH). This catalytic system tolerates α-branched and linear aliphatic and aromatic aldehydes, such as isobutanal, *n*-hexanal, and benzaldehyde, respectively, to achieve excellent yields, diastereo- and enantioselectivities (92–95%, up to >99:1 dr (*syn*/*anti*), 95–97% ee; 88–92%, up to 97:3 dr (*syn*/*anti*), 97–99% ee; and 85–89%, up to 95:5 dr (*syn*/*anti*), 93–98% ee, respectively), providing outstanding generality. Additionally, this catalytic system accepts several glycine Schiff base esters, such as

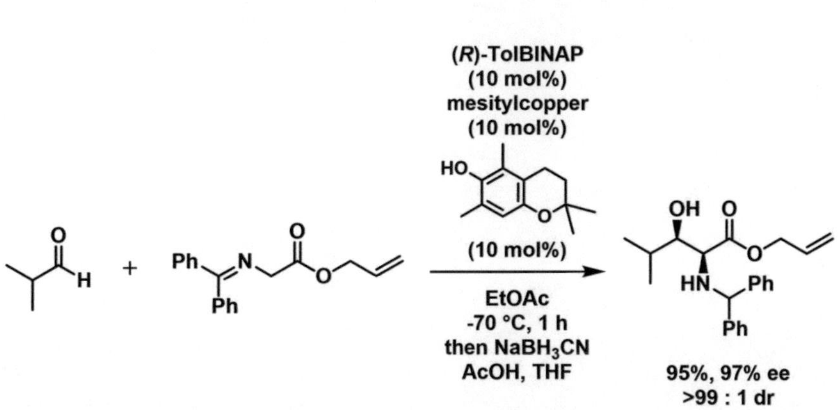

Figure 16.
Synthesis of syn*-β-hydroxy-α-amino esters via the copper-TolBINAP-2,2,5,7,8-pentamethyl-6-chromanol system.*

methyl, allyl, and *tert*-butyl esters, which are widely used as protective ester groups. The versatility of the catalytic system extends beyond its compatibility with various aldehydes and glycine Schiff base esters. This method has demonstrated remarkable efficiency in generating *syn*-β-hydroxy-α-amino esters under mild reaction conditions, making it a valuable tool for synthetic organic chemists. Moreover, the catalytic system tolerates α-chiral aldehydes, affording a high yield of single diastereomers (81–93%, up to >99:1 dr (*syn*/*anti*)). The ability of this catalytic system to tolerate α-chiral aldehydes and consistently yield single diastereomers in high yields is particularly noteworthy. These characteristics significantly broaden the utility for the synthesis of complex chiral molecules and pharmaceutically relevant compounds. Furthermore, the use of commercially available reagents, combined with the system's operational simplicity and robustness, renders it a highly attractive strategy for both academic investigations and industrial-scale applications in asymmetric organic synthesis.

Copper(I) complex **A** coordinates the carbonyl and imine groups of the glycine Schiff base, which increases the α-acidity of the glycine Schiff base and controls the enolate geometry to form complex **B**, as shown in **Figure 17**. [Cu(I)] alkoxide of 2,2,5,7,8-pentamethyl-6-chromanol ([Cu(I)]OAr, **A**) deprotonates the α-proton of the glycine Schiff base to produce *Z*-enolate **C** and HOAr. Then, [Cu(I)] enolate **C** activates the aldehyde by coordination, and the *si*-face of aldehyde approaches from the *si*-face of the enolate, ultimately leading to the preferential formation of the [Cu(I)] alkoxide of (2*S*,3*R*)-β-hydroxy-α-amino ester **F** as a major product through the 6-membered ring transition state **E**. The resulting [Cu(I)] alkoxide of β-hydroxy-α-amino ester **F** deprotonates HOAr to release (2*S*,3*R*)-β-hydroxy-α-amino ester **G** and regenerate the copper(I) complex **A**, thereby continuing the catalytic cycle.

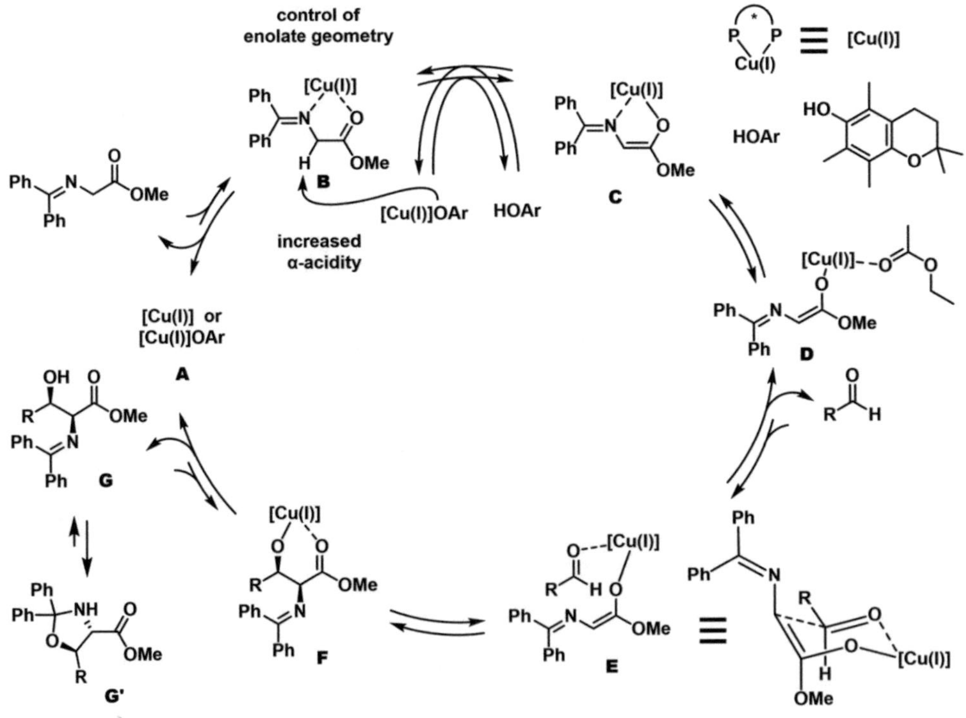

Figure 17.
Plausible reaction cycle of the copper-catalyzed direct asymmetric aldol reaction of glycine Schiff base.

5. Conclusion

β-Hydroxy-α-amino acids are integral components of numerous structurally complex natural products, many of which exhibit potent antibacterial activity. For the total synthesis of these complex structures, efficient and scalable methods for producing large quantities of their building blocks are essential. While various synthetic strategies have been reported, this review covers direct catalytic asymmetric aldol synthesis of β-hydroxy-α-amino esters using glycine Schiff bases. The enantioselective synthesis of β-hydroxy-α-amino acids has been actively pursued for over three decades. Early approaches utilized chiral auxiliaries, which, despite their utility, often suffered from poor atom economy and required harsh conditions for auxiliary removal. In contrast, the direct aldol reaction between a glycine-derived donor and an aldehyde provides a highly atom-economical route for constructing the β-hydroxy-α-amino ester framework in a single step, simultaneously generating adjacent stereocenters. Significant advancements in this methodology have been achieved through the development of both organocatalytic and metal-based catalytic systems. Ongoing research into catalyst design and mechanistic insights will further enhance the scope and efficiency of this approach, facilitating broader access to β-hydroxy-α-amino ester frameworks essential for applications in medicinal and synthetic chemistry.

Acknowledgements

T.T. acknowledges a Grant-in-Aid for Scientific Research (C) from the Japan Society for the Promotion of Science (No. 22 K05341) and (No. 25 K09886), and 2024 Microbial Chemistry Research and Development Fund (No. 24A05).

Conflict of interest

The authors declare no conflict of interest.

Author details

Toshifumi Takeuchi* and Masakatsu Shibasaki*
Institute of Microbial Chemistry (BIKAKEN), Tokyo, Japan

*Address all correspondence to: takeuchit@bikaken.or.jp; mshibasa@bikaken.or.jp

References

[1] Ooi T, Maruoka K. Enantioselective synthesis of α-amino acids by chiral phase-transfer catalysis. Journal of Synthetic Organic Chemistry, Japan. 2003;**61**:1195-1206. DOI: 10.5059/yukigoseikyokaishi.61.1195

[2] Makino K, Hamada Y. Asymmetric synthesis of β-Hydroxy-α-amino acid. Journal of Synthetic Organic Chemistry, Japan. 2005;**63**:1198-1208. DOI: 10.5059/yukigoseikyokaishi.63.1198

[3] Luo Y-C, Zhang H-H, Wang Y, Zu P-F. Synthesis of α-amino acids based on chiral Tricycloiminolactone derived from natural (+)-camphor. Accounts of Chemical Research. 2010;**43**:1317-1330. DOI: 10.1021/ar100050p

[4] Zhang Z, Farrants H, Li X. Adding a functional handle to nature's building blocks: The asymmetric synthesis of β-Hydroxy-α-amino acids. Chemistry – An Asian Journal. 2014;**9**:1752-1764. DOI: 10.1002/asia.201400111

[5] McCormick MH, Stark WM, Pittenger GE, Pit-tenger RC, McGuire JM. Vancomycin, a new antibiotic. I. Chemical and biologic properties. Antibiotics Annual. 1955;**3**:606-611

[6] Review of glycopeptide antibiotics:Nicolaou KC, Boddy CN, Bräse S, Winssinger N. Chemistry, biology, and medicine of the Glycopeptide antibiotics. Angewandte Chemie, International Edition. 1999;**38**:2096-2152. DOI: 10.1002/(SICI)1521-3773(19990802)38:15<2096::AID-ANIE2096>3.0.CO;2-F

[7] Review of glycopeptide antibiotics:Williams DH, Bardsley B. The vancomycin group of antibiotics and the fight against resistant bacteria. Angewandte Chemie, International Edition. 1999;**38**:1172-1193. DOI: 10.1002/(SICI)1521-3773(19990503)38:9<1172::AID-ANIE1172>3.0.CO;2-C

[8] Boger DL. Vancomycin, teicoplanin, and ramoplanin: Synthetic and mechanistic studies. Meddicinal Research Review. 2001;**21**:356-381. DOI: 10.1002/med.1014

[9] Suzuki S, Isono K, Nagatsu J, Mizutani T, Kawashima Y, Mizuno T. A new antibiotic, polyoxin A. The Journal of Antibiotics, Series A. 1965;**18**:131. DOI: 10.11554/antibioticsa.18.3_131

[10] Isano K, Asahi K, Suzuki S. Studies on polyoxins, antifungal antibiotics. XIII. Structure of polyoxins. Journal of the American Chemical Society. 1969;**91**:7490-7502. DOI: 10.1021/ja01054a045

[11] Sakai Y, Yoshida T, Tsujita T, Ochiai K, Agatsuma T, Saitoh Y, et al. GE3, a novel hexadepsipeptide antitumor antibiotic, produced by *Streptomyces* sp. I. Taxonomy, production, isolation, physico-chemical properties, and biological activities. The Journal of Antibiotics. 1997;**50**:659-664. DOI: 10.7164/antibiotics.50.659

[12] Agatsuma T, Sakai Y, Mizukami T, Saitoh Y. GE3, a novel Hexadepsipeptide antitumor antibiotic produced by *Streptomyces* sp. II. Structure determination. The Journal of Antibiotics. 1997;**50**:704. DOI: 10.7164/antibiotics.50.704

[13] Yoshida T, Sakai Y, Mizukami T. Development of novel anticancer

drugs which inhibit E2F transcription factor. Nippon Nogei Kagaku Kaishi. 1997;**71**:516-519

[14] Shoji J, Hinoo H, Hattori T, Hirooka K, Kimura Y, Yoshida T. Isolation and characterization of hypeptin from *pseudomonas* sp. Journal of Antibiotics. 1989;**42**:1460-1464. DOI: 10.7164/antibiotics.42.1460

[15] Shoji J, Hinoo H, Matsumoto K, Hattori T, Yoshida T, Matsuura S, et al. Isolation and characterization of katanosins A and B. Journal of Antibiotics. 1988;**41**:713-718. DOI: 10.7164/antibiotics.41.713

[16] Tymiak AA, McCormick TJ, Unger SE. Structure determination of lysobactin, a macrocyclic peptide lactone antibiotic. The Journal of Organic Chemistry. 1989;**54**:1149-1157. DOI: 10.1021/jo00266a029

[17] Kratena N, Gökler T, Maltrovsky L, Oburger E, Christian SC. A unified approach to Phytosiderophore natural products. Chemistry – A European Journal. 2021;**27**:577-580. DOI: 10.1002/chem.202004004

[18] Kimishima A, Hagimoto D, Honsho M, Sakai K, Honma S, Fuji S, et al. Biosynthesis of the fusarium mycotoxin (−)-Sambutoxin. Organic Letters. 2024;**26**:597-601. DOI: 10.1021/acs.orglett.3c03792

[19] Zhang Y, Borch LA, Fischer NH, Meldal M. Hydrodynamic control of Alzheimer aβ fibrillation with Glucosaminic acid containing click-cyclized β-bodies. Journal of the American Chemical Society. 2024;**146**:2654-2662. DOI: 10.1021/jacs.3c12118

[20] Chen Q-Y, Ratnayake R, Hortigüelac R, Seabra GM, Cameron MD, Díaz JF, et al. Probing a distinct druggable tubulin binding site with gatorbulins 1-7, their metabolic and physicochemical properties, and pharmacological consequences. Bioorganic & Medicinal Chemistry. 2023;**95**:117506. DOI: 10.1016/j.bmc.2023.117506

[21] Medcalfa MR, Kruegera RL, Medcalfa ZT, Rosstona PA, Zhua Y, Kaltenbronnb KM, et al. Building chemical probes based on the natural products YM-254890 and FR900359: Advances toward scalability. Synthesis. 2023;**55**:90-106. DOI: 10.1055/a-1873-6891

[22] Wang H-Y, Yang H, Holm M, Tom H, Oltion K, Al-Khdhairawi AAQ, et al. Synthesis and single-molecule imaging reveal stereospecific enhancement of binding kinetics by the antitumour eEF1A antagonist SR-A3. Nature Chemistry. 2022;**14**:1443-1450. DOI: 10.1038/s41557-022-01039-3

[23] Takashina K, Katsuyama A, Kaguchi R, Yamamoto K, Sato T, Takahashi S, et al. Solid-phase total synthesis of Plusbacin A_3. Organic Letters. 2022;**24**:2253-2257. DOI: 10.1021/acs.orglett.2c00667

[24] Diamandas M, Moreira R, Taylor SD. Solid-phase Total synthesis of Dehydrotryptophan-bearing cyclic peptides Tunicyclin B, Sclerotide a, CDA3a, and CDA4a using a protected β-Hydroxytryptophan building block. Organic Letters. 2021;**23**:3048-3052. DOI: 10.1021/acs.orglett.1c00717

[25] Matthew S, Chen Q-Y, Ratnayake R, Fermaintt CS, Lucena-Agell D, Bonato F, et al. Gatorbulin-1, a distinct cyclodepsipeptide chemotype, targets a seventh tubulin pharmacological site. Proceedings of the National Academy of Sciences of the United States of America.

2021;**118**:e2021847118. DOI: 10.1073/pnas.2021847118

[26] Grab HA, Kirsch VC, Sieber SA, Bach T. Total synthesis of the cyclic Depsipeptide Vioprolide D via its (*Z*)-Diastereoisomer. Angewandte Chemie, International Edition. 2020;**59**:12357-12361. DOI: 10.1002/anie.202002328

[27] Selected reference of conversion to β-lactams:Miller MJ. Hydroxamate approach to the synthesis of β-lactam antibiotics. Accounts of Chemical Research. 1986;**19**:49-56. DOI: 10.1021/ar00122a004

[28] Fugami K, Morizawa Y, Ishima K, Nozaki H. Pd(O) promoted transformation of *N*-tosyl-2-(1,3-butadienyl)-aziridine into *N*-tosyl-2-vinyl-3-pyrroline. Tetrahedron Letters. 1985;**26**:857-860. DOI: 10.1016/S0040-4039(00)61948-2

[29] Review of an enzymatic approach:Dückers N, Baer K, Simon S, Gröger H, Hummel W. Threonine aldolases-screening, properties and applications in the synthesis of non-proteinogenic β-hydroxy-α-amino acids. Applied Microbiology and Biotechnology. 2010;**88**:409-424. DOI: 10.1007/s00253-010-2751-8

[30] Review of an enzymatic approach:Fesko K, Gruber-Khadjawi M. Biocatalytic methods for C–C bond formation. ChemCatChem. 2013;**5**:1248-1272. DOI: 10.1002/cctc.201200709

[31] For a racemic syn-selective aldol reaction to access β-hydroxy-α-amino esters:Seashore-Ludlow B, Torssell S, Somfai P. Addition of Azomethine Ylides to aldehydes: Mechanistic dichotomy of differentially substituted α-Imino esters. European Journal of Organic Chemistry. 2010;**20**:3927-3933. DOI: 10.1002/ejoc.201000377

[32] For a racemic syn-selective aldol reaction to access β-hydroxy-α-amino esters:Lou S, Ramirez A, Conlon DA. Catalytic *syn*-selective direct aldol reactions of benzophenone glycine imine with aromatic, heteroaromatic and aliphatic aldehydes. Advanced Synthesis and Catalysis. 2015;**357**:28-34. DOI: 10.1002/adsc.201400639

[33] Yamada S-I, Oguri T, Shioiri T. Asymmetric synthesis of α-amino-acid derivatives by alkylation of a chiral Schiff base. Journal of the Chemical Society, Chemical Communications. 1976:136-137. DOI: 10.1039/C39760000136

[34] Schöllkopf U, Hartwig W, Groth U. Enantioselective synthesis of α-Methylserines. Angewandte Chemie, International Edition. 1980;**19**:212-213. DOI: 10.1002/anie.198002121

[35] Schöllkopf U. Asymmetric syntheses of amino acids via metalated bis-lactim ethers of 2,5-diketopiperazines. Pure and Applied Chemistry. 1983;**55**:1799-1806. DOI: 10.1351/pac198355111799

[36] Belokon YN, Zel'tzer IE, Loim NM, Tsiryapkin VA, Parnes ZN, Kursanov DN, et al. 1-(*NN*-dimethylaminomethyl)-2-formylcymantrene as a reagent for the asymmetric synthesis and retroracemisation of amino acids. Journal of the Chemical Society, Chemical Communications. 1979;**18**:789-790. DOI: 10.1039/C39790000789

[37] Belokon YN, Kochetkov KA, Ikonnikov NS, Strelkova TV, Harutyunyan SR, Saghiyan AS. A new synthesis of enantiomerically pure *syn*-(*S*)-β-hydroxy-α-amino acids via asymmetric aldol reactions of aldehydes with a homochiral Ni(II)-glycine/(*S*)-BPB Schiff base complex. Tetrahedron: Asymmetry. 2001;**12**:481-485. DOI: 10.1016/s0957-4166(01)00071-4

[38] Eder U, Sauer G, Wiechert R. New type of asymmetric cyclization to optically active steroid CD partial structures. Angewandte Chemie, International Edition. 1971;**10**:496-497. DOI: 10.1002/anie.197104961

[39] Hajos ZG, Parrish DR. Asymmetric synthesis of bicyclic intermediates of natural product chemistry. The Journal of Organic Chemistry. 1974;**39**:1615-1621. DOI: 10.1021/jo00925a003

[40] Yamada YMA, Yoshikawa N, Sasai H, Shibasaki M. Direct catalytic asymmetric aldol reactions of aldehydes with unmodified ketones. Angewandte Chemie, International Edition. 1997;**36**:1871-1873. DOI: 10.1002/anie.199718711

[41] Yoshikawa N, Yamada YMA, Das J, Sasai H, Shibasaki M. The direct catalytic asymmetric aldol reaction. Journal of the American Chemical Society. 1999;**121**:4168-4178. DOI: 10.1021/ja990031y

[42] Shibasaki M, Yoshikawa N. Lanthanide complexes in multifunctional asymmetric catalysis. Chemical Reviews. 2002;**102**:2187-2210. DOI: 10.1021/cr010297z

[43] List B, Lerner RA, Barbas CF. Proline-catalyzed direct asymmetric aldol reactions. Journal of the American Chemical Society. 2000;**122**:2395-2396. DOI: 10.1021/ja994280y

[44] O'Donnell MJ. Benzophenone Schiff bases of glycine derivatives: Versatile starting materials for the synthesis of amino acids and their derivatives. Tetrahedron. 2019;**75**:3667-3696. DOI: 10.1016/j.tet.2019.03.029

[45] Stork G, Leong AYW, Touzin AM. Alkylation and Michael additions of glycine ethyl ester. Use in α-amino acid synthesis and as acyl carbanion equivalent. The Journal of Organic Chemistry. 1976;**41**:3491-3493. DOI: 10.1021/jo00883a044

[46] O'Donnell MJ, Boniece JM, Earp SE. The synthesis of amino acids by phase-transfer reactions. Tetrahedron Letters. 1978;**19**:2641-2644. DOI: 10.1016/S0040-4039(01)91563-1

[47] O'Donnell MJ, Eckrich TM. The synthesis of amino acid derivatives by catalytic phase-transfer alkylations. Tetrahedron Letters. 1978;**19**:4625-4628. DOI: 10.1016/S0040-4039(01)85688-4

[48] O'Donnell MM, Bennett WD, Wu S. The stereoselective synthesis of alpha-amino acids by phase-transfer catalysis. Journal of the American Chemical Society. 1989;**111**:2353-2355. DOI: 10.1021/ja00188a089

[49] O'Donnell MJ, Bennett WD, Bruder WA, Jacobsen WN, Knuth K, LeClef B, et al. Acidities of glycine Schiff bases and alkylation of their conjugate bases. Journal of the American Chemical Society. 1988;**110**:8520-8525. DOI: 10.1021/ja00233a031

[50] Wurtz CA. Sur un aldéhyde-alcool. In: Bulletin de la Société Chimique de Paris. 2nd series. Vol. 17. France: Société Chimique de France; 1872. pp. 436-442

[51] Wurtz CA. Ueber einen Aldehyd-Alkohol. Journal für Praktische Chemie. 1872;**5**:457-464. DOI: 10.1002/prac.18720050148

[52] Wurtz CA. Sur un aldéhyde-alcool. Comptes Rendus de l'Académie des Sciences. 1872;**74**:1361-1367

[53] Kumagai N, Kanai M, Sasai H. A career in catalysis: Masakatsu Shibasaki. ACS Catalysis. 2016;**6**:4699-4709. DOI: 10.1021/acscatal.6b01227

[54] Yoshikawa N, Shibasaki M. Catalytic asymmetric synthesis of β-hydroxy-α-amino acid esters by direct aldol reaction of glycinate Schiff bases. Tetrahedron. 2002;**58**:8289-8298. DOI: 10.1016/S0040-4020(02)00979-1

[55] Rahmani R, Matsumoto M, Yamashita Y, Kobayashi S. Direct-type aldol reactions of Fluorenylidene-protected/activated glycine esters with aldehydes for the synthesis of β-Hydroxy-α-amino acid derivatives. Chemistry – An Asian Journal. 2012;**7**:1191-1194. DOI: 10.1002/asia.201200081

[56] Gasparski CM, Miller MJ. Synthesis of β-Hydroxy-α-amino acids by aldol condensation using a chiral phase transfer catalyst. Tetrahedron. 1991;**47**:5367-5378. DOI: 10.1016/S0040-4020(01)80971-6

[57] O'Donnell MJ. The enantioselective synthesis of α-amino acids by phase-transfer catalysis with achiral Schiff Base esters. Accounts of Chemical Research. 2004;**37**:506-517. DOI: 10.1021/ar0300625

[58] Mettah S, Srikanth GSC, Dangerfield BS, Castle SL. Asymmetric synthesis of β-Hydroxy amino acids via aldol reactions catalyzed by chiral ammonium salts. The Journal of Organic Chemistry. 2004;**69**:6489-6492. DOI: 10.1021/jo049283u

[59] Ma B, Parkinson JL, Castle SL. Novel *cinchona* alkaloid derived ammonium salts as catalysts for the asymmetric synthesis of β-hydroxy α-amino acids via aldol reactions. Tetrahedron Letters. 2007;**48**:2083-2086. DOI: 10.1016/j.tetlet.2007.01.132

[60] Ooi T, Taniguchi M, Kameda M, Maruoka K. Direct asymmetric aldol reactions of glycine Schiff Base with aldehydes catalyzed by chiral quaternary ammonium salts. Angewandte Chemie, International Edition. 2002;**41**:4542-4544. DOI: 10.1002/1521-3773(20021202)41:23<4542::AID-ANIE4542>3.0.CO;2-3

[61] Ooi T, Kameda M, Taniguchi M, Maruoka K. Development of highly Diastereo- and enantioselective direct asymmetric aldol reaction of a Glycinate Schiff Base with aldehydes catalyzed by chiral quaternary ammonium salts. Journal of the American Chemical Society. 2004;**126**:9685-9694. DOI: 10.1021/ja048865q

[62] Kano T, Lan Q, Wang X, Maruoka K. Effects of aromatic substituents on Binaphthyl-based chiral Spiro-type ammonium salts in asymmetric phase-transfer reactions. Advanced Synthesis and Catalysis. 2007;**349**:556-560. DOI: 10.1002/adsc.200600576

[63] Kitamura M, Shirakawa S, Arimura Y, Wang X, Maruoka K. Combinatorial design of simplified high-performance chiral phase-transfer catalysts for practical asymmetric synthesis of α-alkyl- and α,α-Dialkyl-α-amino acids. Asian Journal of Organic Chemistry. 2008;**3**:1702-1714. DOI: 10.1002/asia.200800107

[64] Vera S, Vázquez A, Rodriguez R, del Pozo S, Urruzuno I, de Cózar A, et al. Synthesis of β-Hydroxy α-amino acids through Brønsted Base-catalyzed *syn*-selective direct aldol reaction of Schiff bases of glycine *o*-Nitroanilide. The Journal of Organic Chemistry. 2021;**86**:7757-7772. DOI: 10.1021/acs.joc.1c00406

[65] Trost BM, Miege F. Development of pro phenol ligands for the Diastereo- and enantioselective synthesis of β-Hydroxy-α-amino esters. Journal of the American Chemical Society. 2014;**136**:3016-3019. DOI: 10.1021/ja4129394

[66] Trost BM, Bartlett MJ. Pro phenol-catalyzed asymmetric additions by spontaneously assembled Dinuclear main group metal complexes. Accounts of Chemical Research. 2015;**48**:688-701. DOI: 10.1021/ar500374r

[67] Takeuchi T, Shibasaki M. Copper-catalyzed direct asymmetric aldol reaction of glycine Schiff bases to access *syn*-β-Hydroxy-α-amino esters. Organic Letters. 2024;**26**:7981-7986. DOI: 10.1021/acs.orglett.4c03085

Chapter 4

Recent Applications of TMS Imines in Organic Synthesis

Plato A. Magriotis

Abstract

Recent applications of TMS imines in organic synthesis, from the author's laboratory as well as from the recent literature, are described and analyzed in this chapter. These include novel syntheses of saturated and unsaturated heterocyclic rings such as β-lactams, piperazines, aziridines, and morpholines, as well as phenanthridines and pyridines. The manifestation description of an operating β-silicon effect in some of these applications, such as the syntheses of piperazines and morpholines, is also included. Possible applications of these heterocycles in medicinal chemistry in terms of novel drug discovery are described. Finally, an additional section is included, describing the amphiphilic electronic nature of TMS imines as indicated by the recent literature.

Keywords: β-lactams, piperazines, morpholines, TMS imines, β-silicon effect, aziridines

1. Introduction

Trimethylsilyl (TMS) imines are monomeric compounds that are stable under anhydrous conditions (their silicon-nitrogen bond is readily hydrolyzed). Therefore, TMS imines may be considered a protected form of the corresponding elusive imines of ammonia that are known to be highly unstable, trimerizing to triazines. TMS imines hold a unique position among Schiff bases owing to their dual reactivity, which renders them electrophilic at carbon and nucleophilic on the nitrogen atom. *N*-TMS imines are considerably more Lewis basic and less electrophilic than *N*-acylimines [1]. Since their synthesis and use were both comprehensively reviewed toward the end of the previous century [2], only recent developments regarding their employment that were reported after the turn of this century will be covered in this chapter.

2. TMS imines as nucleophiles

Although the Diels-Alder reaction is without a doubt the most powerful synthesis tool for the construction of six-membered cyclic structures, the related, well-documented, dehydro-Diels-Alder reaction (DDAR) had been rather less developed until the Barluenga group reported the first example of an intermolecular metal-catalyzed hetero-dehydro-Diels-Alder (HDDA) reaction [3]. Specifically, nonactivated nitriles regioselectively attack the metal-complexed triple bond of esters **1,** leading, after cyclization, to tetrasubstituted pyridines **2** (Eq. (1)).

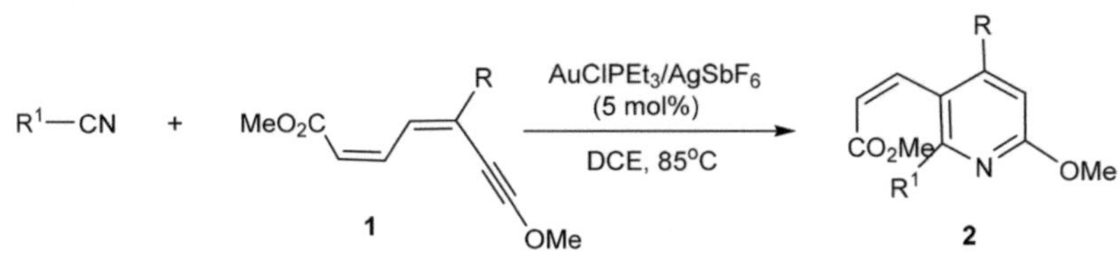

(1)

Recently, the group of Aguilar extended the scope of this HDDA reaction with respect to the nucleophile to include imines and especially TMS imines **3**, as nucleophiles, for the production of *N*-unsubstituted dihydropyridones **4** (Eq. (2)) [4].

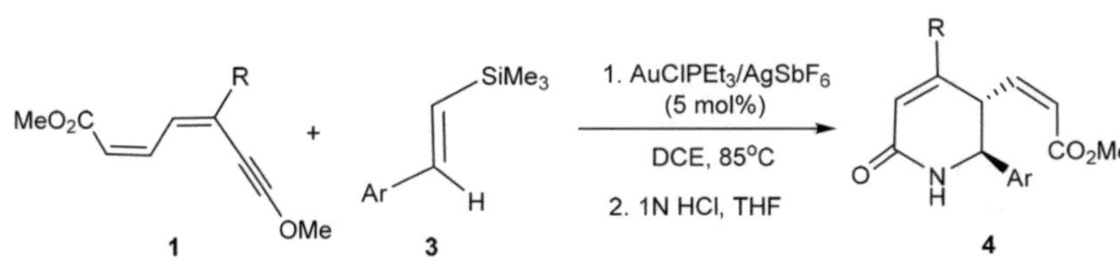

(2)

During ongoing research in this laboratory, it was considered that TMS imines may actually be effective dienophiles in inverse-electron demand formal dihetero Diels-Alder reactions toward an efficient, *uncatalyzed* synthesis of carbon-substituted piperazines, as shown in Eq. (3).

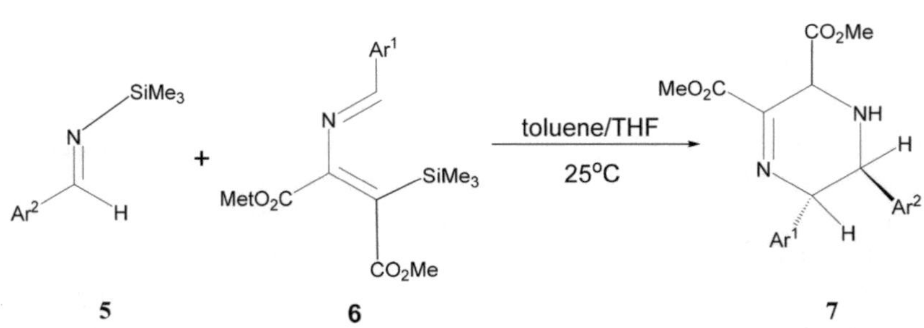

(3)

This premise has been found to be the case, and as a result, a manuscript was submitted for publication [5]. Thus, TMS imine **5** served as an excellent dienophile since its reaction with diene **6** in an inverse-electron demand diaza-Diels-Alder reaction (DADAR) [6], at ambient temperature, provided dehydropiperazine **7** (Eq. (3)). It is rather interesting, in this regard, to note two issues: First, that a TMS imine **8** does also react as a nucleophile adding to dimethyl acetylenedicarboxylate (DMAD) to furnish the required electron-deficient 2-azadiene **6** (Eq. (4)) [7]; and second that the ambient temperature at which the DADAR occurs, can be attributed to the operation of the β-silicon effect

DOI: http://dx.doi.org/10.5772/intechopen.1011940

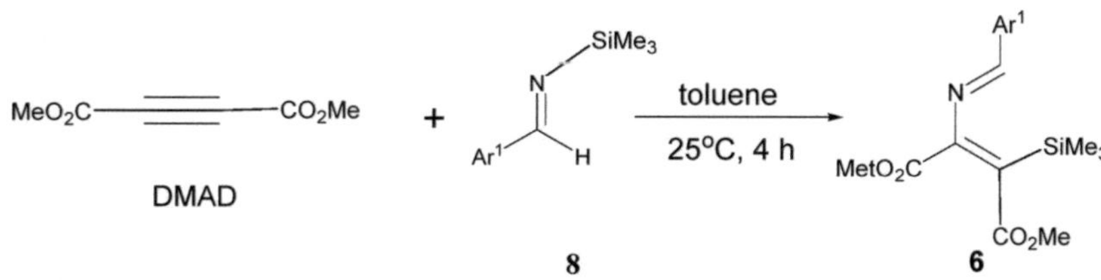

(4)

from the diene, accelerating the DADAR through intermediate **9** as indicated in **Figure 1**.

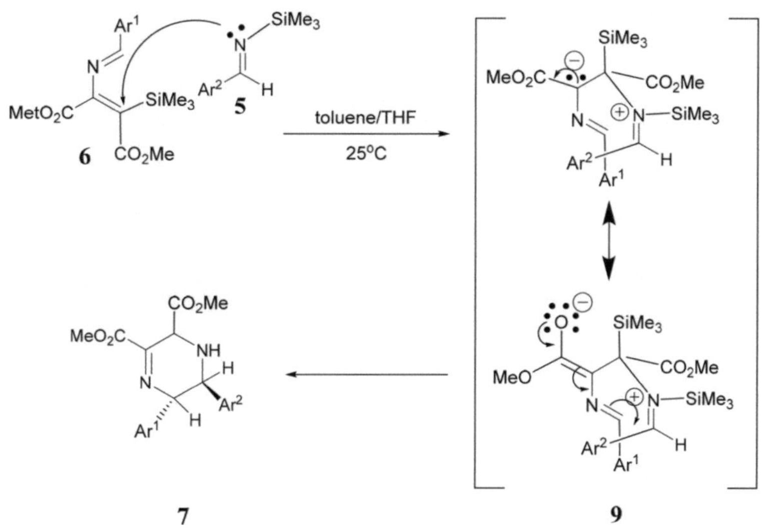

Figure 1.
Manifestation of the β-silicon effect from the diene ***6****, accelerating the DADAR through intermediate* ***9****.*

The phenanthridine aromatic heterocycle is an important structural moiety found in natural products and biologically interesting compounds. Thus, significant effort has been devoted to the synthesis of such derivatives to study structure-activity relationships and as a result, discover new analogs with better pharmacological profiles. An additional synthetic approach to phananthridines **10**, therefore, was reported by the Lautens group based on a palladium-catalyzed Domino arylation/*N*-arylation of aryl iodide **11** and TMS imine **12** (X = Cl, Br) as indicated in Eq. (5) [8].

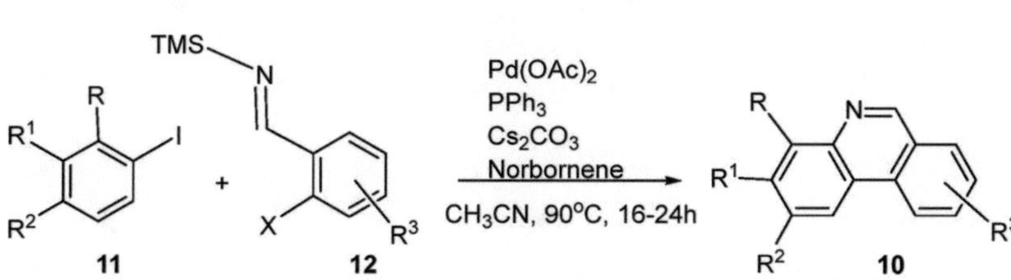

(5)

A simplified catalytic cycle was proposed based on mechanistic work performed independently by Catellani et al., Hartwig et al., and Barluenga et al. [8]. This report represents the first application of TMS imine derivatives in palladium-catalyzed cross-coupling processes.

Nucleophilic TMS imines such as **3** (Eq. (2)) have also been used by the Ghosez group for the construction of electron-rich 2-azadienes **13** (Eq. (6)), which participate

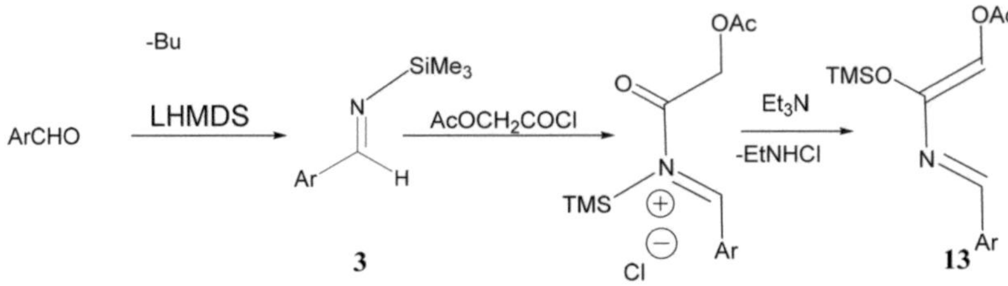

(6)

in normal-electron demand aza-Diels-Alder reaction (ADAR) at ambient temperature to produce complex piperidine scaffolds **14,** as shown in Eq. (7) [9].

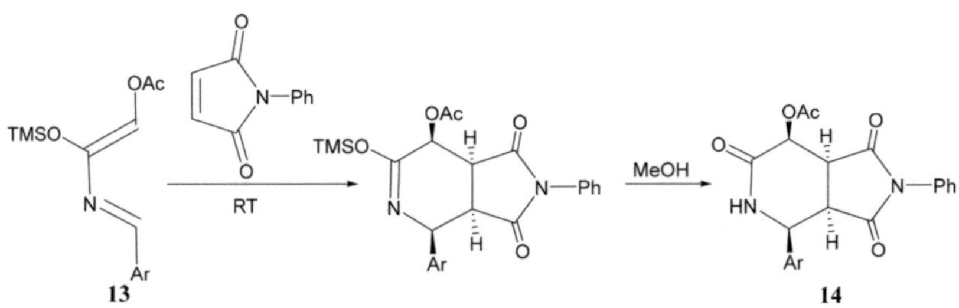

(7)

This technology for the synthesis of 2-azadienes, such as **13**, advanced by the Ghosez group toward the end of the past century [10], was utilized by the Panunzio group to demonstrate that these compounds can be considered stable forms of the zwitterionic intermediates involved in the classical Staudinger synthesis of β-lactams [11]. Accordingly, thiosubstituted-2-azadienes such as **15** were prepared (Eq. (8)), and their thermal electrocyclic ring closure to diastereomeric *trans* β-lactams **16** and **17** was reported (Eq. (9)) [11].

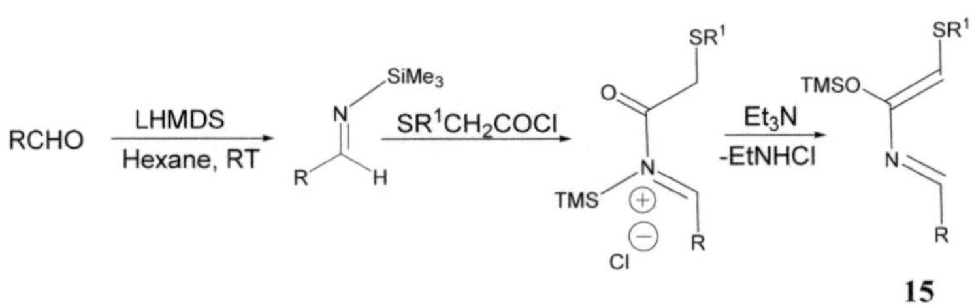

(8)

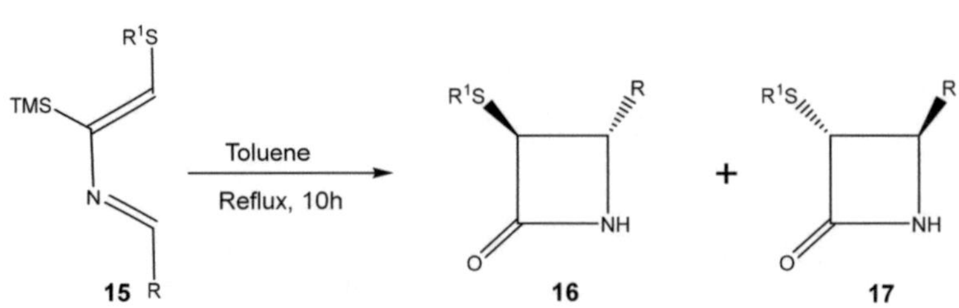

(9)

The Lewis-acid electrocyclization of 2-aza-1,3-butadienes like **15** was also studied, and stereochemical differences from the uncatalyzed cyclization were discussed [12].

Interestingly, the Lewis acid-catalyzed dihetero Diels-Alder reaction of a related 2-azadiene **18** with an aldehyde gave rise to two diastereomeric 5-Phenylthio-1,3-oxazinan-4-ones **19** and **20** that were converted to racemic Prozac **21** (Eq. (10)) [13].

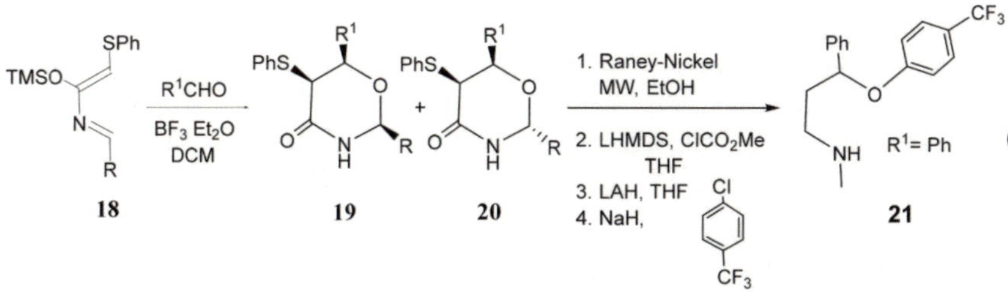

18 19 20 21 (10)

A subsequent publication from the same laboratory [14] described the application of the latter finding to the synthesis of enantiomerically pure (*S*)- and (*R*)- fluoxetine (Prozac®) [14].

3. TMS imines as electrophiles

The Mannich reaction has become an important method for C-C bond formation in synthetic organic chemistry. Despite the prevalence of quaternary centers in complex synthetic targets, the stereoselective Mannich reaction employing non-stabilized enolates and forming such quaternary centers remains largely underexplored. Specifically, a stereoselective Mannich reaction using unstabilized α-substituted lactams as pronucleophiles to form quaternary centers has not been reported. In fact, the first report of this kind appeared recently in the literature by the Stoltz group, describing the potassium *tert*-butoxide-mediated stereoselective/direct Mannich reaction of α-substituted γ-lactams with *in situ* generated aryl *N*-silyl imines from aryl nitriles [15]. Namely, boron-catalyzed hydrosilylation of nitrile furnished trialkylsilylimine *in situ* that reacted with α-substituted γ-lactam **22** to produce the quaternary Mannich base **23** in excellent yield and stereoselectivity as well as modest enantioselectivity (72%, Eq. (11)) [15]. This *in situ* formation of the *N*-silyl imines was deemed necessary due to the water-instability and challenging isolation of *N*-silyl imines [15].

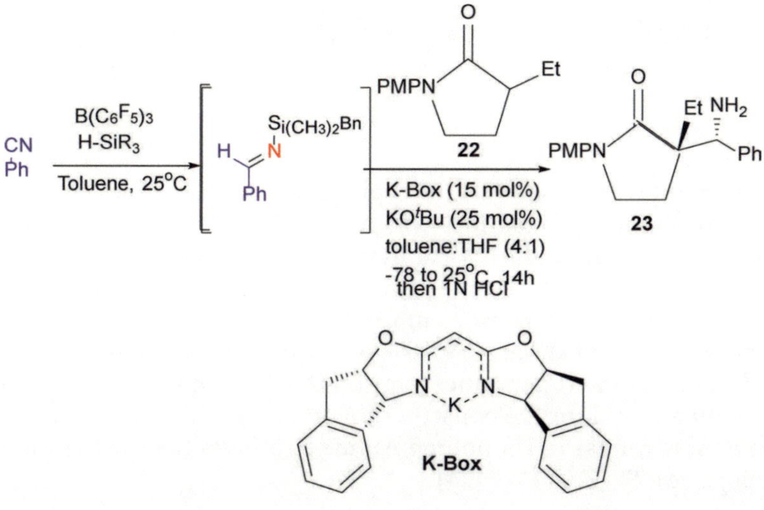

(11)

Besides the Staudinger synthesis of β-lactams, mentioned in the previous section, the ester enolate-imine condensation route το β-lactams that was communicated by Gilman and Speeter in 1943 [16] and comprehensively reviewed by Hart and Ha in 1989 [17], was shown by the group of Hart to become an extremely efficient process when TMS imines were employed as the imine (Schiff base) component. Mindful of this pioneering work by Hart using lithium enolates [18], it was reasoned that trimethylsilyl ketene acetals might be productive partners in this Gilman-Speeter process [19]. It was also thought that the *in situ* production of one equivalent of trimethylsilyloxy lithium upon formation of the TMS imine would obviate the need for catalysis [20]. Accordingly, addition of a 1.0 M solution of lithium hexamethyldisilazide in THF to benzaldehyde in THF at 0°C, followed by stirring at ambient temperature for 0.5 h, addition of TMS-ketene acetal **24**, and heating at THF reflux for 1.5 h, produced β-lactam **25** as a white solid after extractive isolation and recrystallization (Eq. (12)) [19].

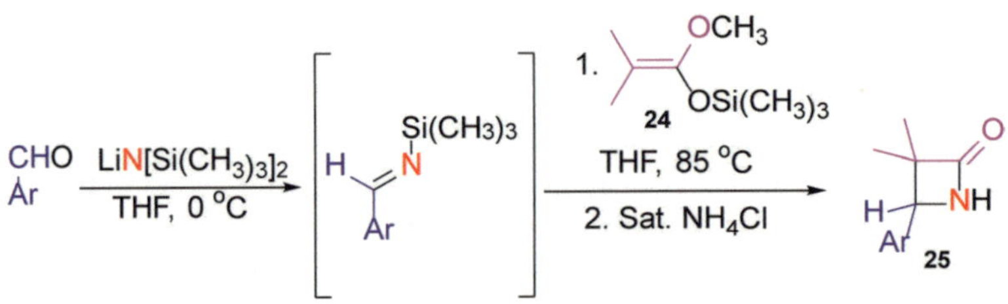

(12)

The first catalytic enantioselective reaction of TMS imines in this century was recorded by the group of Morken, who described the synthesis of nonracemic α-amino boronates **26** [21] of which the corresponding acids have found widespread use as protease inhibitors (Eq. (13)) [22].

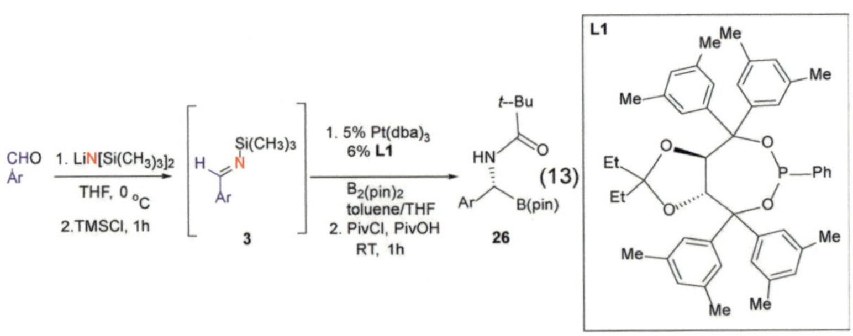

(13)

Another very interesting and useful application of TMS imines electrophiles in catalytic enantioselective reactions was achieved by the group of Johnston, namely in the Bis(AMidine, PBAM)-catalyzed addition of bromonitromethane to TMS imines with a high degree of enantioselection (Eq. (14)) [23]. Thus, the addition of bromonitromethane in 20 mol% increments evenly over 10 h led to excellent enantioselection at the 93/91% level with 10 mol% catalyst loading to ensure low bromonitromethane concentration throughout the reaction. This important result allowed for the construction of a range of protected α-bromo nitroalkane donors for use in Umpolung Amide Synthesis (UmAS), as indicated in structure (peptide) **27** (Eq. (14)) [24].

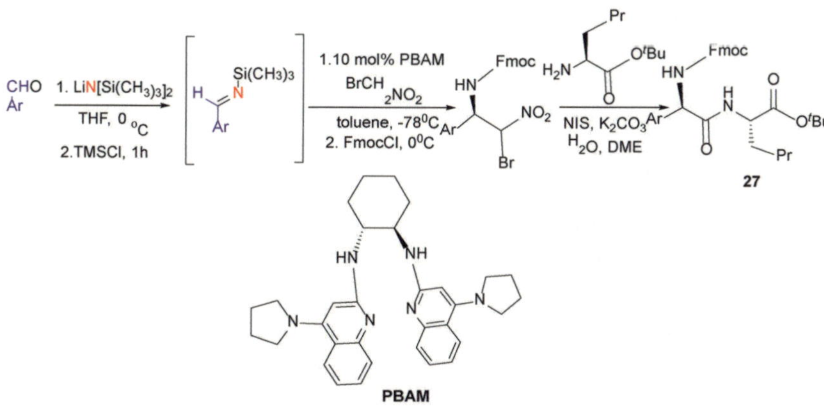

(14)

Among saturated *N*-heterocycles, piperazines and morpholines are considered the most valuable and useful as far as medicinal chemistry is concerned [25]. In fact, morpholine represents the most useful scaffold for the development of CNS drug candidates [26], due to its well-balanced lipophilic-hydrophilic profile, the reduced p*K*a value, and the chair-like flexible conformation, all of which render this important *N*, *O*-heterocyclic scaffold blood-brain barrier (BBB) penetrant [27]. As a result, it was considered whether a [3 + 3]-cycloaddition reaction involving an azaoxyallyl cation, generated from bromoamide **28**, and styrene oxide would lead to morpholinone **29** as the product (Eq. (15)) [28]. After extensive experimentation, it was found to be the case, as **29** was isolated and characterized [29].

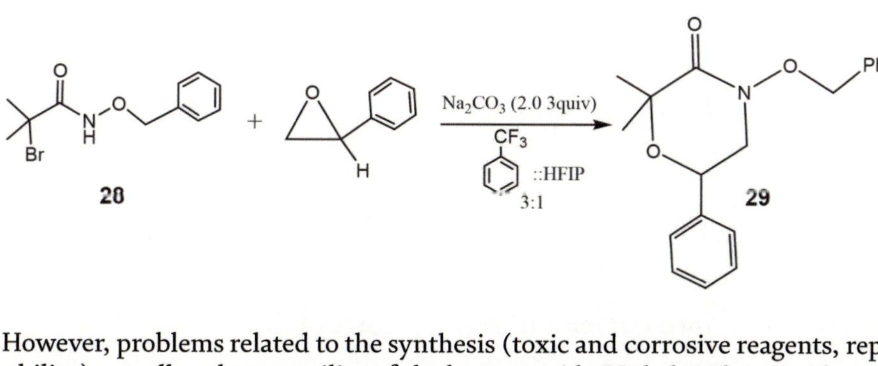

(15)

However, problems related to the synthesis (toxic and corrosive reagents, reproducibility) as well as the versatility of the bromoamide **28**, led to the consideration of an alternative as far as the generation of azaoxyallyl cations is concerned. Thus, the more stable (right, full octets at both atoms) resonance structure of carbon monoxide (**Figure 2**) points out the high electrophilicity of the carbon that, in turn, calls for the potential organocatalytic activation of CO.

Accordingly, nucleophilic attack of an organocatalyst should generate an acyl anion equivalent **30,** which will react with the electrophilic TMS imine **3** to generate zwitterion **31.** Collapse of this intermediate will result in unstable *N*-TMS α-lactam **32** [30], which will open up to the azaoxyallyl cation shown in brackets, stabilized at the central carbon by the β-silicon effect (**Figure 2**). This premise was found to be the case in a reaction that included formylsaccharin (5.0 equiv) as a benign source of CO, 4A molecular sieves, the TMS imine of benzaldehyde, sodium carbonate, and benzotetramisole for CO release and activation, to finally provide the free morpholinone **33** (Eq. (16)) that was characterized [31].

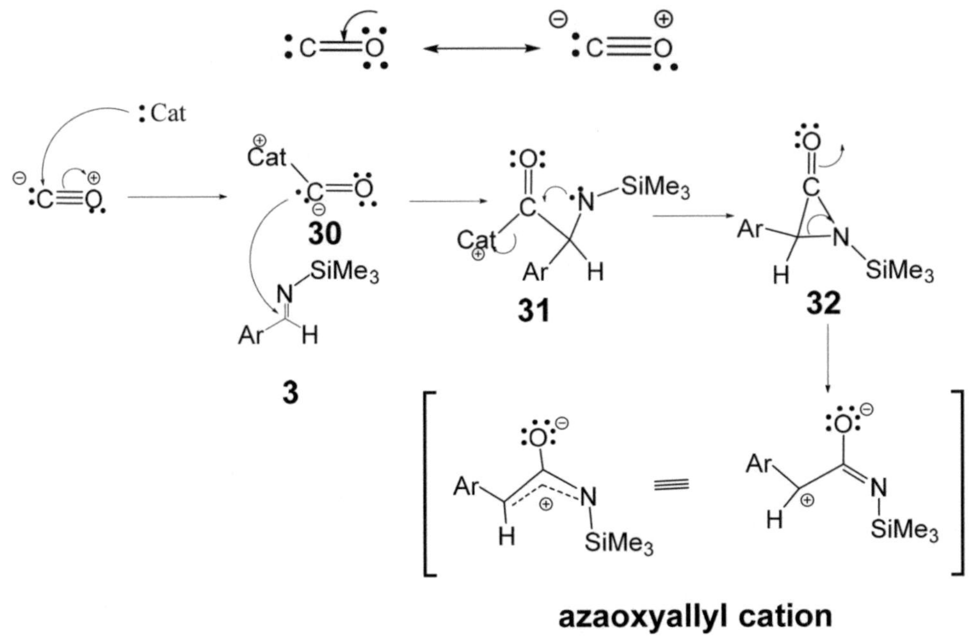

Figure 2.
Manifestation of the β-silicon effect within the N*-silyl azaoxyallyl cation (central carbon) generated by organocatalytic activation of CO and TMS imine* 3.

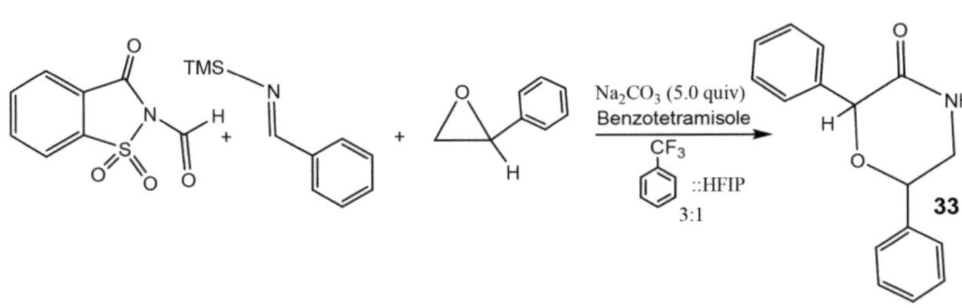

(16)

4. TMS imines as both nucleophiles and electrophiles

The synthesis of optically active homoallylic amines represents an important task in organic synthesis, since such amines make excellent building blocks for the synthesis of a large number of nitrogen-containing natural products. Consequently, the Ramachandran group studied both the allylboration and the crotylboration as well as the alkoxyallylboration of TMS imines employing α-pinene-based reagents for asymmetric reactions. For instance, alkoxy-allylboration of the TMS imine derived from benzaldehyde provided β-alkoxy homoallylic amine **34** in high yield and stereoselectivity (de and ee, Eq. (17)) [32]. The mechanism of this reaction involves displacement of methoxy anion by the TMS imine nitrogen, followed by desilylation with methanol and intramolecular alkoxyallylation, showcasing both nucleophilic and electrophilic properties of the TMS imine. It is assumed that the allylboration of TMS imines may proceed in a similar fashion to the allylboration of aldehydes *via* a six-membered chair-like transition state, with the stereochemical outcome determined by the isopinocampheyl auxiliary [32].

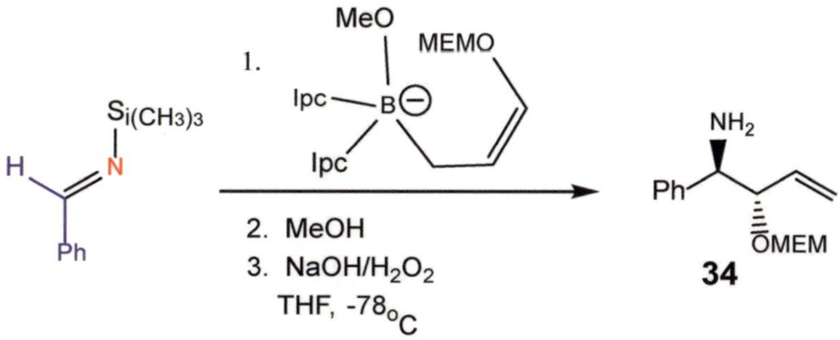

(17)

Aziridines are relatively reactive, synthetically useful, three-membered heterocycles that are commonly employed in the synthesis of other heterocyclic entities. Another example of the amphiphilic electronic character of TMS imines is provided by the report that TMS imine **35** reacts with ethyldiazoacetate (EDA), organocatalyzed by *N*-fluoropyridinium triflate **36** to produce cis-aziridine **37** (Eq. (18)) [33].

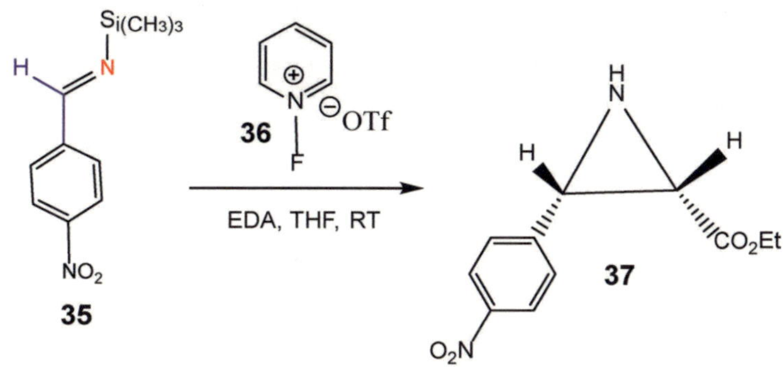

(18)

Finally, chiral, nonracemic 1,2-diamines have been used extensively in asymmetric synthesis. Accordingly, the stereoselective synthesis of these compounds has become a very active area of research. The reductive coupling of imines to give 1,2-diamines naturally represents one of the most versatile methods for the synthesis of symmetrical 1,2-diamines due to the ready availability of imines from aldehydes. Many reducing agents have been used in this coupling, including Ti, Zr, Sm, Zn, Al, In, and other metals. However, mixtures of *d*, *l*, and meso diastereomers are usually obtained. The most *d*,*l*-selective imine coupling reaction to give 1,2-diphenyl-1,2-ethanediamine analogs is the coupling of *N*-TMS imines by low valent niobium reagents reported by Pederson in 1987, and applied, by the Denmark group, to aromatic TMS imines **3** with different substitution patterns furnishing the desired *d*, *l* diamines **38** in good yields with high diastereoselectivity (Eq. (19)) [34]. The individual enantiomers were then obtained by resolution employing the most practical for each case chiral nonracemic reagent, e.g. (−)-menthyl chloroformate [34].

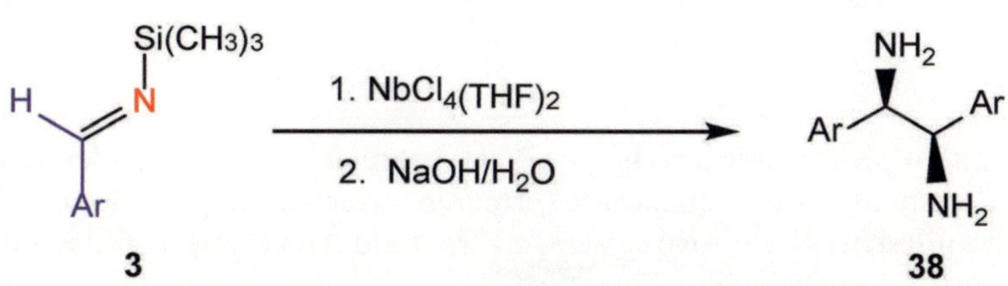

(19)

Author details

Plato A. Magriotis
Department of Pharmacy, Laaboratory of Medicinal Chemistry, University of Patras, Rio, Greece

*Address all correspondence to: pmagriotis@upatras.gr

References

[1] Cainelli G, Panunzio M, Andreoli P, Martelli G, Spunta G, Giacomini D, et al. Metallo-imines: Useful reagents in organic synthesis. Pure and Applied Chemistry. 1990;**62**:605-612

[2] Panunzio M, Zarantonello P. Synthesis and use of *N*-(Trimethylsilyl) imines. Organic Process Research & Development. 1998;**2**:49-59

[3] Barluenga J, Fernández-Rodriguez MA, Garcia-Garcia P, Aguilar E. Gold-catalyzed intermolecular hetero-dehydro diels-alder cycloaddition of captodative dienynes with nitriles: A new reaction and regioselective direct access to pyridines. Journal of the American Chemical Society. 2008;**130**:2764-2765

[4] Fernández-Garcia JM, Fernández-Rodriguez MA, Aguilar E. Catalytic intermolecular hetero dehydro-diels-alder cycloadditions: Regio- and diastrereoselective synthesis of 5,6-dihydropyridinones. Organic Letters. 2011;**13**:5172-5175

[5] Sehou M, Tsigilaras A, Asimakopoulos A, Magriotis P, Inverse electron demand formal dihetrero diels-alder reactions: Application to the synthesis of C-substituted piperazines and morpholines chemistry select 2025; Manuscript submitted for publication

[6] Magriotis P. Synthetic approaches to the Stereochemically complex antitumor drug Ecteinascidin-743: A marine natural product by the name Yondelis® or Trabectidin. In: Chang H, Maryanoff C, Miller B, Schmidt D, editors. The Legacy of Ernest L. Eliel. Vol. 2. Washington, DC: American Chemical Society; 2017. pp. 61-78

[7] Barluenga J, Tomás M, Ballesteros A, Gotor V. An easy synthesis of electron-withdrawing substituted 2-Aza-1,3-dienes and their 1,4-cycloaddition with enamines. Journal of the Chemical Society, Chemical Communications. 1987:1195-1196

[8] Candito D, Lautens M. Palladium-catalyzed domino direct arylation/N-arylation: Convenient synthesis of Phenanthridines. Angewandte Chemie, International Edition. 2009;**48**:6713-6716

[9] Zhu W, Mena M, Jnoff E, Sun N, Pasau P, Ghosez L. Multicomponent reactions for the synthesis of complex Piperidine scaffolds. Angewandte Chemie International Edition. 2009;**48**:5880-5883

[10] Ghosez L. Stereoselective synthesis with and without organometallics. Pure and Applied Chemistry. 1996;**68**:15-22

[11] Panunzio M, Bongini A, Tamanini E, Campana E, Martelli G, Vincennati P, et al. Synthesis of *NH*-3-phenylsulfanyl- and *NH*-3-benzylsulfanyl-azetidinones from 1-phenylsulfanyl- or 1-benzylsulfanyl-3-aza-1,3-dienes. Tetrahedron. 2003;**59**:9577-9582

[12] Bongini A, Panunzio M, Tamanini E, Martelli G, Vicennati P, Monari M. Lewis acid-catalyzed electrocyclization of 2-aza-1,3-butadienes to *NH*-β-lactams. Tetrahedron: Asymmetry. 2003;**14**:993-998

[13] Panunzio M, Tamanini E, Bandini E, Campana E, D'Aurizio A, Vincennati P. 5-Phenylthio-1,3-oxazinan-4-ones via hetero Diels-Alder reactions: Synthesis of (*R*)- and (*S*)- duloxetines and fluoxetines. Tetrahedron. 2006;**62**:12270-12280

[14] Panunzio M, Rossi K, Tamanini E, Campana E, Martelli G. Synthesis of enantiomerically pure (*S*)- and (*R*)-fluoxetine (Prozac®) via a hetero Diels-Alder strategy. Tetrahedron: Asymmetry. 2004;**15**:3489-3493

[15] Casselman T, Madhusudhanan M, Mai BK, Liu P, Stoltz B. Potassium *tert*-Butoxide mediated stereoselective/direct Mannich reaction of α-substituted γ-lactams with *in situ* generated aryl *N*-silyl imines. Chemical Science. 2025;**16**:9863-9871

[16] Gilman H, Speeter M. The reformatsky reaction with Bezalaniline. Journal of the American Chemical Society. 1943;**65**:2255-2256

[17] Hart D, Ha D-C. The Ester-Enolate-imine condensation route to β-lactams. Chemical Reviews. 1989;**89**:1447-1465

[18] Hart D, Kanai K-I, Thomas D, Yang T-K. Preparation of primary amines and 2-Azetidinones via *N*-Trimethylsilyl Imi;nes. The Journal of Organic Chemistry. 1983;**48**:289-294

[19] Panagiotou M, Demos V, Magriotis P. Direct and practical of 3,4-trisubstituted β-lactams via the Thorpe-Ingold effect Tetrahedron Letters. 2020;**61**. DOI: 10.1016/j.tetlet.2020.152375

[20] Colvin E, McGarry D, Nugent M. Silicon assisted synthesis of β-lactams. Tetrahedron. 1988;**44**:4157-4172

[21] Hong K, Morken J. Catalytic enantioselective one-pot aminoborylation of aldehydes: A strategy for construction of nonracemic α–amino boronates. Journal of the American Chemical Society. 2013;**135**:9252-9254

[22] Baker S, Tomsho J, Benkovic S. Boron-containing inhibitors of synthatases. Chemical Society Reviews. 2011;**40**:4279-4285

[23] Makley D, Johnston J. Silyl imine electrophiles in enantioselective catalysis: A Rosetta stone for peptide homologation, enabling diverse *N*-protected aryl glycines from aldehydes in three steps. Organic Letters. 2014;**16**:3146-3149

[24] Shen B, Makley D, Johnston J. *Umpolung* reactivity in amide and peptide synthesis. Nature. 2010;**465**:1027-1033

[25] Marshall C, Federice J, Bell C, Cox P, Njardason J. An Update on the nitrogen heterrocycle compositions and properties of U.S. FDA-approved pharmaceuticals. Journal of Medicinal Chemistry. 2024;**67**:11622-11655

[26] Lenci E, Calugi L, Trabocchi A. Occurance of morpholine in central nervous system drug discovery. ACS Chemical Neuroscience. 2021;**12**:378-390

[27] Tzara A, Xanthopoulos D, Kourounakis A. Morpholine as a scaffold in medicinal chemistry: An update on synthetic strategies. ChemMed Chem. 2020;**15**:392-403

[28] Xuan J, Cao X, Cheng X. Advances in heterocycle synthesis *via* [3+m]-cycloaddition reactions involving an azaoxyallyl cation as the key intermediate. Chemical Communications. 2018;**54**:5154-5163

[29] Lambrou A. Development of a novel [3+3] cycloaddition of azaoxyallyl cations with epoxides for the synthesis of morpholinones [MSc thesis]. Greece: University of Patras; 2023

[30] Lengyel I, Sheehan J. β-Lactams (Aziridinones). Angewandte Chemie, International Edition. 1968;**7**:25-36

[31] Magriotis P, Lambrou A, Konidaris E, Asimakopoulos A. Manuscript to be submitted to org. Lett. for publication

[32] Ramachandran P, Burghardt T. Highly diastereoselective and enantioselective preparation of Hmoallylic amines: Application for the synthesis of, β-amino acids and γ-lactams. Chemistry - A European Journal. 2005;**11**:4387-4395

[33] Bew S, Fairhurst S, Hughes D, Legentil L, Liddle J, Pesce P, et al. Organocatalytic aziridine synthesis using F^+ salts. Organic Letters. 2009;**11**:4552-4555

[34] Denmark S, Yutaka X, Coe N, Wong K-T, Winter S, Choi J. Synthesis of phosphoramides for the Lewis base-csatalyzed allylation and aldol addition reactions. The Journal of Organic Chemistry. 1999;**64**:1958-1967

Chapter 5

Density Functional Theory as a Tool for Assessing the Therapeutic Activity of New Compounds

Maria Marinescu and Claudia-Valentina Popa

Abstract

Research over the past 10 years has shown that a comprehensive assessment is needed for the predictive assessment of the therapeutic activity of soft synthesized organic compounds, especially in the field of heterocyclic compounds. Current research confirms that density functional theory is an absolutely indispensable tool for the predictive study of biological activity, of almost any type, antibacterial, antiviral, anticancer, anti-inflammatory, antidiabetic, etc. This chapter aims to review a series of results of the applicability of density functional theory in the study of the therapeutic properties of newly synthesized compounds, by correlating the various DFT properties of the compounds, i.e., chemical reactivity indices, E_{HOMO}, E_{LUMO}, E_{gap}, etc., with the determined biological characteristics of the compounds, minimum inhibitory concentrations (MIC), minimum biofilm inhibition concentrations (MBEC), IC_{50}, etc. The interactions with specific receptors anticipated by DFT theory will also be discussed in this review.

Keywords: DFT, heterocycles, antimicrobial activity, anticancer, HOMO-LUMO orbitals, chemical reactivity indices

1. Introduction

Over the past 10 years, medicinal chemistry has produced a wide range of new molecules with a wide variety of structures from all classes of compounds, both organic and inorganic, materials, formulations of nanocompounds or nanocompounds with different metallic nanoparticles or derivatives, such as Ag, Au, ZnO, gels, and encapsulated compounds, all of which are meant to improve the therapeutic activity of simple synthesized compounds [1]. Simple heterocycles, which are organic compounds with a single heterocyclic ring that is either aromatic or aliphatic, or heterocyclic hybrids, which have two or more heterocycles and typically exhibit the synergism of the two nuclei, are notable among organic compounds as novel molecules with pharmacological properties [2–6]. Closely related to these are their intricate combinations with different metals, such as Ag, Cu, Ni, Fe, Cr, and Mo, which have amazing therapeutic qualities [7]. Numerous chemical compounds from different classes, including steroids, carbohydrates, peptides, and multifunctional compounds, are also known to have unique therapeutic properties [8]. This encourages study in these areas to discover new compounds with exceptional medical properties [9].

The question of what new structures should be created in order to outperform the newly discovered molecules due to their medical qualities is closely tied to the development of new medicinal compounds [10]. Naturally, molecular mechanics algorithms play an overwhelming contribution in this situation. They are able to predict the biological characteristics of newly synthesized compounds and, more importantly, guide the creation of molecules with therapeutic qualities that are better than those now understood [11].

The remarkable importance that density functional theory plays in identifying the geometric, energetic, and interactional aspects of newly synthesized compounds—properties that define and determine their unique therapeutic capabilities—is evident in these new algorithms [12]. Current research also focuses on the way a molecule interacts with a particular protein, enzyme, or other biomolecule. This interaction determines the molecule's mode of interaction and the mechanism by which it binds to the protein "pocket." The target molecule's desired biological action is determined by the conformational changes, favoring energy, and overall stereochemistry [13].

The following are some parameters that the functional density theory identified to describe the therapeutic compounds under study and that positively and significantly contribute to good biological activity:

a. The first parameter examined is the energy differential between the molecular orbitals HOMO (highest occupied molecular orbital) and LUMO (lowest unoccupied molecular orbital), which is frequently represented in the literature as E_{gap} [14].

$$E_{gap} = E_{HOMO} - E_{LUMO} \tag{1}$$

A significant energy difference indicates a very stable molecule that interacts very strongly, making it more difficult for the molecule to exhibit its therapeutic effects via interacting with the body's biomolecules [15]. Conversely, a molecule's susceptibility to biological activity is determined by a minor energy difference, E_{gap}, which allows it to interact with the proteins that cause its biological activity due to the proximity of its HOMO-LUMO orbitals [16].

b. The second factor is a compound's "chemical hardness," which may be computed using the following formula and indicates a compound's stability and reactivity [17]:

$$\eta = \frac{E_{LUMO} - E_{HOMO}}{2} \tag{2}$$

A low value of chemical hardness defines a reactive compound, more therapeutically active than other compounds with higher values of the same parameter [18].

c. The third factor taken into account is the electronegativity with negative value, which is a molecule's electronic chemical potential as established by the equation [19]:

$$\mu - \frac{E_{HOMO} + E_{LUMO}}{2} \tag{3}$$

Thus, the bigger the electronic chemical potential (μ) is, the more the compound looks to be less stable or more reactive [20].

d. The global electrophilicity index (ω), which is derived from the electronic chemical potential and chemical hardness values, is an additional measure that Parr introduced [21].

$$\omega = \frac{\mu^2}{2\eta} \tag{4}$$

One molecule's stability after absorbing an electrical charge from its surroundings is gauged by this index. A more reactive molecule that is more therapeutically available for interactions in the body is indicated by a lower value of this index [22].

e. Atomic charges of the new compounds represent another parameter that defines the reactivity of biologically active compounds [23]. Atomic charges are connected to the ideas of electrophilicity and nucleophilicity. DFT theory is used to compute Mulliken charges on atoms [24].

It has been discovered that compounds with strongly nucleophilic groups—such as halogens, including F, Cl, Br, I pseudohalogens, CN, SCN, and groups like OH, SH, OCH_3, SH, NH_2, NHR, COOH, COH, and COR—have better therapeutic activity strictly related to the presence of these groups on the molecules [25].

The correlation between DFT parameters and the biological properties of various compounds reported in the specialized literature with various therapeutic properties, such as antibacterial, antifungal, antiviral, anti-inflammatory, analgesic, anticancer, antidiabetic, antioxidant, and so on, will be thoroughly discussed in the following analysis of various molecular structures optimized by density functional theory [26].

2. Density functional theory applied to medicinally active compounds

2.1 Correlation of DFT properties with structural, energetic, and electronic parameters of antimicrobial compounds

As shown in **Figure 1** [27], Grilot et al. reported the synthesis of second-generation antibacterial benzimidazole-pyrazoles **1** from 2,3,6-trifluorobenzenamine in seven steps. With a 3-pyridine moiety at C-5 that keeps a hydrogen bond with Arg136, as a result from DFT studies, compound **1a** demonstrated a respectable minimum inhibitory concentration (MIC) against *S. aureus* (0.25 μg/mL). As previously reported [28], the addition of a fluorine atom at position C-6 on pyrazole **1b** did not improve its antibacterial potency (0.25 μg/mL against *S. aureus*) or affinity for Gyrase B and Topoisomerase IV. However, it did improve oral exposure by two-fold, which led to the study of C-6-fluorobenzimidazole pyrazoles exclusively. As shown in **Table 1**, the authors also investigated how the addition of the substituent in the "5" position affected the polarity of molecules and therefore, the variation of MIC (minimum inhibitory concentration).

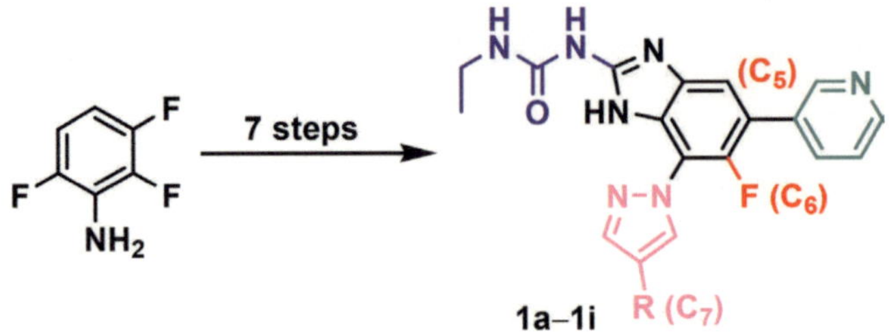

Figure 1.
Synthesis of pyrazoles grafted benzimidazole **1a–1i***.*

Compounds	C-5	C-6	C-7	MIC (μg/mL)	
				Sa	Sa + HS
1a		H	H	0.25	4
1b		F	H	0.25	2
1c	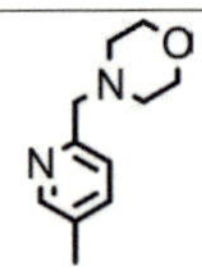	F	H	0.25	1
1d	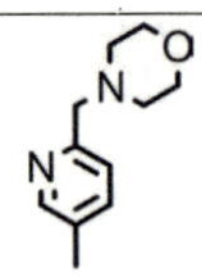	F	Me	0.063	0.5
1e	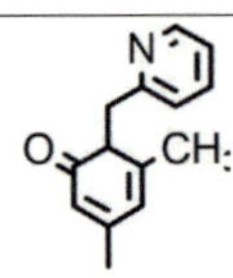	F	Me	0.016	0.125
1f	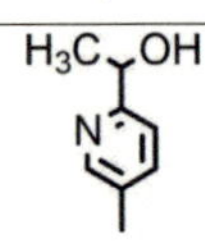	F	H	0.125	3
1g	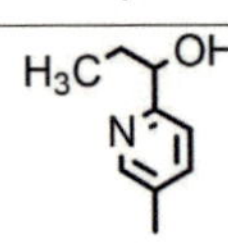	F	H	0.063	2
1h	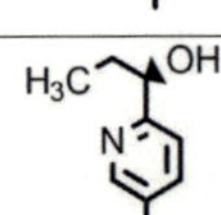	F	H	0.25	4
1i	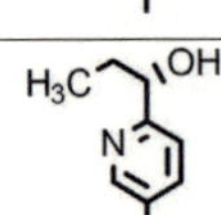	F	H	0.063	1

Sa = S. aureus; Sa + HS = S. aureus + 50% human serum.

Table 1.
Structure-activity relationship (SAR) investigations and MIC values for pyrazole series **1a–1i***.*

The addition of a methyl group to the C-5 substituent, as in compound **1 g**, could somewhat improve the antimicrobial properties of compound **1f**, which had a MIC against *S. aureus* of 0.125 μg/mL. Compounds **1 h** and **1i** were the result of its resolution. While compound **1 h** showed a large serum shift (16-fold), the (*S*)-isomer **1i** was 4-fold more effective against *S. aureus* and demonstrated acceptable oral exposure. Marinescu et al. synthesized a series of benzimidazole compounds and used DFT to optimize their structures at the M11/ktzvp level of theory (**Figure 2**). The calculated DFT parameters of the compounds are given in **Table 2**. Considering this data, it can be said that: i. The order of decreasing the reactivity from the E_{gap} value is: **2b** > **2c** > **2d** > **2e** > **1b** > **2f**, therefore a difference in reactivity for the compounds **2a** and **2e**. The chemical hardness (η) values of the benzimidazoles under study indicate that compound **2b** should be the most reactive due to its lower hardness, followed by compounds **2c**, **2d**, **2e**, and **2a**. Compound **2f** should be the least reactive. Additionally, it is evident that the data from compound

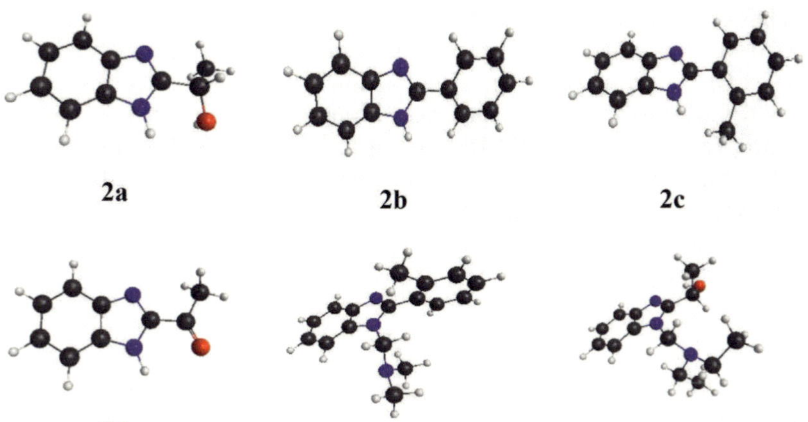

2a 2b 2c

2d 2e 2f

Figure 2.
Compounds **2a–2f** *with optimal structures (black—C, gray—H, red—O, blue—N).*

Energy (Hartree)		Benzimidazole compounds					
		2a	2b	2c	2d	2e	2f
E_{HOMO}		−0.322	−0.310	−0.311	−0.337	−0.315	−0.325
E_{LUMO}		0.048	0.016	0.023	−0.001	0.039	0.046
E_{gap}		−0.370	−0.326	−0.334	−0.336	−0.354	−0.371
η		0.185	0.163	0.167	0.168	0.177	0.1855
μ		−0.137	−0.152	−0.144	−0.169	−0.138	−0.1395
ω		0.050	0.070	0.062	0.085	0.053	0.052
Antimicrobial activity	Qualitative	44.44%	44.44%	55.55%	22.22%	11.11%	22.22%
	Quantitative (MIC)	33.33%	44.44%	33.33%	—	—	33.33%
	Quantitative (MBEC)	44.44%	33.33%	22.22%	22.22%	11.11%	44.44%

Table 2.
Indices of chemical reactivity for benzimidazoles **2a–2f**.

1b does not fit all. It is well known that a compound's stability or reactivity increases with its electronic chemical potential (μ). **Table 2** shows that the chemicals' reactivity decreases in the following order: **2c** > **2b** > **2d** *via* **2a** > **2e** > **2 f**. With the exception of **2e**, all chemicals correlate well. The global electrophilicity index ω has a decreasing order that best matches the reactivity of all compounds: **2a** > **2f** > **2e** > **2c** > **2b** > **2d**. The strongest nucleophile is compound **2a**, and the strongest electrophile is compound **2d**. With the exception of compound **2e**, it was found that the antibacterial activity and electronic characteristics correlate well. Using an environmentally friendly synthesis from aromatic aldehydes, 2-aminobenzothiazole, and dimedone with thiamine hydrochloride (Vitamine B_1) as an organocatalyst, Sethiya et al. produced pyrimidine derivatives **3a–3g** (**Figure 3**). The DFT program was used to conduct molecular docking experiments on compounds **3a–3g** utilizing DNA gyrase (1KZN) and *Staphylococcus aureus* dihydropteroate synthase (saDHPS) (6CLV) proteins. Against both proteins, 1KZN and 6CLV, compound **3e** was the most effective, exhibiting excellent binding interactions and the highest docking score [29]. Dihydropyrimidinones **4** and **5** were synthesized by Ramachandran et al. utilizing a solvent-free grindstone chemistry technique, which was catalyzed by HCl and $CuCl_2{\bullet}2H_2O$ (**Figure 4**) [30]. Significant action was shown by all of the produced compounds against the harmful pathogens *Staphylococcus aureus* and *Salmonella typhi*. These dihydropyrimidinones (DHPMs) also target the bacterial ribosomal A site RNA. Additionally, a number of DFT docking studies were conducted using the human 40S rRNA as the target. Compared to pyrimidine **5e**, which showed comparatively low binding affinity toward the human rRNA site, Amikacin medication was discovered to have a higher binding affinity.

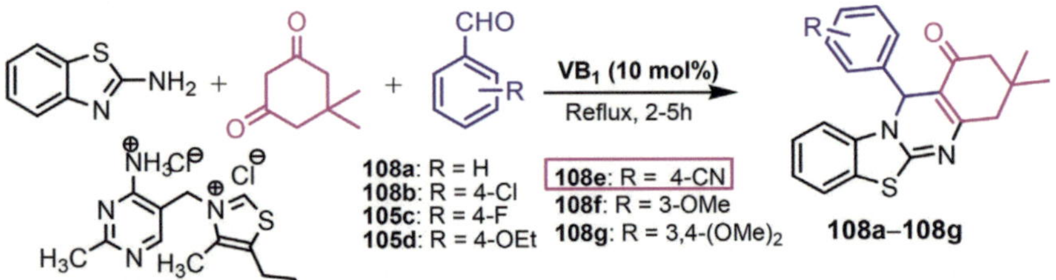

Figure 3.
Synthesis of compounds 3a–3g.

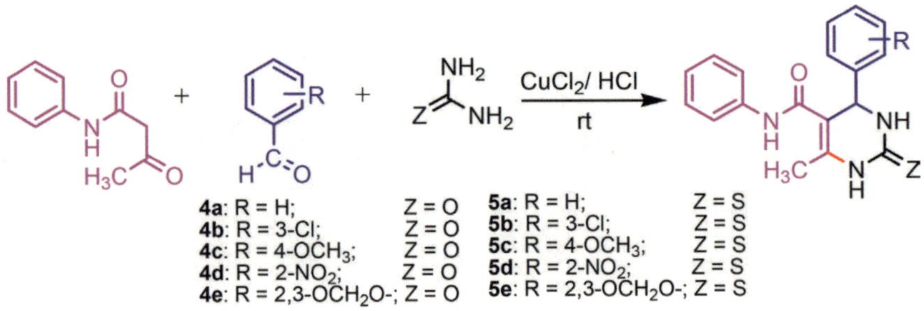

Figure 4.
Synthesis of compounds 4a–4e *and* 5a–5e.

A new family of antibacterial drugs, 3,4-dihydropyrimidinone-coumarin analogues **6** and 7 (**Figure 5**), was synthesized and published by Naik et al. [31]. The *in vitro* antibacterial investigations of the 3,4-dihydropyrimidinone grafted coumarins against *Pseudomonas aeruginosa, Bacillus subtilis, Escherichia coli*, and *Staphylococcus aureus* were assessed using the broth microdilution method (**Table 3**). All of the compounds showed good antibacterial activity against *S. aureus*, as indicated by bold values (MIC = 0.2–6.25 μg/mL). However, all of the compounds showed reduced activity against *B. subtilis.* According to the DFT experiments, the effectiveness of the substituent at position C-6 declined in the following order: $-CH_3$ > −7,8-Benzo > −Cl > $-OCH_3$ (b > e > c > a). In a similar vein, compounds **7a**, **7c**, and **7e** had considerable action against *E. coli,* while the remaining compounds exhibited notably lower activity. Octahydroacridines **8–11**, a class of tetra-substituted pyridines, were synthesized, and their structures were optimized at the M11/ktzvp level using DFT theory. The frontier orbitals' contour plots, including the lowest unoccupied molecular orbital (LUMO) and the highest occupied molecular orbital (HOMO), were also obtained using DFT at the M11/ktzvp level of theory [32, 33]. The energies of the HOMO and LUMO orbitals, E_{HOMO}, E_{LUMO}, E_{gap}, dipole moment (DM), and hydration energy (HE) for **8–11** were also determined (**Table 4**). The lowest E_{gap} for **8** indicates a better reactivity for it,

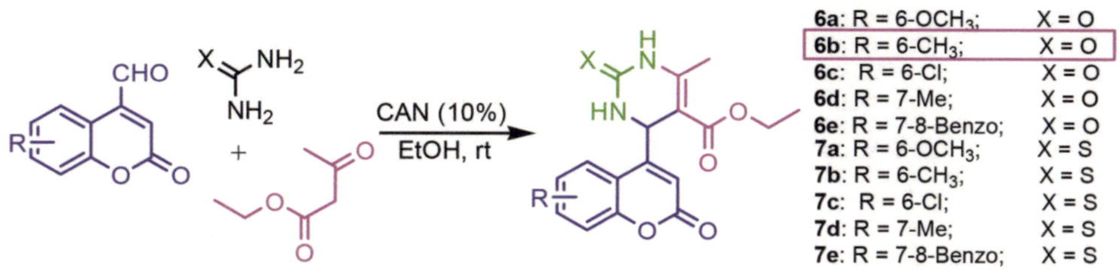

Figure 5.
Synthesis of compounds **6a–6e** *and* **7a–7e**.

Compound	Minimum inhibitory concentrations (MIC) (μg/mL)			
	S. aureus	B. subtilis	E. coli	P. aeruginosa
6a	**0.4**	12.5	50	100
6b	**0.8**	100	50	25
6c	3.12	100	25	50
6d	**0.2**	50	12.5	100
6e	**0.4**	100	25	25
7a	**0.4**	100	**1.6**	100
7b	**0.8**	50	25	100
7c	**0.8**	50	**0.4**	100
7d	6.25	50	25	100
7e	**0.4**	100	**0.4**	100
Ciprofloxacin	2	2	2	<4

Table 3.
Antibacterial activity of compounds **6a–6e** *and* **7a–7e** *under* in vitro *conditions.*

Properties/compound	Octahydroacridine-N-oxide			
	8	**9**	**10**	**11**
HOMO/eV	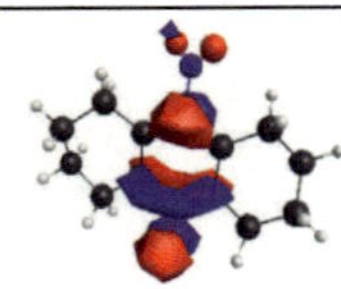	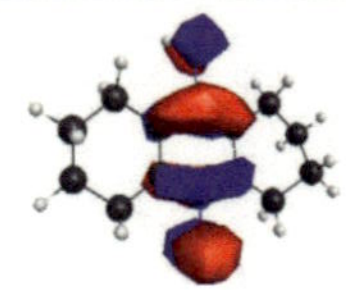	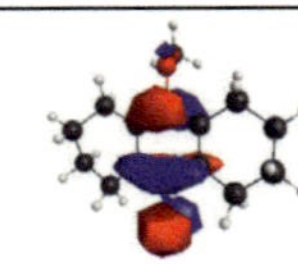	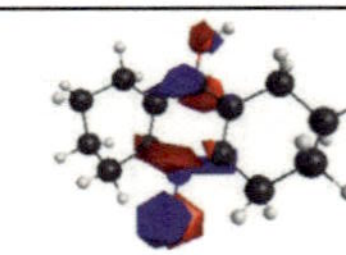
LUMO/eV	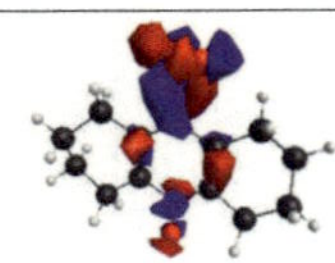	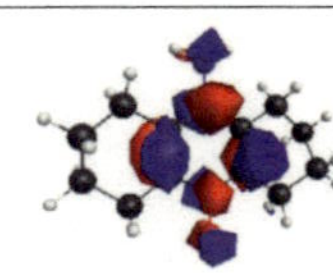		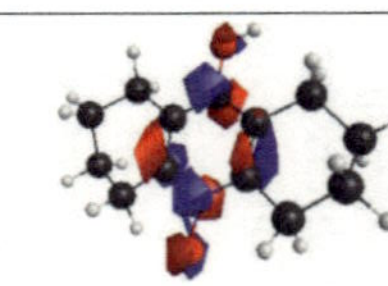
E_{total} (kcal/Mol)	−839.941	−690.785	−749.940	−710.661
E_{HOMO}/eV	−8.870	−7.428	−8.054	−7.809
E_{LUMO}/eV	−0.489	1.659	1.224	1.442
E_{GAP}/eV	−8.381	−9.087	**−9.278**	−9.251
DM (D)	0.674	5.682	4.223	4.442
HE (Kcal/Mol)	−3.14	−1.72	−0.59	−3.76
Antimicrobial activity	41.6%	33.3%	Weak	50%

Table 4.
The HOMO and LUMO frontier orbital contour plots, their energies, E_{GAP} (eV), dipole moments, hydration energies, and antibacterial activity for octahydroacridine-N-oxides **8–11**.

which concretizes in good qualitative antibacterial activity as well as the best minimum inhibitory concentrations; the largest E_{gap} for **10**, which is the minimum reactivity of all chemicals, correlates with the worst antimicrobial activity [34].

The results are summarized in **Figure 6**, which shows that compound **8** has the strongest antibacterial activity with a minimum E_{gap} of −8.381 eV and compound **9** has the weakest antimicrobial activity with a maximum E_{gap} of −9.278 eV [32, 33, 35].

Kalinowska et al. correlated the antimicrobial activities of a series of natural acids with the DFT parameters determined for the compounds (**Table 5**) [34]. It was found that the compounds under test had no effect on the viability of *E. coli* within the studied concentration range (0.98–500 μM). This indicates that the compounds do not promote the growth of a potentially harmful bacterium that is naturally found in the human microbiome [36]. Furthermore, at the concentration range of 0.98–500 μM, **GA**, **CA**, **5-CQA**, ***p*-CA**, and **RA** (**Figure 7**) demonstrated prebiotic qualities toward *L. rhamnosus*. The prebiotic effect was seen for **TA** at lower concentrations between 0.98 and 15.63 μM [37]. The phenolic compounds' antibacterial activity against *L. rhamnosus* was associated with their lipophilicity, as indicated by the logP parameter [38].

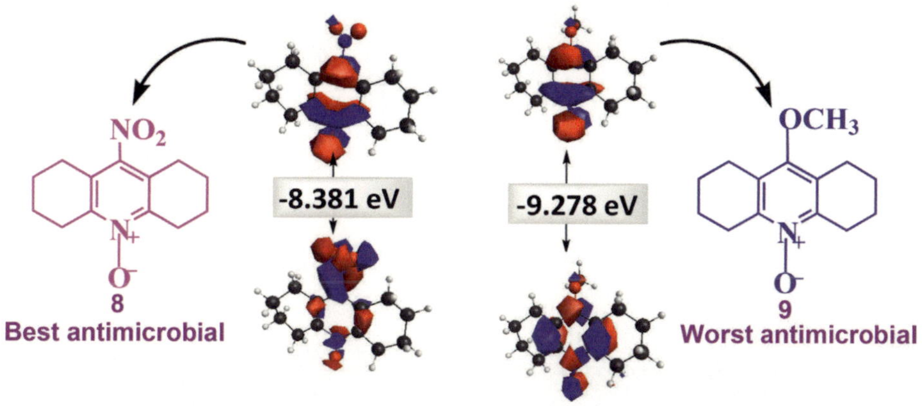

Figure 6.
*Compounds **8** and **9** with their HOMO-LUMO orbitals, and E_{gap}.*

Descriptors	10 *p*-CA	11 CA	12 RA	13 5-CQA	14 GA	15 TA
E_{HOMO} (eV)	−6.431	−6.252	−5.796	−6.281	−6.344	−8.554
E_{LUMO} (eV)	−2.148	−2.114	−2.251	−2.198	−1.613	−1.805
Energy gap (eV)	4.283	4.138	3.545	4.083	4.731	6.749
Ionization potential, $I = -E_{HOMO}$	6.431	6.252	5.796	6.281	6.344	8.554
Electron affinity, $A = -E_{LUMO}$	2.148	2.114	2.251	2.198	1.613	1.805
Electronegativity, $\chi = I + A/2$	4.289	4.183	4.023	4.240	3.978	5.179
Chemical potential, $\mu = —I + A/2$	−4.289	−4.183	−4.023	−4.240	−3.978	−5.179
Chemical hardness, $\eta = I-A/2$	2.141	2.069	1.773	2.042	2.366	3.374
Chemical softness, $S = 1/2\eta$	0.234	0.242	0.282	0.245	0.211	0.148
Electrophilicity index, $\omega = \mu^2/2\eta$	4.296	4.228	4.566	4.402	3.345	3.975

Table 5.
Energy of HOMO/LUMO orbitals and other reactivity descriptors.

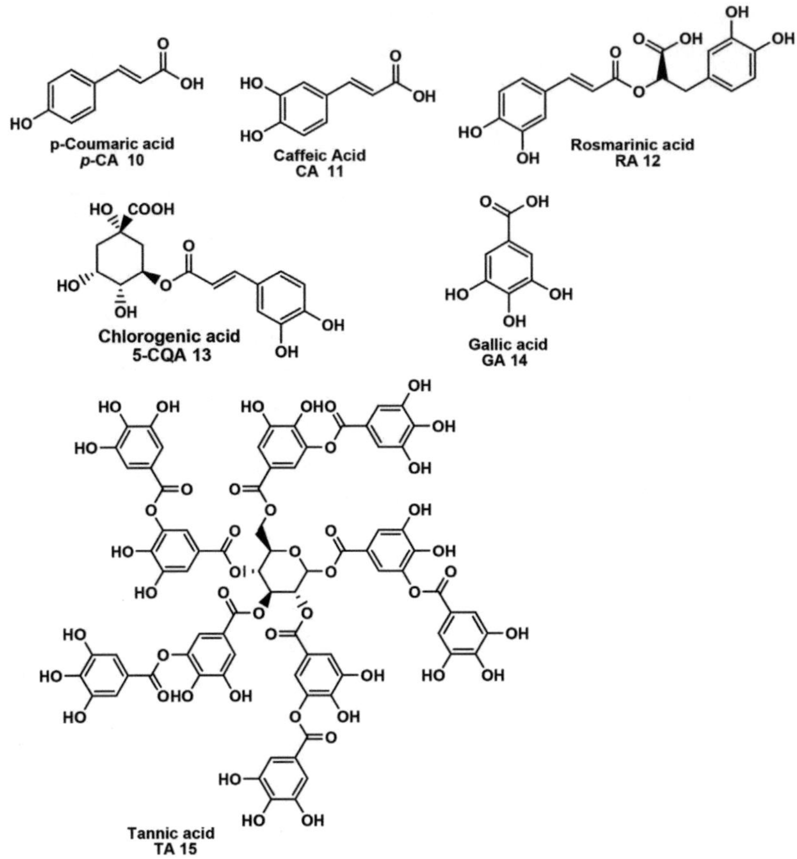

Figure 7.
Natural compounds studied for their descriptors.

At lower concentration levels, the compounds (**TA**, *p*-**CA**) with the highest logP values demonstrated an inhibitory effect on microbial growth compared to the phenolic compounds with greater logP [39]. It is interesting to note that the computed HOMO orbital energy represented variations in these compounds' lipophilicity and can be utilized as an extra metric to characterize the biological activity of phenolic compounds against *L. rhamnosus* [40]. Phenolic acids' cytotoxic properties were examined within the same concentration range as the antibiotic [41].

3. Conclusions

In this chapter of the book, a literature search was attempted to reflect the correlations between the electronic parameters of organic molecules and their antimicrobial properties. A variety of compounds studied in this regard were found, especially heterocyclic compounds with one or more identical or different rings, but also natural compounds, such as polyphenols, flavonoids, coumarins, and so on.

Among the important parameters that correlate well with the antimicrobial activity of the compounds, we can note: HOMO-LUMO energy difference (E_{gap}), chemical hardness (η), electronegativity (μ), global electrophilicity index (ω), and atomic charges on the electronegative elements in the molecules [42].

The examples illustrated in this chapter, as well as others, which were only referred to in the literature, are eloquent and reflect very well the importance of the DFT study in the prediction of antimicrobial activities [43], and not only, antiviral [44], antitumor [45], and so on.

More and more scientific journals refer to the correlations between determined antimicrobial activities and calculated DFT properties, correlations that have become useful tools in predicting the biological activities of compounds proposed for synthesis, and their evaluation as new therapeutic treatment agents.

We hope that this chapter will be a useful tool in the synthesis of new antimicrobial compounds, which will facilitate the work of young researchers.

Author details

Maria Marinescu[1*] and Claudia-Valentina Popa[1,2*]

1 Faculty of Chemistry, University of Bucharest, Bucharest, Romania

2 Cantacuzino National Military Medical Institute for Research and Development, Bucharest, Romania

*Address all correspondence to: maria.marinescu@chimie.unibuc.ro
and claudia.popa@cantacuzino.ro

References

[1] Yılmaz GE, Göktürk I, Ovezova M, Yılmaz F, Kılıç S, Denizli A. Antimicrobial nanomaterials: A review. Hygiene. 2023;**3**(3):269-290. DOI: 10.3390/hygiene3030020

[2] Marinescu M. Biginelli reaction mediated synthesis of antimicrobial pyrimidine derivatives and their therapeutic properties. Molecules. 2021;**26**(19):6022. DOI: 10.3390/molecules26196022

[3] Potmischil F, Marinescu M, Nicolescu A, Deleanu C, Hillebrand M. Hydroacridines: Part 29. 15N NMR chemical shifts of 9-substituted 1,2,3,4,5,6,7,8-octahydroacridines and their N-oxides—Taft, swain–Lupton, and other types of linear correlations. Magnetic Resonance in Chemistry. 2008;**46**(12):1141-1147. DOI: 10.1002/mrc.2335

[4] Marinescu M. Bisindole compounds—Synthesis and medicinal properties. Antibiotics. 2024;**13**(12):1212. DOI: 10.3390/antibiotics13121212

[5] Zalaru C, Dumitrascu F, Draghici C, Ferbinteanu M, Tarcomnicu I, Marinescu M, et al. Synthesis, spectroscopic characterization, structural analysis, and evaluation of anti-tumor, antimicrobial, and antibiofilm activities of halogenoaminopyrazoles derivatives. Antibiotics. 2024;**13**(12):1119. DOI: 10.3390/antibiotics13121119

[6] Potmischil F, Deleanu C, Marinescu M, Gheorghiu MD. Hydroacridines: Part 23.† 1H and 13C NMR spectra of sym-octahydroacridine, its 9-(3-pyridyl) and 9-(4-pyridyl) derivatives and the corresponding N(10)-oxides. An experimental approach to the diamagnetic anisotropy of the pyridine nucleus. Magnetic Resonance in Chemistry. 2002;**40**(3):237-240. DOI: 10.1002/mrc.996

[7] Islam T, Rahaman MM, Mia MN, Ara I, Islam MT, Alam Riaz T, et al. Therapeutic perspectives of metal nanoformulations. Drugs and Drug Candidates. 2023;**2**(2):232-278. DOI: 10.3390/ddc2020014

[8] Joseph TM, Kar Mahapatra D, Esmaeili A, Piszczyk Ł, Hasanin MS, Kattali M, et al. Nanoparticles: Taking a unique position in medicine. Nanomaterials. 2023;**13**(3):574. DOI: 10.3390/nano13030574

[9] Visan AI, Negut I. Integrating artificial intelligence for drug discovery in the context of revolutionizing drug delivery. Life. 2024;**14**(2):233. DOI: 10.3390/life14020233

[10] Serrano DR, Luciano FC, Anaya BJ, Ongoren B, Kara A, Molina G, et al. Artificial intelligence (AI) applications in drug discovery and drug delivery: Revolutionizing personalized medicine. Pharmaceutics. 2024;**16**(10):1328. DOI: 10.3390/pharmaceutics16101328

[11] Blanco-González A, Cabezón A, Seco-González A, Conde-Torres D, Antelo-Riveiro P, Piñeiro Á, et al. The role of AI in drug discovery: Challenges, opportunities, and strategies. Pharmaceuticals. 2023;**16**(6):891. DOI: 10.3390/ph16060891

[12] Zois KP, Tzeli D. A critical look at density functional theory in chemistry: Untangling its strengths and weaknesses. Atoms. 2024;**12**(12):65. DOI: 10.3390/atoms12120065

[13] Akther T, McFadden WM, Zhang H, Kirby KA, Sarafianos SG, Wang Z. Design and synthesis of new GS-6207 subtypes for targeting HIV-1 capsid protein. International Journal of Molecular Sciences. 2024;**25**(7):3734. DOI: 10.3390/ijms25073734

[14] Mubarik A, Rasool N, Hashmi MA, Mansha A, Zubair M, Shaik MR, et al. Computational study of structural, molecular orbitals, optical and thermodynamic parameters of thiophene sulfonamide derivatives. Crystals. 2021;**11**(2):211. DOI: 10.3390/cryst11020211

[15] Róg T, Girych M, Bunker A. Mechanistic understanding from molecular dynamics in pharmaceutical research 2: Lipid membrane in drug design. Pharmaceuticals. 2021;**14**(10):1062. DOI: 10.3390/ph14101062

[16] Lefi N, Kazachenko AS, Raja M, Issaoui N, Kazachenko AS. Molecular structure, spectral analysis, molecular docking and physicochemical studies of 3-Bromo-2-hydroxypyridine monomer and dimer as bromodomain inhibitors. Molecules. 2023;**28**(6):2669. DOI: 10.3390/molecules28062669

[17] Mutahir S, Khan MA, Mushtaq M, Deng H, Naglah AM, Almehizia AA, et al. Investigations of electronic, structural, and in silico anticancer potential of persuasive phytoestrogenic isoflavene-based Mannich bases. Molecules. 2023;**28**(15):5911. DOI: 10.3390/molecules28155911

[18] Khalid Z, Shafqat SS, Ahmad HA, Munawar MA, Mutahir S, Elkholi SM, et al. A combined experimental and computational study of novel benzotriazinone carboxamides as alpha-glucosidase inhibitors. Molecules. 2023;**28**(18):6623. DOI: 10.3390/molecules28186623

[19] Geerlings P. From density functional theory to conceptual density functional theory and biosystems. Pharmaceuticals. 2022;**15**(9):1112. DOI: 10.3390/ph15091112

[20] Pratiwi R, Ibrahim S, Tjahjono DH. Reactivity and stability of metalloporphyrin complex formation: DFT and experimental study. Molecules. 2020;**25**(18):4221. DOI: 10.3390/molecules25184221

[21] Elshakre ME, Noamaan MA, Moustafa H, Butt H. Density functional theory, chemical reactivity, pharmacological potential and molecular docking of dihydrothiouracil-indenopyridopyrimidines with human-DNA topoisomerase II. International Journal of Molecular Sciences. 2020;**21**(4):1253. DOI: 10.3390/ijms21041253

[22] Bramki A, Benouchenne D, Salvatore MM, Benslama O, Andolfi A, Rahim N, et al. In vitro and in silico biological activities investigation of ethyl acetate extract of *Rubus ulmifolius* Schott leaves collected in Algeria. Plants. 2024;**13**(23):3425. DOI: 10.3390/plants13233425

[23] Shanthakumar DC, Nagarajappa LT, Isamura BK, Leoma MB, Mokgopa KP, Anandalwar SM, et al. The thiadiazole ring (THD) is a building block for potential inhibitors of the SARS-CoV-2 main protease (Mpro): Theoretical look into the structure, reactivity, and binding profile of three 1,3,4-THD derivatives toward Mpro. Engineering Proceedings. 2023;**59**(1):94. DOI: 10.3390/engproc2023059094

[24] Deghady AM, Hussein RK, Alhamzani AG, Mera A. Density functional theory and molecular docking investigations of the chemical and antibacterial activities for

1-(4-hydroxyphenyl)-3-phenylprop-2-en-1-one. Molecules. 2021;**26**(12):3631. DOI: 10.3390/molecules26123631

[25] Marinescu M. Synthesis of antimicrobial benzimidazole–pyrazole compounds and their biological activities. Antibiotics. 2021;**10**(8):1002. DOI: 10.3390/antibiotics10081002

[26] Chouchène N, Toumi A, Boudriga S, Edziri H, Sobeh M, Abdelfattah MAO, et al. Antimicrobial activity and DFT studies of a novel set of spiropyrrolidines tethered with thiochroman-4-one/chroman-4-one scaffolds. Molecules. 2022;**27**(3):582. DOI: 10.3390/molecules27030582

[27] Marinescu M. Benzimidazole-triazole hybrids as antimicrobial and antiviral agents: A systematic review. Antibiotics. 2023;**12**(7):1220. DOI: 10.3390/antibiotics12071220

[28] Khwaza V, Mlala S, Aderibigbe BA. Advancements in synthetic strategies and biological effects of ciprofloxacin derivatives: A review. International Journal of Molecular Sciences. 2024;**25**(9):4919. DOI: 10.3390/ijms25094919

[29] Sethiya A, Soni J, Manhas A, Prakash CJ, Shikha A. Green and highly efficient MCR strategy for the synthesis of pyrimidine analogs in water via C–C and C–N bond formation and docking studies. Research on Chemical Intermediates. 2021;**47**:4477-4496. DOI: 10.1007/s11164-021-04529-0

[30] Ramachandran V, Arumugasamy K, Singh SK, Naushad E, Ramesh P, Kamaraj SK. Synthesis, antibacterial studies, and molecular modeling studies of 3,4-dihydropyrimidinone compounds. Journal of Chemical Biology. 2016;**9**:31-40. DOI: 10.1007/s12154-015-0142-4

[31] Naik NS, Shastri LA, Joshi SD, Dixit SR, Chougala BM, Samundeeswari S, et al. 3,4-Dihydropyrimidinone-coumarin analogues as a new class of selective agent against S. aureus: Synthesis, biological evaluation and molecular modelling study. Bioorganic Medicinal Chemistry. 2017;**25**:1413-1422. DOI: 10.1016/j.bmc.2017.01.001

[32] Marinescu M, Cinteză LO, Marton GI, Chifiric MC, Popa M, Stanculescu I, et al. Synthesis, density functional theory study and in vitro antimicrobial evaluation of new benzimidazole Mannich bases. BMC Chemistry. 2020;**14**:45. DOI: 10.1186/s13065-020-00697-z

[33] Marinescu M, Cinteza LO, Marton G, Măruțescu L, Chifiriuc MC, Constantinescu C. Density functional theory molecular modeling and antimicrobial behaviour of selected 1,2,3,4,5,6,7,8-octahydroacridine-N(10)-oxides. Journal of Molecular Structure. 2017;**1144**:14-23. DOI: 10.1016/j.molstruc.2017.05.003

[34] Kalinowska M, Świsłocka R, Wołejko E, Jabłońska-Trypuć A, Wydro U, Kozłowski M, et al. Structural characterization and evaluation of antimicrobial and cytotoxic activity of six plant phenolic acids. PLoS One. 2023;**19**(6):e0299372. DOI: 10.1371/journal.pone.0299372

[35] Marinescu M, Tudorache DG, Marton G, Zalaru CM, Popa M, Chifiriuc MC, et al. Density functional theory molecular modeling, chemical synthesis, and antimicrobial behaviour of selected benzimidazole derivatives. Journal of Molecular Structure. 2017;**1130**:463-471. DOI: 10.1016/j.molstruc.2016.10.066

[36] Thapa K, Julianingsih D, Tung C-W, Phan A, Hashmi MA, Bleich K,

et al. Berry pomace extracts as a natural washing aid to mitigate enterohaemorrhagic *E. coli* in fresh produce. Food. 2024;**13**(17):2746. DOI: 10.3390/foods13172746

[37] Li S, Li Y, Sui D, Ren Q, Ai C, Li M, et al. Anti-inflammatory effects of novel probiotic lactobacillus rhamnosus RL-H3-005 and pedicoccus acidilactici RP-H3-006: In vivo and in vitro evidence. Food. 2024;**13**(22):3676. DOI: 10.3390/foods13223676

[38] Wróblewska B, Kuliga A, Wnorowska K. Bioactive dairy-fermented products and phenolic compounds: Together or apart. Molecules. 2023;**28**(24):8081. DOI: 10.3390/molecules28248081

[39] Kauffmann AC, Castro VS. Phenolic compounds in bacterial inactivation: A perspective from Brazil. Antibiotics. 2023;**12**(4):645. DOI: 10.3390/antibiotics12040645

[40] Stefaniu A, Pirvu LC, Pintilie L, Godeanu SC. Bioavailability computations for natural phenolic derivatives for druglikeness assessment. Chemistry Proceedings. 2023;**13**(1):26. DOI: 10.3390/chemproc2023013026

[41] Krzemińska M, Owczarek A, Gonciarz W, Chmiela M, Olszewska MA, Grzegorczyk-Karolak I. The antioxidant, cytotoxic and antimicrobial potential of phenolic acids-enriched extract of elicited hairy roots of salvia bulleyana. Molecules. 2022;**27**(3):992. DOI: 10.3390/molecules27030992

[42] Świsłocka R, Regulska E, Karpińska J, Świderski G, Lewandowski W. Molecular structure and antioxidant properties of alkali metal salts of rosmarinic acid. Experimental and DFT Studies Molecules. 2019;**24**(14):2645. DOI: 10.3390/molecules24142645

[43] El-Shamy NT, Alkaoud AM, Hussein RK, Ibrahim MA, Alhamzani AG, Abou-Krisha MM. DFT, ADMET and molecular docking investigations for the antimicrobial activity of 6,6′-diamino-1,1′,3,3′-tetramethyl-5,5′-(4-chlorobenzylidene) bis[pyrimidine-2,4(1H,3H)-dione]. Molecules. 2022;**27**(3):620. DOI: 10.3390/molecules2703062

[44] Laref S, Harrou F, Sun Y, Gao X, Gojobori T. Exploring antiviral drugs on monolayer black phosphorene: Atomistic theory and explainable machine learning-assisted platform. International Journal of Molecular Sciences. 2024;**25**(9):4897. DOI: 10.3390/ijms25094897

[45] Dhahri M, Khan FA, Emwas A-H, Alnoman RB, Jaremko M, Rezki N, et al. Synthesis, DFT molecular geometry and anticancer activity of symmetrical 2,2′-(2-oxo-1H-benzo[d] imidazole-1,3(2H)-diyl) diacetate and its arylideneacetohydrazide derivatives. Materials. 2022;**15**(7):2544. DOI: 10.3390/ma15072544

Section 3

Biomedical and Therapeutic Applications

Chapter 6

Molecular Dynamics Insights into Novel 1,2,4-Triazole-Based Schiff Base Compounds as Dual-EGFR/Tubulin Inhibitor for the Treatment of Cancer Patient

Ahmed A. Elrashedy and Asmaa A. Magd El-Din

Abstract

A significant obstacle in cancer therapy is multidrug resistance (MDR), where tumors become unresponsive to both traditional and advanced chemotherapeutic agents. One emerging approach to combat MDR involves simultaneously targeting EGFR and Tubulin polymerization, disrupting critical pathways that drive uncontrolled cancer cell proliferation. Recent research has highlighted **Compound 1** as a potent dual inhibitor of EGFR and Tubulin, yet its exact binding mechanisms and selectivity remain incompletely understood. To elucidate its inhibitory effects, molecular dynamics (MD) simulations were performed, uncovering critical interactions between the compound and key residues in both targets. To explore its inhibitory effects, molecular dynamics (MD) simulations were conducted, revealing key interactions between the compound and specific residues in the binding pockets of both targets. For EGFR, **Compound 1** selectively binds to catalytic site residues, including Leu 23 (−2.097 kcal/mol), Val 31 (−1.697 kcal/mol), Ala 48 (−1.28 kcal/mol), Lys 50 (−3.365 kcal/mol), and Leu 149 (−1.775 kcal/mol). In contrast, for Tubulin, it interacts strongly with Lys 251 (−2.067 kcal/mol), Leu 252 (−3.037 kcal/mol), Cys 238 (−1.18 kcal/mol), and Leu 245 (−0.571 kcal/mol). These findings provide crucial structural insights for rational drug design, paving the way for next-generation dual inhibitors with enhanced selectivity and efficacy against resistant cancers. Such advancements could offer new therapeutic options for patients who no longer respond to existing treatments.

Keywords: 1,2,4 triazole, EGFR/tubulin inhibitor, cancer, molecular dynamic, MDR

1. Introduction

Over the past two decades, significant research efforts have focused on understanding the molecular mechanisms underlying metastatic cell transformation [1]. Scientists are working to develop novel therapeutic agents capable of suppressing cancer cell proliferation by blocking the conversion of normal cells into metastatic cells, thereby reducing reliance on conventional cytotoxic treatments that indiscriminately damage both healthy and cancerous cells [2]. A key approach involves identifying tumor-specific therapies by analyzing the biological differences between normal and malignant cells [3]. Such targeted treatments could minimize the nonspecific toxic effects of chemotherapy by selectively disrupting cellular processes critical for tumor growth, replication, or malignant progression. These therapies may exert either cytotoxic or cytostatic effects on cancer cells [4].

Considerable research attention has been directed toward the epidermal growth factor receptor (EGFR) due to its role in cancer progression [5]. Upon binding to its ligands, EGFR undergoes autophosphorylation, initiating a signaling cascade that promotes cellular proliferation [6]. Elevated EGFR expression in tumor cells is associated with poor prognosis and enhanced metastatic growth [7]. Consequently, pharmacological research has prioritized the development of small-molecule EGFR inhibitors over the last 20 years [8]. Inhibiting dual-EGFR and Tubulin polymerization represents a unique chemotherapeutic approach to limit the uncontrolled proliferation of several cancer cell types [9]. These inhibitors have shown promise in enhancing the efficacy of traditional anticancer treatments, such as chemotherapy and radiation, leading to significant tumor shrinkage in some cases [10].

Additionally, 1,2,4-triazole derivatives have attracted interest due to their diverse biological properties, including anti-inflammatory, analgesic, antimicrobial, and anticancer effects [9, 11]. Microtubules also represent an important therapeutic target, as they play vital roles in cell division, intracellular transport, and maintaining cellular structure. Both natural and synthetic compounds that disrupt microtubule function—particularly those interfering with mitotic spindle formation—are being explored as potential anticancer agents [12].

Vorozole, letrozole, and anastrozole are prominent anticancer drugs featuring the 1,2,4-triazole moiety, functioning as aromatase inhibitors in breast cancer treatment (**Figure 1**) [13]. Beyond their hormonal activity, the 1,2,4-triazole scaffold has also exhibited inhibitory effects on EGFR and Tubulin, as documented in prior research [14].

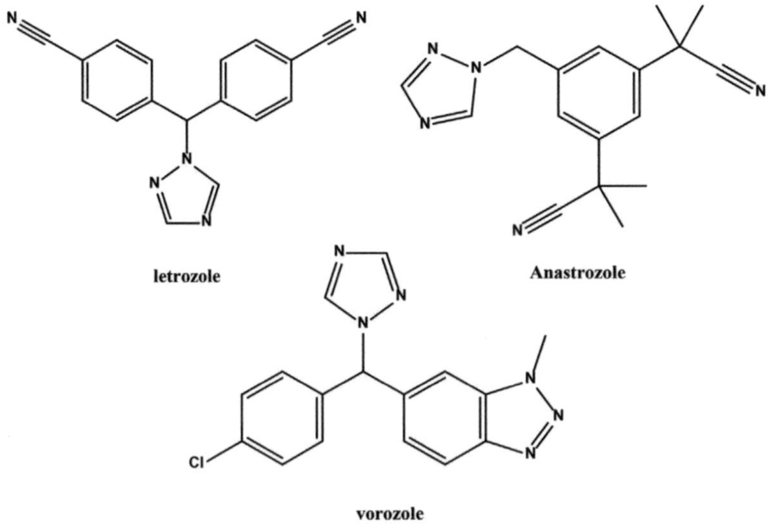

Figure 1.
1,2,4-triazoles as anticancer drugs.

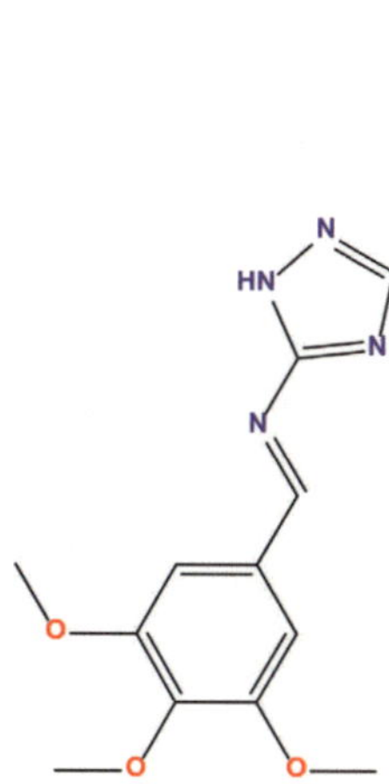

Figure 2.
2D structure of ***compound 1.***

Schiff bases represent another pharmacologically significant class of compounds, demonstrating diverse biological activities such as antimicrobial, anti-inflammatory, antiviral, and antioxidant effects [15–21]. Additionally, they exhibit anticancer potential by targeting key enzymes, proteins, and receptors involved in cell proliferation [22]. A promising strategy in anticancer drug development involves combining two pharmacophores with distinct mechanisms of action [23, 24].

A recent study by Mohamed *et al.* has experimentally investigated the potential of inhibiting Tubulin and EGFR at their source to develop an effective therapeutic method for cancer patients [25].

Tubulin and epidermal growth factor receptor (EGFR) were therapeutically targeted by **compound 1 (Figure 2)** [26]. To address this challenge, a combined *in silico* and *in vitro* approach was utilized, leading to the discovery of **Compound 1** as a multi-target inhibitor capable of simultaneously blocking both Tubulin polymerization and epidermal growth factor receptor (EGFR) activity. Computational modeling and docking studies suggested that this compound selectively binds to both targets, though further investigation is needed to fully understand its inhibitory mechanisms—particularly at the catalytic binding sites of Tubulin and EGFR.

To gain deeper insights, molecular dynamics (MD) simulations were employed to analyze the interaction dynamics, binding affinity, and selectivity of **Compound 1** with these two enzymes. These findings are expected to contribute to the structure-based design of novel, highly selective dual-target inhibitors that can effectively disrupt both Tubulin and EGFR in cancer therapy. This research could pave the way for more efficient multi-target drugs, offering a promising strategy to overcome multidrug resistance (MDR) in cancer treatment.

2. Computational methods

2.1 System preparation and MD simulations

The crystal structures of the Human Epidermal Growth Factor Receptor (EGFR) tyrosine kinase domain and Tubulin (PDB IDs: 1 M17 and 5LYJ, respectively) [27, 28] were retrieved from the RCSB Protein Data Bank [29]. The structures were prepared for molecular dynamics (MD) simulations using UCSF Chimera [30] and Molegro Molecular Viewer (MMV) [31]. To reduce computational costs, the monomeric form of each protein was used. Missing residues were modeled using MODELER 9.19 integrated into Chimera [32]. The ligand (**Compound 1**) was obtained from PubChem [33], and hydrogen atoms were added while nonpolar hydrogens were removed from the receptor.

2.2 Molecular dynamic simulations

All the molecular dynamic simulation were carried out using the GPU amber 14 software package [34]. The methodology section of this research was previously reported [35].

3. Results and discussion

3.1 Molecular dynamic and system stability

The stability of both unbound (apo) and bound protein-ligand complexes is crucial for guaranteeing reliable simulation results and avoiding artifacts during the 20 ns molecular dynamics production run. We assessed system stability by monitoring the root mean square deviation (RMSD) of the protein backbone atoms throughout the simulation trajectory.

The average RMSD values across all simulation frames were calculated as 3.29 ± 0.89 Å for apo-EGFR and 2.67 ± 0.58 Å for the EGFR-**compound 1** complex (**Figure 3A**). Similarly, Tubulin systems showed average RMSD values of 0.97 ± 0.09 Å for apo-Tubulin and 0.93 ± 0.11 Å for the Tubulin-**compound 1** complex (**Figure 3B**).

These results demonstrate that **compound 1** binding contributes to enhanced conformational stability in both protein systems compared to their apo forms. Notably, the lower RMSD fluctuations in the bound states suggest that **compound 1** forms

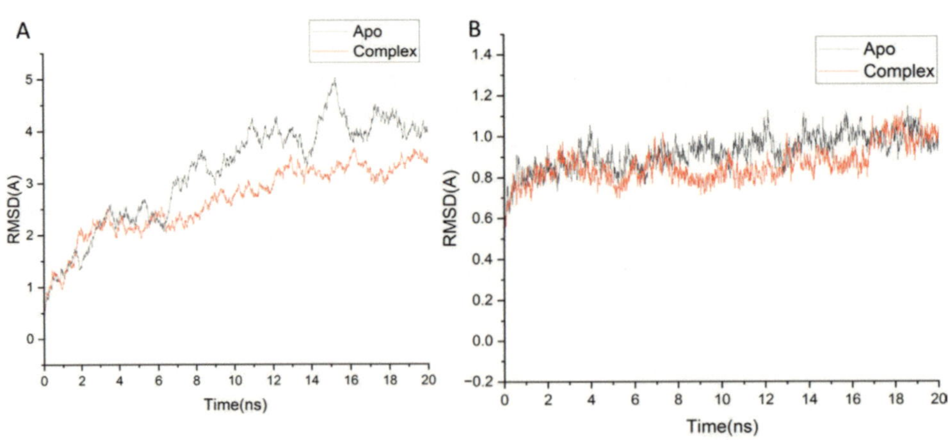

Figure 3.
*RMSD of Cα atoms of the protein backbone atoms showing the degree of stability upon **compound 1** binding [A] EGFR [B] Tubulin.*

stable interactions with both EGFR and Tubulin, maintaining structural integrity throughout the simulation period. The observed stability supports the reliability of subsequent analyses of binding interactions and dynamics.

Furthermore, the binding of **compound 1** to both target proteins induced distinct structural perturbations, as evidenced by increased backbone atom deviations compared to their unbound counterparts. Root mean square fluctuation (RMSF) analysis revealed average atomic fluctuations of 1.63 ± 0.24 Å for apo-EGFR versus 1.50 ± 0.41 Å for the EGFR-**compound 1** complex (**Figure 4A**). Similarly, Tubulin systems showed fluctuations of 0.83 ± 0.35 Å for apo-Tubulin and 0.79 ± 0.31 Å for the Tubulin-**compound 1** complex (**Figure 4B**). The data indicated that the **compound 1** attached to the protein complex system exhibits reduced residue variation compared to the other systems.

To assess the l structural compactness of EGFR and Tubulin upon ligand binding, we calculated the radius of gyration (Rg), which measures the mass-weighted root-mean-square distance of atoms from the protein's center of mass [36, 37]. The average ROG values were 22.81 ± 0.20 Å, 22.00 ± 0.32, 21.07 ± 0.04 Å, and 20.51 ± 0.04 Å for EGFR-apo, and EGFR- **compound 1**, Tubulin -Apo, Tubulin – **compound 1** systems, respectively, as shown in **Figure 5A** and **B**. Based on the observed behavior, **compound 1** exhibits a highly rigid structure in relation to the catalytic binding site of the target receptor.

To further elucidate the structural effects of ligand binding, we examined the solvent-accessible surface area (SASA) to quantify protein-solvent interactions and assess alterations in the compactness of the hydrophobic core. SASA measures the protein surface area accessible to solvent molecules, offering essential insights into bimolecular stability and folding kinetics [38]. The computed average SASA values were 17065.32, 16801.53, 17987.09, and 17496.26 Å for EGFR-apo, and EGFR- **compound 1**, Tubulin -Apo, Tubulin – **compound 1** systems, respectively, as shown in **Figure 6A** and **B**. The SASA results, along with data from RSMD, RMSF, and ROG calculations, further confirm that EGFR and Tubulin demonstrate enhanced structural stability when bound to **compound 1**.

The disruption of backbone atoms in EGFR and Tubulin may elucidate the mechanistic inhibitory activity of **compound 1**, since an induced loss of structural integrity in a protein correlates with a loss of functionality.

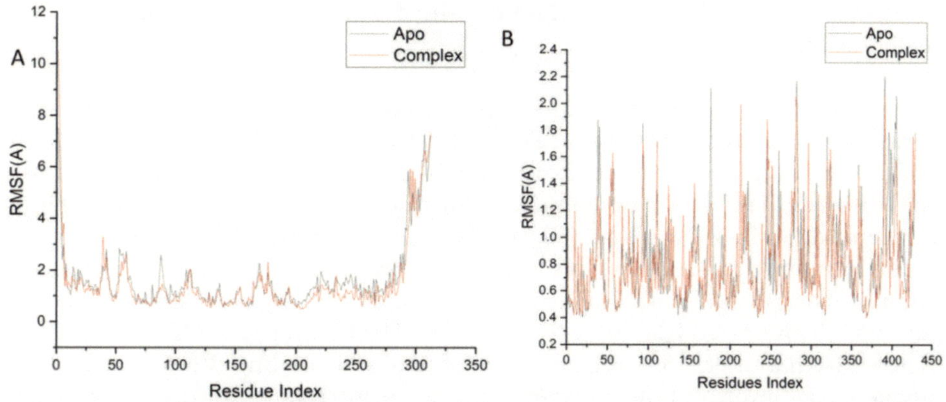

Figure 4.
RMSF of each residue of the protein backbone Cα atoms indicates the extent of flexibility upon the binding of ***compound 1*** *to [A] EGFR and [B] Tubulin.*

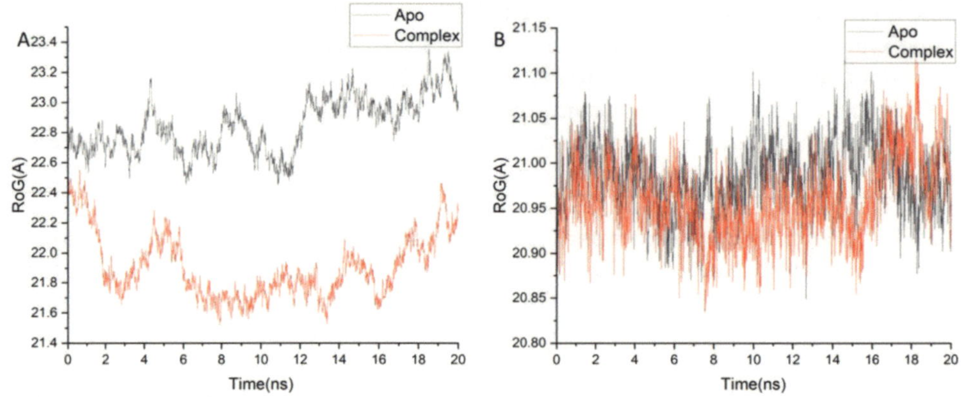

Figure 5.
Radius of Gyration (ROG) of Cα atoms in protein residues indicates the level of compactness upon the binding of ***compound 1*** *to [A] EGFR and [B] Tubulin.*

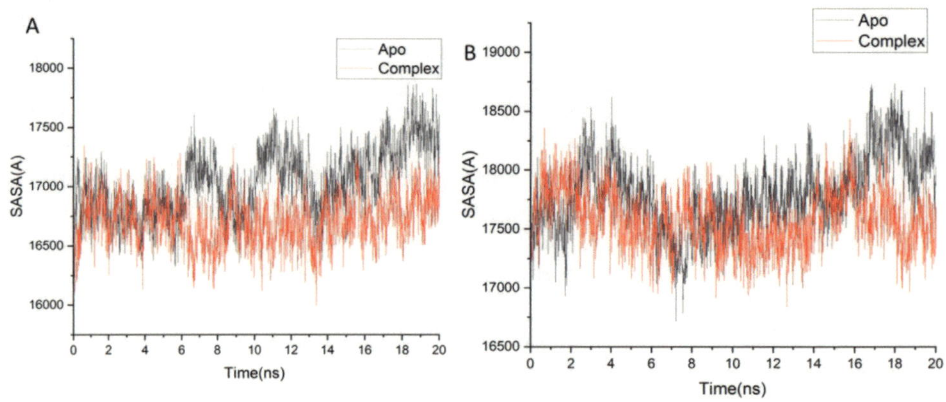

Figure 6.
Solvent accessible surface area (SASA) of the backbone atoms relative to the starting minimized over 20 ns for [A] EGFR [B] Tubulin systems.

3.2 Mechanism of binding interactions obtained from binding free energy calculations

The total binding free energy was computed to elucidate the binding energetics of **compound 1** as a dual inhibitor of EGFR and Tubulin. The MM-GBSA program in AMBER14 was employed to calculate binding free energies by extracting snapshots from the trajectories of the compounds, as illustrated in **Table 1**. The binding energy of **compound 1** to EGFR was −44.58 kcal/mol, in contrast to −38.95 kcal/mol for Tubulin, indicating a more favorable binding of **compound 1** to EGFR compared to Tubulin. The calculated binding energies exhibited a strong correlation with the experimentally reported IC_{50} value for EGFR and % of inhibition of Tubulin [25].

The MM-GBSA method's disaggregation of total free binding energy into distinct components has improved the understanding of the intricate binding process. In all systems, Van der Waals interaction energies are shown to promote favorable binding free energies, whereas polar solvation energy terms adversely impact the binding of the inhibitors.

Energy Components (kcal/mol)						
Complex	**ΔE_{vdW}**	**ΔE_{elec}**	**ΔG_{gas}**	**ΔG_{solv}**	**ΔG_{bind}**	**Biological activity**
EGFR-**compound 1**	-56.47 ± 0.13	-25.40 ± 0.61	-81.88 ± 0.68	37.29 ± 0.59	-44.58 ± 0.31	IC_{50} = 0.135 μM
Tubulin-**compound 1**	-47.66 ± 0.39	-20.39 ± 0.50	-68.05 ± 0.75	29.09 ± 0.37	-38.95 ± 0.55	%inhibition = 73.40

Table 1.
*Lowest energy of the EGFR-**compound 1**/Tubulin -**compound 1** obtained during MD simulation.*

3.3 Identification of the key residues responsible for inhibitor binding

To provide deeper insight into the critical residues implicated in the inhibitory process, the total binding energy of **compound 1** toward EGFR and Tubulin was deconstructed to assess the contribution of each site residue. **Figure 7** illustrates that the primary favorable contributions of **compound 1** to EGFR binding are predominantly attributed to residues Leu 23 (−2.097 kcal/mol), Val 31 (−1.697 kcal/mol), Ala 48 (−1.28 kcal/mol), Lys 50 (−3.365 kcal/mol), and Leu 149 (−1.775 kcal/mol). In contrast, for Tubulin, the contributions are due to residues Lys 251 (−2.067 kcal/mol), Leu 252 (−3.037 kcal/mol), Cys 238 (−1.18 kcal/mol), and Leu 245 (−0.571 kcal/mol).

3.4 Ligand: Residue interaction network profiles

One objective of drugs design is to implement structural alterations that enhance bioavailability, diminish toxicity, and improve pharmacokinetics of pharmaceutical compounds [39, 40]. The drug action mechanism entails the contact between receptor-specific active site residues and distinct groups within the drug molecule, leading to signal transduction and the subsequent activation of a specific reaction recognized as drug action [41]. Ligand–residue profiling analyzes drug-receptor interactions to elucidate the precise functions and forms of interaction associated with residues. It has been discovered that the majority of active site residues engage in hydrophobic interactions with **compound 1** in both the EGFR and Tubulin active site residue systems.

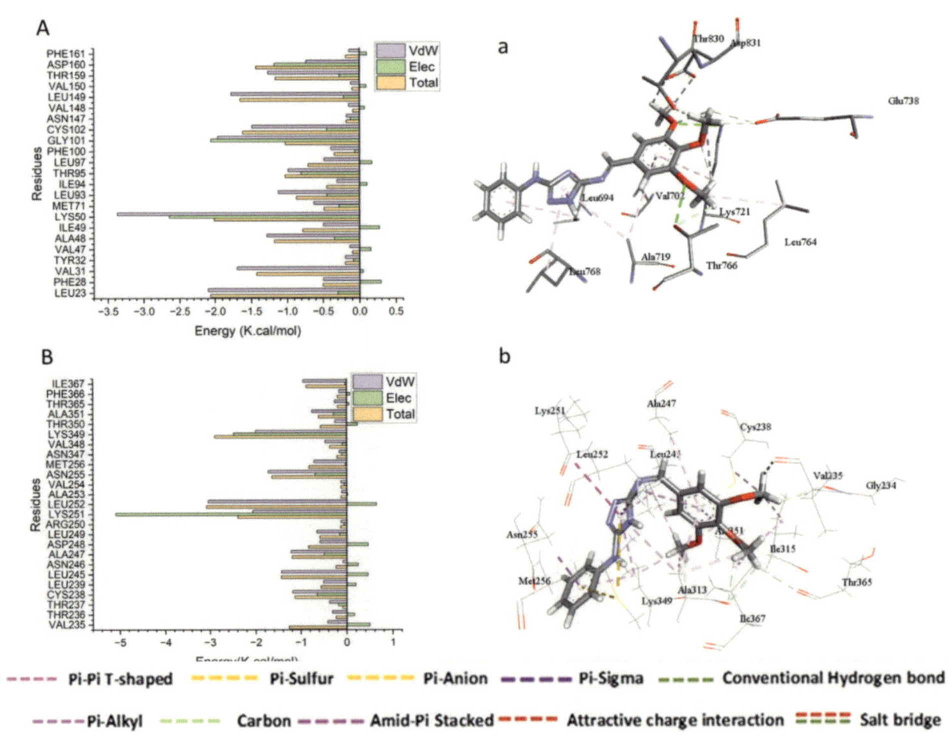

Figure 7.
Graphs depicting the per-residue energy contributions to the binding and stability of ***compound 1*** *at the catalytic contacts of [A] EGFR and [B] Tubulin. Intermolecular interactions are depicted in [a] EGFR and [b] Tubulin.*

Figure 8A demonstrates that the ATP binding site residues Thr95 and Lys50 of EGFR establish stable hydrogen bonds with the oxygen of the trimethoxy group. Furthermore, API alkyl interactions have been noted between **compound 1** and Leu 23, Val 31, Ala 48, Leu 93, and Leu 97. The observed interaction (**Figure 8A**) thus confirmed the attraction between the positively charged residue and the negatively charged group. Conversely, the residues in the Tubulin ATP binding site, namely, Ala351, Ile367, Ile315, Ala247, Leu252, Leu245, and Lys251, have established Pi-alkyl interactions with **compound 1** (**Figure 7B**). It is noteworthy that the hotspot residues Ala 313 and Lys 349 have established Pi-donor hydrogen bonding and Pi-alkyl interactions with **compound 1** (**Figure 8B**).

3.5 Free energy landscape (FEL) analysis

The stability of protein conformations correlates with reduced Gibbs free energy values; thus, a PCA-based free energy landscape (FEL) was conducted to assess the stability of protein conformations for the simulated complexes illustrated in **Figure 6**.

The FEL analysis revealed that the simulated complex achieved a stable structure with low Gibb's free energy (blue and violet, **Figure 9**).

3.6 Probability density function (PDF) analysis

The analysis of the probability density function (PDF) indicates the probability of protein trajectory occurrences using kernel density estimation (KDE) [42]. **Figure 9A** and **B** present the radius of gyration (Rg) and RMSD-based PDF plots for the EGFR-**compound 1** complex and the Tubulin-**compound 1** complex, respectively. The PDF analysis for the EGFR-**compound 1** complex indicated that the most crowded conformations exhibit an Rg value of 22.59 Å and an RMSD value of 2.48 Å (**Figure 10A**). The PDF analysis of the Tubulin-**compound 1** complex indicated that the most populated conformations exhibit a Rg value of 20.97 Å and an RMSD value of 0.99 Å (**Figure 10B**).

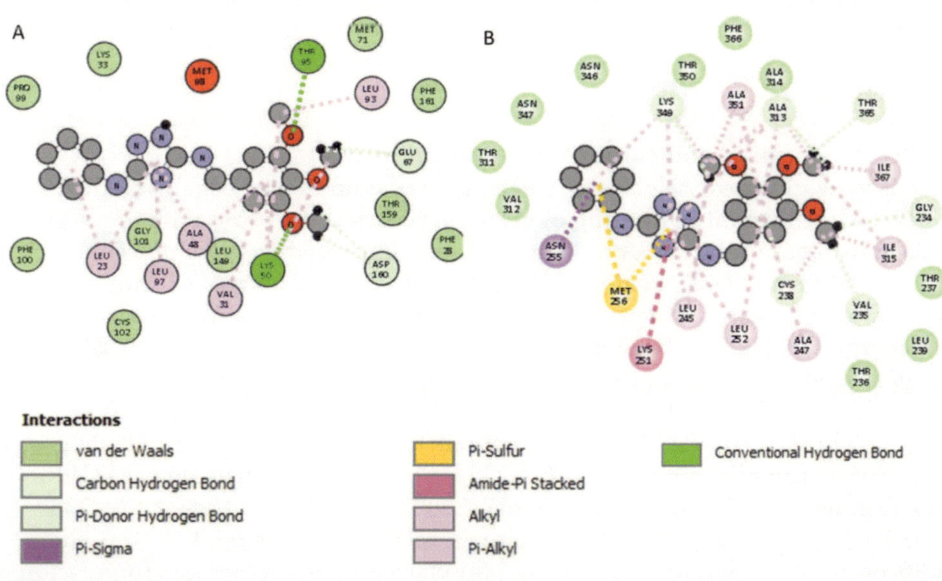

Figure 8.
The interacting residues of EGFR [A] and Tubulin [B] complexed with ***compound 1***.

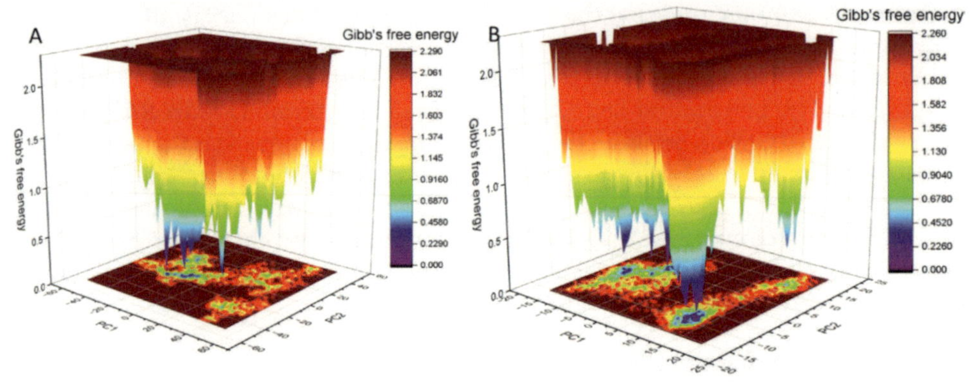

Figure 9.
Free energy landscape (FEL) of ***compound 1*** *with [A] EGFR [B] Tubulin.*

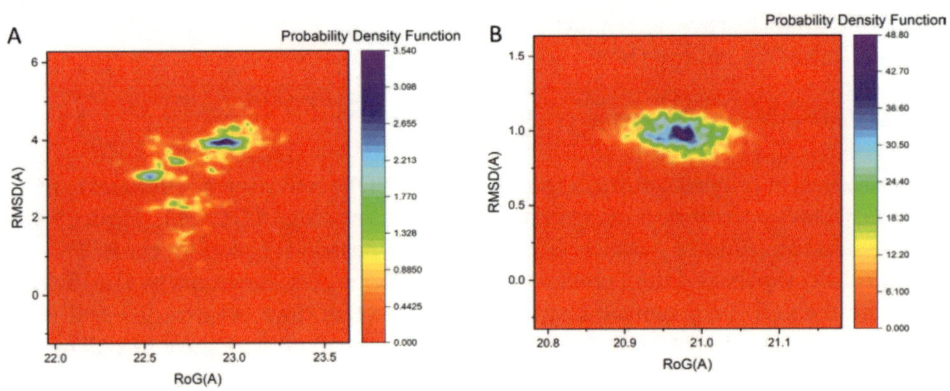

Figure 10.
Probability density function (PDF) of ***compound 1*** *with EGFR [A] Tubulin [B] complex the least (red) and the most (blue) populated conformations.*

4. Conclusion

This work utilized comparative MD simulation and binding free energy analysis to examine the selectivity mechanism of **compound 1** against EGFR and Tubulin. The differential binding of **compound 1** to these protein targets was assessed using the MM/GBSA technique, which indicated positive interactions with ΔG values of -44.58 kcal/mol (EGFR) and - 38.95 kcal/mol (Tubulin). The examination of binding free energy components indicates that the predominant energy factor contributing to this synergistic impact is the van der Waals energy component. The analysis of total energy contributions from the active site residues of EGFR and Tubulin indicated that the amino acid residues Leu 23 (-2.097 kcal/mol), Val 31 (-1.697 kcal/mol), Ala 48 (-1.28 kcal/mol), Lys 50 (-3.365 kcal/mol), and Leu 149 (-1.775 kcal/mol) are significant for EGFR, while residues Lys 251 (-2.067 kcal/mol), Leu 252 (-3.037 kcal/mol), Cys 238 (-1.18 kcal/mol), and Leu 245 (-0.571 kcal/mol) are notable for Tubulin. The results of our investigation are essential for clarifying the molecular foundation of the activity difference between **compound 1** against EGFR and Tubulin, as well as for the advancement of more potent selective inhibitors.

Acknowledgements

The authors would like to express their sincere appreciation to National Research Centre (NRC) Egypt for its financial support of this research project (No., 13010142).

Conflict of interest

The authors declare no conflict of interest.

Author details

Ahmed A. Elrashedy[1,2*] and Asmaa A. Magd El-Din[1]

1 Chemistry of Natural and Microbial Products Department, National Research Center (NRC), Egypt

2 Department Organic and Medicinal Chemistry, Faculty of Pharmacy, University of Sadat City, Menoufia, Egypt

*Address all correspondence to: ahmedelrashedy45@gmail.com

References

[1] ElSherief HAM, Abdel-Aziz M, Abdel-Rahman HA. Synthesis and evaluation of the antioxidant activity of 1,2,4-Triazole derivatives. Journal of Advanced Biomedical and Pharmaceutical Sciences. 2018;**1**:1-5

[2] Al-Sanea MM, Parambi DGT, Shaker ME, Elsherif HAM, Elshemy HAH, Bakr RB, et al. Design, synthesis, and in vitro cytotoxic activity of certain 2-[3-phenyl-4-(pyrimidin-4-Yl)-1H-pyrazol1-Yl] acetamide derivatives. Russian Journal of Organic Chemistry. 2020;**56**:514-520. DOI: 10.1134/S1070428020030239

[3] Qazi S, Khanna K, Raza K. Dihydroquercetin (DHQ) has the potential to promote apoptosis in ovarian cancer cells: An in silico and in vitro study. Journal of Molecular Structure. 2023;**1271**:134093. DOI: 10.1016/J.MOLSTRUC.2022.134093

[4] Peralta-Zaragoza O, Bermúdez-Morales VH, Pérez-Plasencia C, Salazar-León J, Gómez-Cerón C, Madrid-Marina V. Targeted treatments for cervical cancer: A review. OncoTargets and Therapy. 2012;**5**:315-328. DOI: 10.2147/OTT.S25123

[5] Bogdan S, Klämbt C. Epidermal growth factor receptor signaling. Current Biology. 2001;**11**(8):R292-R295. DOI: 10.1016/S0960-9822(01)00167-1

[6] Mitsudomi T, Yatabe Y. Epidermal growth factor receptor in relation to tumor development: EGFR gene and cancer. The FEBS Journal. 2010;**277**:301-308. DOI: 10.1111/J.1742-4658.2009.07448.X

[7] Prasad S, Yadav VR, Sung B, Reuter S, Kannappan R, Deorukhkar A, et al. Ursolic acid inhibits growth and metastasis of human colorectal cancer in an orthotopic nude mouse model by targeting multiple cell Signaling pathways: Chemosensitization with capecitabine. Clinical Cancer Research. 2012;**18**:4942-4953. DOI: 10.1158/1078-0432.CCR-11-2805

[8] Roskoski R. A historical overview of protein kinases and their targeted small molecule inhibitors. Pharmacological Research. 2015;**100**:1-23. DOI: 10.1016/J.PHRS.2015.07.010

[9] Kharb R, Sharma PC, Yar MS. Pharmacological significance of Triazole scaffold. Journal of Enzyme Inhibition and Medicinal Chemistry. 2011;**26**:1-21. DOI: 10.3109/14756360903524304

[10] Kemp JA, Shim MS, Heo CY, Kwon YJ. "Combo" nanomedicine: Co-delivery of multi-modal therapeutics for efficient, targeted, and safe cancer therapy. Advanced Drug Delivery Reviews. 2016;**98**:3-18. DOI: 10.1016/J.ADDR.2015.10.019

[11] Dixit D, Verma PK, Marwaha RK. A review on 'Triazoles': Their chemistry, synthesis and pharmacological potentials. Journal of the Iranian Chemical Society. 2021;**18**:2535-2565. DOI: 10.1007/S13738-021-02231-X/TABLES/23

[12] El-Sherief HAM, Youssif BGM, Bukhari SNA, Abdel-Aziz M, Abdel-Rahman HM. Novel 1,2,4-Triazole derivatives as potential anticancer agents: Design, synthesis, molecular docking and mechanistic studies. Bioorganic Chemistry. 2018;**76**:314-325. DOI: 10.1016/J.BIOORG.2017.12.013

[13] Rashdan HRM, Shehadi IA. Triazoles synthesis and Applications

as nonsteroidal aromatase inhibitors for hormone-dependent breast cancer treatment. Heteroatom Chemistry. 2022;**2022**:5349279. DOI: 10.1155/2022/5349279

[14] El-Sherief HAM, Youssif BGM, Abdelazeem AH, Abdel-Aziz M, Abdel-Rahman HM. Design, synthesis and antiproliferative evaluation of novel 1,2,4-Triazole/Schiff Base hybrids with EGFR and B-RAF inhibitory activities. Anti-Cancer Agents in Medicinal Chemistry. 2019;**19**:697-706. DOI: 10.2174/1871520619666181224115346

[15] Kapila A, Kaur M, Kaur H. Organotin (IV) complexes of tridentate (O,N,O) Schiff base ligand: Computational, spectroscopic and biological studies. Materials Today Proceedings. 2021;**40**:S102-S106. DOI: 10.1016/J.MATPR.2020.04.080

[16] Wang M, Wang LF, Li YZ, Li QX, Xu ZD, Qu DM. Antitumour activity of transition metal complexes with the thiosemicarbazone derived from 3-Acetylumbelliferone. Transition Metal Chemistry. 2001;**26**:307-310. DOI: 10.1023/A:1007159301849/METRICS

[17] Bharti SK, Nath G, Tilak R, Singh SK. Synthesis, anti-bacterial and anti-fungal activities of some novel Schiff bases containing 2,4-disubstituted Thiazole ring. European Journal of Medicinal Chemistry. 2010;**45**:651-660. DOI: 10.1016/J.EJMECH.2009.11.008

[18] Prakash CR, Raja S, Panneer Selvam T, Saravanan G, Karthick V, Dinesh Kumar P. Synthesis and anti-microbial activities of some novel Schiff bases of 5-substituted isatin derivatives. Rasayan Journal of Chemistry. 2009;**2**:960-968

[19] Youssif BGM, Mohamed YAM, Salim MTA, Inagaki F, Mukai C, Abdu-Allah HHM. Synthesis of some benzimidazole derivatives endowed with 1,2,3-Triazole as potential inhibitors of hepatitis C virus. Acta Pharmaceutica. 2016;**66**:219-231. DOI: 10.1515/ACPH-2016-0014

[20] Romagnoli R, Baraldi PG, Salvador MK, Prencipe F, Bertolasi V, Cancellieri M, et al. Synthesis, antimitotic and antivascular activity of 1-(3′,4′,5′-trimethoxybenzoyl)-3-arylamino-5-amino-1,2,4-triazoles. Journal of Medicinal Chemistry. 2014;**57**:6795. DOI: 10.1021/JM5008193

[21] Yapar G, Demir N, Kiraz A, Özkat GY, Yıldız M. Synthesis, biological activities, antioxidant properties, and molecular docking studies of novel Bis-Schiff base podands as responsive chemosensors for anions. Journal of Molecular Structure. 2022;**1266**:133530. DOI: 10.1016/J.MOLSTRUC.2022.133530

[22] Manojkumar Y, Ambika S, Arulkumar R, Gowdhami B, Balaji P, Vignesh G, et al. Synthesis, DNA and BSA binding, in vitro anti-proliferative and in vivo anti-angiogenic properties of some cobalt(III) Schiff base complexes. New Journal of Chemistry. 2019;**43**:11391-11407. DOI: 10.1039/C9NJ01269A

[23] Demirci S, Doıan A, Başak N, Telci D, Dede B, Orhan C, et al. A Schiff Base derivative for effective treatment of diethylnitrosamine-induced liver cancer in vivo. Anti-Cancer Drugs. 2015;**26**:555-564. DOI: 10.1097/CAD.0000000000000221

[24] Elbastawesy MAI, Ramadan M, El-Shaier YAMM, Aly AA, Abuo-Rahma GEDA. Arylidenes of quinolin-2-one scaffold as erlotinib analogues with activities against leukemia through

inhibition of EGFR TK/STAT-3 pathways. Bioorganic Chemistry. 2020;**96**:1-13. DOI: 10.1016/J.BIOORG.2020.103628

[25] Mohamed MS, Ibrahim NA, Gouda AM, Badr M, El-Sherief HAM. Design, synthesis and molecular docking of 1,2,4-triazole Schiff base hybrids as tubulin, EGFR inhibitors and apoptosis-inducers. Journal of Molecular Structure. 2023;**1286**:135621. DOI: 10.1016/J.MOLSTRUC.2023.135621

[26] Xu J, Wang J-X, Zhou J-M, Xu C-L, Huang B, Xing Y, et al. A novel protein kinase inhibitor IMB-YH-8 with anti-tuberculosis activity. Scientific Reports. 2017;**7**:5093. DOI: 10.1038/s41598-017-04108-7

[27] Stamos J, Sliwkowski MX, Eigenbrot C. Structure of the epidermal growth factor receptor kinase domain alone and in complex with a 4-anilinoquinazoline inhibitor. The Journal of Biological Chemistry. 2002;**277**:46265-46272. DOI: 10.1074/JBC.M207135200

[28] Gaspari R, Prota AE, Bargsten K, Cavalli A, Steinmetz MO. Structural basis of cis- and trans-combretastatin binding to tubulin. Chem. 2017;**2**:102-113. DOI: 10.1016/J.CHEMPR.2016.12.005/ATTACHMENT/8C14FA77-1ED7-4DE3-89E3-7A49E8EBA993/MMC2.PDF

[29] Kollman P, A, Massova I, Reyes C, Kuhn B, Huo S, Chong L, et al. Calculating structures and free energies of complex molecules: Combining molecular mechanics and continuum models. 2000;**33**(12):889-897. DOI: 10.1021/AR000033J

[30] Pettersen EF, Goddard TD, Huang CC, Couch GS, Greenblatt DM, Meng EC, et al. UCSF Chimera–A visualization system for exploratory research and analysis. Journal of Computational Chemistry. 2004;**25**:1605-1612. DOI: 10.1002/jcc.20084

[31] Kusumaningrum S, Budianto E, Kosela S, Sumaryono W, Juniarti F. The molecular docking of 1,4-naphthoquinone derivatives as inhibitors of polo-like kinase 1 using Molegro virtual Docker. Journal of Applied Pharmaceutical Science. 2014;**4**:47-53. DOI: 10.7324/JAPS.2014.4119

[32] Webb B, Sali A. Comparative protein structure Modeling using MODELLER. In: Current Protocols in Protein Science. Vol. 86. Hoboken, NJ, USA: John Wiley & Sons, Inc; 2016. pp. 1-37

[33] Kim S, Thiessen PA, Bolton EE, Chen J, Fu G, Gindulyte A, et al. PubChem substance and compound databases. Nucleic Acids Research. 2016;**44**:1202-1213

[34] Case DA, Cheatham TE, Darden T, Gohlke H, Luo R, Merz KM, et al. The Amber biomolecular simulation programs. Journal of Computational Chemistry. 2005;**26**:1668-1688. DOI: 10.1002/jcc.20290

[35] Elrashedy AA. Targeting Inhibitor of *Enterococcus Faecalis*: Insights from Comparative Molecular Dynamics and Binding Free EnergyAnalyses. 2024. DOI: 10.5772/INTECHOPEN.114329

[36] Pan L, Patterson JC, Deshpande A, Cole G, Frautschy S. Molecular dynamics study of Zn(aβ) and Zn(aβ)2. PLoS One. 2013;**8**:70681-70688. DOI: 10.1371/journal.pone.0070681

[37] Wijffels G, Dalrymple B, Kongsuwan K, Dixon N. Conservation of eubacterial replicases. IUBMB Life. 2005;**57**:413-419. DOI: 10.1080/15216540500138246

[38] Richmond TJ. Solvent accessible surface area and excluded volume in proteins. Analytical equations for overlapping spheres and implications for the hydrophobic effect. Journal of Molecular Biology. 1984;**178**:63-89

[39] Nassar AEF, Kamel AM, Clarimont C. Improving the decision-making process in the structural modification of drug candidates: Enhancing metabolic stability. Drug Discovery Today. 2004;**9**:1020-1028. DOI: 10.1016/S1359-6446(04)03280-5

[40] Ejalonibu MA, Elrashedy AA, Lawal MM, Soliman ME, Sosibo SC, Kumalo HM, et al. Dual targeting approach for mycobacterium tuberculosis drug discovery: Insights from DFT calculations and molecular dynamics simulations. Structural Chemistry. 2019;**31**:1-15

[41] Mori G, Chiarelli LR, Esposito M, Makarov V, Bellinzoni M, Hartkoorn RC, et al. Thiophenecarboxamide derivatives activated by EthA kill mycobacterium tuberculosis by inhibiting the CTP synthetase PyrG. Chemistry & Biology. 2015;**22**:917-927. DOI: 10.1016/j.chembiol.2015.05.016

[42] Lanka G, Banerjee S, Regula S, Adhikari N, Ghosh B. Pharmacophore modeling, 3D-QSAR, and MD simulation-based overture for the discovery of new potential HDAC1 inhibitors. Journal of Biomolecular Structure & Dynamics. 25 Nov 2024:1-24. DOI: 10.1080/07391102.2024.2429020

Chapter 7

Recent Applications of Schiff Bases in Biomedical Sciences

Ruijuan Zhang, Yali Yao and Xinying Liu

Abstract

Schiff base compounds exhibit multidimensional application potential in the biomedical field due to their unique imine bond (C=N) structure and modifiability. This chapter systematically reviews the research progress of Schiff bases and their metal complexes in antibacterial, antiviral, antifungal, antiparasitic, antitumor applications, as well as drug delivery and sensing/imaging. Structural design optimization strategies, such as introducing functional groups, metal coordination, and hybridization with natural products, significantly enhance their bioactivity and targeting specificity. These advancements have not only elucidated the dynamic reversible nature of imine bonds and the molecular mechanisms underlying the synergistic effects of metal coordination, laid a solid interdisciplinary theoretical foundation for developing environmentally responsive smart theranostic systems, but also provide theoretical foundations and technical references for the molecular design and interdisciplinary applications of Schiff bases.

Keywords: Schiff bases, anti-pathogen, antitumor, drug delivery systems, sensor/biological imaging

1. Introduction

Schiff bases are a class of nitrogen-containing organic compounds characterized by an imine structure, with the general chemical formula R−CH=N−R'. In this formula, R and R' may represent alkyl, aryl, or substituted derivatives thereof, where the substituents can adopt linear or cyclic configurations. Structural functionalization of these compounds can be achieved through diverse chemical modification strategies [1–3]. First reported in 1860 by the German chemist Hugo Schiff, these compounds have emerged as a research hotspot in interdisciplinary chemistry and biology due to their facile synthesis, structural tunability, and multifunctionality [1, 2, 4, 5]. The electron push-pull effect of the imine bond enables Schiff bases to form stable complexes with various metal ions while allowing precise modulation of molecular polarity, lipophilicity, and spatial configuration through structural modifications, endowing them with unique advantages in drug development, smart material design, and environmental monitoring [6, 7]. In biomedicine, their multi-target mechanisms underpin applications in antibacterial, antiviral, and antitumor therapies.

For instance, synergistic strategies—such as targeting pathogen-specific enzymes, disrupting membrane integrity, and regulating signaling pathways—demonstrate potential to overcome drug resistance. In drug delivery systems, the reversibility of dynamic Schiff base bonds inspires innovative designs for smart responsive materials, including pH-sensitive hydrogels and targeted drug carriers [8].

Despite significant advancements in Schiff bases's research, their biomedical applications still face multiple challenges. For example, key issues in current research include how to further optimize the bioavailability and targeting specificity of Schiff bases through molecular design, how to balance their antibacterial or antitumor activity with toxicity, and how to develop more efficient and intelligent Schiff base-based materials. This chapter systematically explores the latest research progress of Schiff base compounds in antibacterial, antiviral, antifungal, antiparasitic, and antitumor applications, as well as drug delivery systems and sensing/imaging technologies. It aims to provide theoretical foundations and practical guidelines for future molecular design, functional optimization, and applications of Schiff bases in biomedicine and materials science.

2. Applications of Schiff bases in antimicrobial activity

Schiff bases, with their unique structural designability and multitarget activity, have become a cutting-edge direction in the development of antimicrobial agents. The imine bond (C=N) in Schiff base molecules not only endows them with excellent coordination ability to form stable complexes with metal ions but also allows flexible modulation of electronic distribution and spatial configuration through structural modification, thereby enhancing their permeability through bacterial cell membranes, as well as their inhibitory effects on nucleic acid synthesis and key metabolic enzymes.

2.1 Antibacterial activity

Introducing different substituents modulates electronic effects to optimize charge distribution and enhances target protein binding affinity. Santu et al. [9] construct four types of novel Schiff base compounds (3a–3d) (**Figure 1a**) based on the 3-acetyl coumarin, achieving diverse structural designs through the precise introduction of

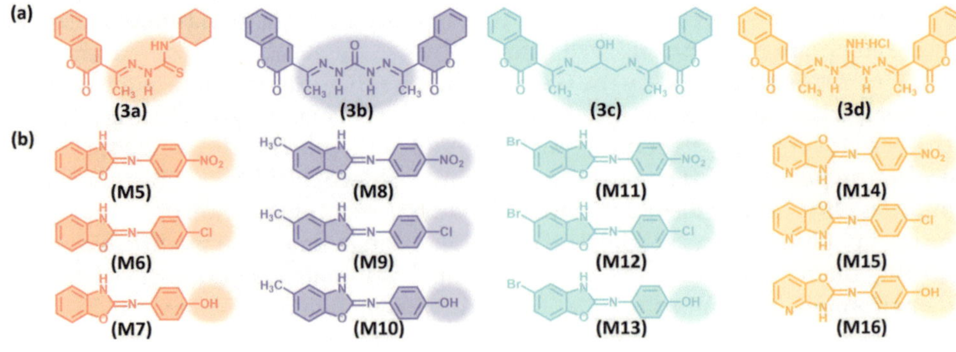

Figure 1.
Molecular structures of Schiff base complexes. (a) 3a–3d, (b) M5–M16.

N4-cyclohexyl thiosemicarbazide (3a), carbohydrazide (3b), 1, 3-diamino-2-propanol (3c), and 1, 3-diaminoguanidine hydrochloride (3d). Screening for *in vitro* antibacterial activity shows that these compounds exhibit broad-spectrum inhibitory effects against both gram-positive (*Staphylococcus aureus*) and gram-negative (*Escherichia coli*). Molecular docking and dynamics simulations reveal that the compounds exert their inhibitory effects by targeting the active center of dihydrofolate reductase (DHFR); the diaminoguanidine dihydrazone 3d has the highest binding energy to DHFR, thus demonstrating the most strongest antibacterial potential. Similarly, Maria et al. [10] construct a series of Schiff bases (M5–M16) (**Figure 1b**) by condensing BOA derivatives with substituted anilines and elucidate the regulatory laws of antibacterial activity by substituent electronic effects through quantum chemical calculations. Molecular docking reveals that nitro derivatives, due to their strong electron-withdrawing properties, optimize the molecular charge distribution and form multiple hydrogen bonds and hydrophobic interactions with the target protein of *Escherichia coli*, significantly enhancing binding affinity. Activity assessment confirms that nitrogen-containing/nitro derivatives have the best antibacterial efficacy, with in vitro antibacterial experiments highly consistent with theoretical predictions, verifying the feasibility of precisely designing antibacterial Schiff bases based on electronic effects.

Natural products serve as a vital source of bioactive compounds, and structural derivatization studies have emerged as an effective strategy for developing novel functional molecules [11–13]. Recent researches demonstrate that Schiff base metal coordination complexes derived from natural molecular frameworks can significantly enhance biological activities [14–16]. Pinheiro et al. [17] successfully prepare a vanillin-based Schiff base ligand (VTRIS) and its Zn (II) complex $[Zn(VTRIS)_2](CH_3COO)_2$ (**Figure 2a**) through a coordination-driven self-assembly strategy. Antibacterial assays reveal that the ligand VTRIS exhibits a minimum inhibitory concentration (MIC) >1024 μg/mL against *Staphylococcus aureus*, while $[Zn(VTRIS)_2](CH_3COO)_2$ shows substantially improved antimicrobial activity with an MIC of 512 μg/mL. Mechanistic investigations indicate that the coordination effect enhances antibacterial performance through two primary pathways: (1) Specific adsorption of Zn(II) on microbial cell wall surfaces disrupts cellular respiration chains, thereby interfering with energy metabolism and subsequent inhibition of protein biosynthesis; (2) The metal-ligand synergistic effects induced by coordination markedly modify molecular properties, where partial positive charges on Zn(II) become effectively delocalized through electron sharing with Schiff base donor groups. Concurrently, orbital overlap within the complex further reduces the system's overall polarity, enhancing membrane penetration capability. This study confirms that the

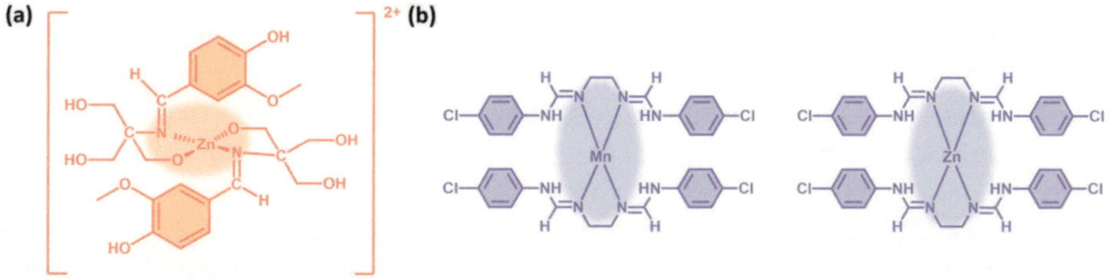

Figure 2.
Molecular structures of Schiff base complexes. (a) $[Zn(VTRIS)_2](CH_3COO)_2$*, (b)* $[M(C_{16}H_{16}Cl_2N_4)_2(OAc)_2]$*, M = Mn, Zn.*

synergistic coordination between metal centers and ligands plays a decisive role in optimizing biological activities.

The Schiff bases with bidentate coordination, due to their excellent coordination ability, structural diversity, and potential for functionalization, hold significant importance in the biological field. Juyal et al. [18] design and synthesize a bidentate Schiff base ligand ($C_{16}H_{16}Cl_2N_4$) and its Mn (II)/Zn (II) complex [M($C_{16}H_{16}Cl_2N_4$)$_2$(OAc)$_2$] (**Figure 2b**). They systematically evaluate the antibacterial activity of these compounds against gram-positive bacteria (*Staphylococcus aureus*, *Bacillus subtilis*) and gram-negative bacteria (*Escherichia coli*, *Salmonella typhi*). Experiments show that the antibacterial efficacy of the [M($C_{16}H_{16}Cl_2N_4$)$_2$(OAc)$_2$] is significantly better than that of the free ligand. This can be attributed to Tweedy's chelation theory: chelation enhances the bactericidal ability of the metal complex. The sharing of positive charges between the metal and the ligand leads to electron delocalization across the entire complex, imparting lipophilicity, which facilitates easier penetration through the cell membrane [19]. In addition, it has been found that [M($C_{16}H_{16}Cl_2N_4$)$_2$(OAc)$_2$] exhibits stronger activity against gram-positive bacteria. This may be related to the cell membrane structure of gram-positive bacteria, which lacks an outer membrane, making it easier for molecules to diffuse [20]. Through molecular docking studies, it finds that [M($C_{16}H_{16}Cl_2N_4$)$_2$(OAc)$_2$] has the best 3D binding interaction with tyrosyl–tRNA synthetase of *Staphylococcus aureus* and the OmpF complex of *Salmonella typhi*. The negative value of the binding energy indicates stronger interactions between the ligand and the target receptor, thereby generating a more favorable free binding energy [21].

2.2 Antiviral activity

Schiff bases, as versatile pharmacophores, have an imine group in their molecules that forms hydrogen bonds with the active sites of receptors and enzymes, playing a crucial role in modulating biological activities. This unique capability allows these compounds to intervene in cellular processes [22–24]. Azzouzi et al. [25], targeting the viral reverse transcriptase, design and synthesize a series of Imidazo(1,2-a) pyridine-Schiff base derivatives (**Figure 3a**). They evaluate these compounds' inhibitory activity against HIV-1 and HIV-2 using the MT-4 cell model. Results show that the compounds' solubility significantly impacts their antiviral performance. 4d in **Figure 3a**, with the highest solubility (125 μg/mL), shows good antiviral activity with an half maximal effective concentration (EC_{50}) of 17.91 mg/mL. In contrast, 4a and 4b, with lower solubility (25 μg/mL), have limited antiviral effects, and compound 4c, with very low solubility (<1 μg/mL), shows no antiviral activity. Molecular docking studies reveal that 4d forms various non-covalent interactions (e.g., Pi–Pi stacked, T shaped and Pi-alkyl interactions) with key aromatic residues of HIV-1 reverse transcriptase and additional electrostatic stabilization via Pi-sigma interactions with residues like LEU A:100. This inhibits the reverse transcriptase's activity, preventing viral genome integration into host cells. These findings indicate that optimizing solubility and bioavailability is crucial for enhancing the antiviral potential of future derivatives.

Abbas et al. [26] targeting the catalytic mechanism of SARS-CoV-2's main protease (6 LU7) design and synthesize four new ferrocene Schiff bases (**Figure 3b**: L1–L4). L1 shows remarkable inhibitory activity against 6 LU7, outperforming current anti-COVID-19 drugs. Molecular docking reveals that L1 forms hydrogen bonds with key amino acids like GLY143 and CYS145, inhibiting the protease and blocking viral

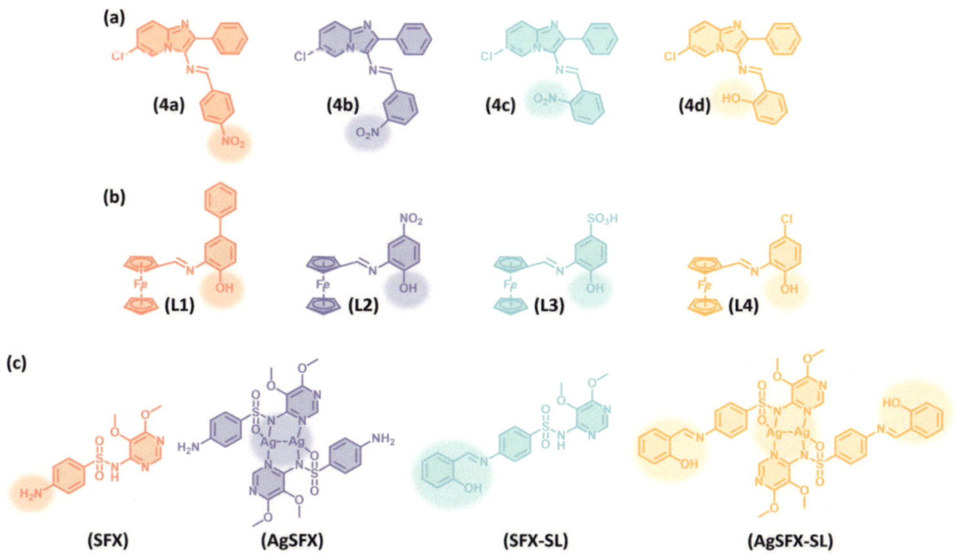

Figure 3.
Molecular structures of Schiff base complexes. (a) 4a–4b, (b) L1–L2, (c) SFX, AgSFX, SFX-SL, and AgSFX-SL.

replication. Quantum chemical calculations indicate that L1 has favorable molecular reactivity for effective electron transfer with viral targets. Its electronegativity and chemical hardness further suggest high reactivity, enabling efficient virus inhibition via electron transfer. These traits give L1 excellent antiviral potential.

In response to the lack of specific therapeutic drugs against *Alphavirus chikungunya* (CHIKV), Martins et al. [27] employ a drug repurposing strategy to systematically evaluate the antiviral activity of sulfadoxine (SFX) and its metal/Schiff base-modified derivatives (**Figure 3c**). The study demonstrates that SFX alone inhibits 34% of CHIKV replication, while its derivatives AgSFX, SFX-SL, and AgSFX-SL exhibit significantly enhanced antiviral efficacy with inhibition rates of 84, 89, and 95%, respectively. Mechanistically, SFX and its derivatives block viral entry into host cells by specifically binding to the viral envelope (E) protein and the host receptor MXRA8. Notably, the Ag(I) in AgSFX and AgSFX-SL improve compound lipophilicity and structural stability, facilitating direct interaction with viral particles and thereby strengthening the inhibition of viral entry. Additionally, SFX suppresses viral RNA synthesis through inhibition of dihydropteroate synthase (dhps), which interferes with the folate metabolism pathway.

2.3 Antifungal activity

Schiff bases provide a unique molecular platform for developing highly efficient, low-toxicity, and resistance-breaking antifungal therapeutics by enabling precise modulation of structure–activity-mechanism relationships through chemical tunability, synergized with metal coordination and natural molecular hybridization. Hao et al. [28] construct a 3D-QSAR model that successfully predicts 30 1,2,4-triazole Schiff base derivatives (**Figure 4**) with high inhibitory activity, and seven of these compounds are synthesized and tested to confirm the model's reliability. DES-3 and DES-5 show higher inhibitory activity than standard triazole fungicides, with EC_{50} values of 9.915 and 9.384 mg/L, lower than the standard drug's 10.820 mg/L.

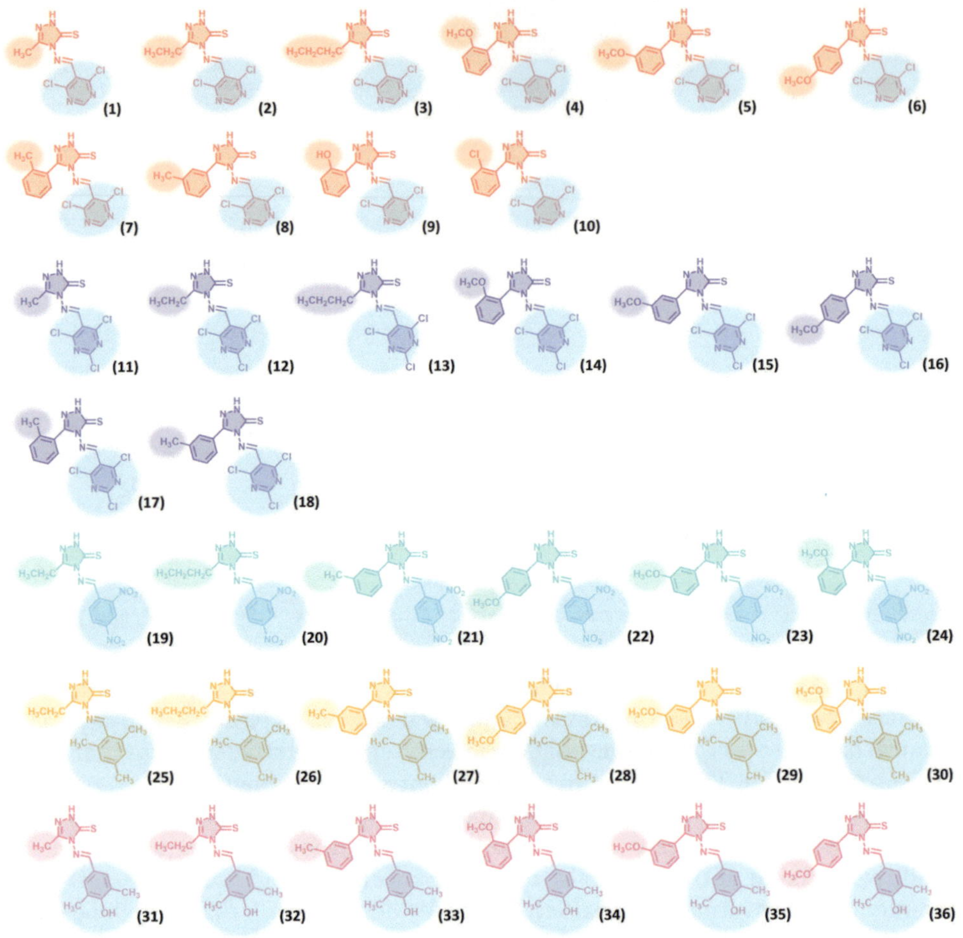

Figure 4.
Molecular structures of Schiff base complexes: DES1–36.

Molecular docking studies reveal that these compounds inhibit enzyme activity and block fungal growth by forming hydrogen bonds with key amino acid residues of isocitrate lyase (ICL), such as MET-218, HIS-217, LYS-406, and SER-436. Notably, DES-5, with the lowest binding energy to ICL in its single crystal structure, shows the strongest inhibitory activity. These findings confirm the 3D-QSAR model's accuracy and provide a significant basis for developing novel, efficient fungicides.

Buran et al. [29] employ a strategy targeting the disruption of fungal cell membrane integrity to innovatively design and construct a sulfamethazine and benzoin Schiff base ligand (L) and its corresponding Cu(II), Co(II), Ni(II), and Cd(II) metal complexes (1–4) (**Figure 5a**). Systematic evaluation using the microdilution method against *Candida albicans* ATCC 10231 demonstrates that the metal complexes exhibit significantly enhanced antifungal activity compared to the parent ligand L, with the Cu(II) complex 1 showing optimal inhibitory efficacy (MIC = 257.81 μg/mL). Molecular orbital theory analysis reveals the electronic communication characteristics of ligand L: the highest occupied molecular orbital (HOMO) primarily localizes on the sulfamethoxazole moiety, while the lowest unoccupied molecular orbital (LUMO) resides in the benzaldehyde unit. This intramolecular charge transfer characteristic

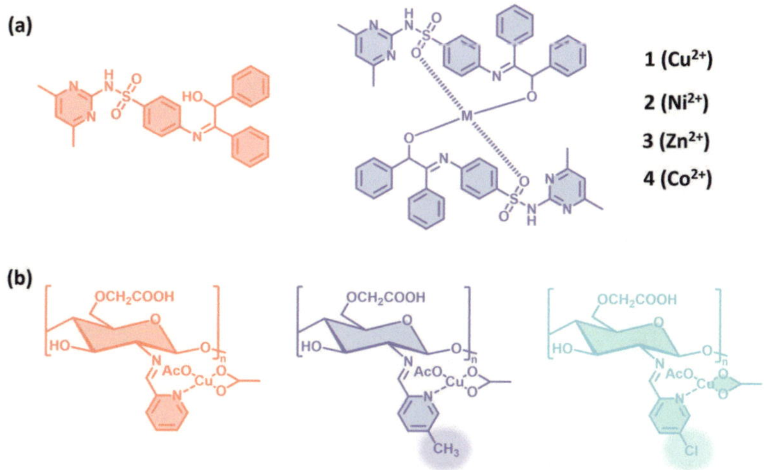

Figure 5.
Molecular structures of Schiff base complexes. (a) Sulfamethazine and its corresponding Cu(II), Co(II), Ni(II), and Cd(II) metal complexes. (b) CPSx-Cu complexes.

effectively strengthens the compound's binding capacity to fungal targets. Notably, the introduction of metal ions enhances the interaction between the complexes and microbial cell membranes, thereby further improving antifungal efficacy. This work provides key theoretical foundations and molecular design paradigms for developing novel antifungal agents with dual functionalities of membrane targeting and enzyme inhibition.

Similarly, Liu et al. [30] successfully synthesize chitooligosaccharide pyridine Schiff base copper complexes (CPSx-Cu) (**Figure 5b**) through Schiff base reactions and ionic complexation. These complexes exhibit significant antifungal activity in both in vitro and in vivo experiments. The study proposes that the antifungal efficacy of CPSx-Cu complexes primarily stems from the action of copper ions, which disrupt fungal cell wall synthesis and interfere with membrane functionality, thereby inhibiting fungal growth. Meanwhile, the inherent bioactivity of chitooligosaccharide may synergize with copper ions through a cooperative interaction, further enhancing the antifungal performance of the complexes.

2.4 Antiparasitic activity

Parasitic diseases are a major global public health concern, posing a serious threat to human health and livestock development. Although existing drugs have achieved some success in controlling infections, drug resistance is becoming an increasingly serious issue. For instance, resistance to artemisinin in *Plasmodium falciparum* has emerged in several countries, leading to reduced treatment efficacy. In response to this challenge, researchers are actively exploring new drug targets and therapeutic strategies. Schiff bases, known for their structural diversity and dynamic coordination ability, have emerged as promising candidates for the development of antiparasitic drugs [31].

Aggarwal et al. [32] successfully prepare 18 novel pyrazole Schiff base hybrids (**Figure 6**) through multi-step organic synthesis reactions. These compounds feature a core structure containing a pyrazole ring linked to a benzothiazole moiety,

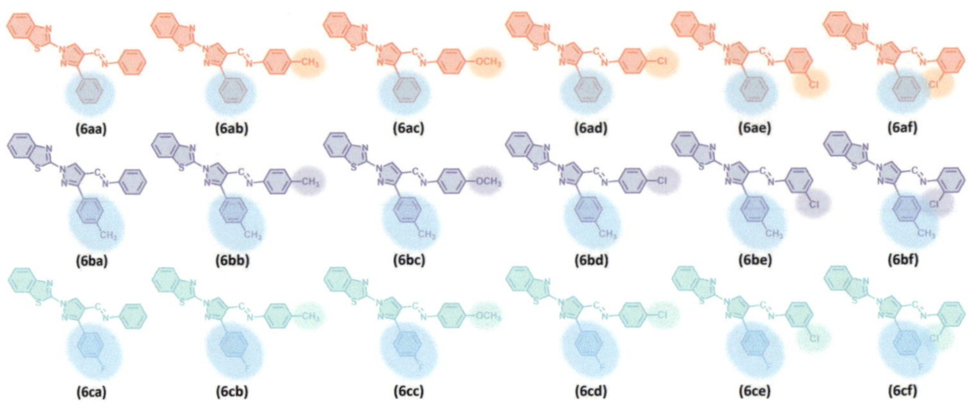

Figure 6.
Molecular structures of pyrazole Schiff base hybrid.

with aromatic side chains modified by various substituents (e.g., chlorine, methyl, methoxy, and fluorine). The research team systematically evaluates the in vitro antiplasmodial activity of these compounds against the asexual blood stage of *Plasmodium falciparum*. Experimental results demonstrate that compounds 6bf and 6bd exhibit the strongest antimalarial efficacy, with EC_{50} of 1.95 and 1.98 μg/mL, respectively. Structure–activity relationship (SAR) analysis reveals the following patterns: *Electronic effects*: Methyl groups (electron-donating) at position 3 of the pyrazole ring and halogen substituents (electron-withdrawing) at position 4 significantly enhance activity. *Substitution position*: Meta-chlorine (*m*-Cl) substitutions show superior activity compared to ortho-(*o*-Cl) and para-(*p*-Cl) substitutions. The pyrazole Schiff base derivatives inhibit malaria parasite development by forming stable interactions with the active site of *PfFP*-2, a key therapeutic target. Molecular docking studies reveal binding energies ranging from −7.9 to −8.6 kcal/mol, with critical π-π stacking interactions involving residues such as tryptophan TRP206. This molecular recognition pattern disrupts enzymatic function, serving as the key mechanism for suppressing parasite growth and schizont maturation.

Rani et al. [33] design and synthesize two novel Schiff base ligands and their Co(II), Ni(II), Cu(II), and Zn(II) metal complexes (**Figure 7**) using a coordination chemistry strategy. They systematically evaluate the biological activity of these compounds. In vitro pharmacological studies demonstrate that the Co(II) and

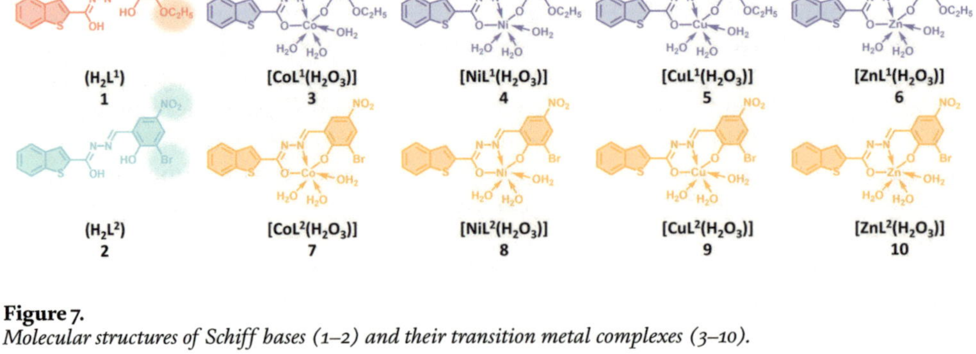

Figure 7.
Molecular structures of Schiff bases (1–2) and their transition metal complexes (3–10).

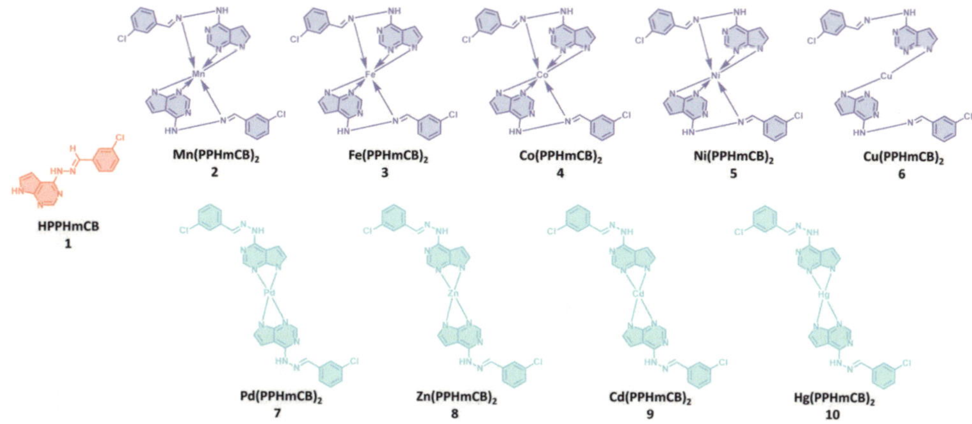

Figure 8.
HPPHmCB ligand (1) and proposed structures of their metal (II) complexes (2–10).

Zn(II) complexes (9 and 10) exhibit significant antimalarial efficacy, with half-maximal inhibitory concentrations (IC_{50}) of 1.08 ± 0.09 and 1.18 ± 0.04 μM against *Plasmodium falciparum*, respectively. These values outperform the frontline clinical drug chloroquine (IC_{50} = 1.29 μM). The study further reveals that the compounds inhibit parasite development by interacting with *P. falciparum* dihydroorotate dehydrogenase, thereby suppressing the parasite's metabolic processes.

Bagul et al. [34] employ a strategy targeting *Plasmodium* mitochondrial energy metabolism to design and synthesize a pyrrolopyrimidine hydrazone-3-chlorobenzaldehyde hybrid ligand (HPPHmCB, 1) and its nine transition metal complexes (2–10) (**Figure 8**) based on a three-dimensional pharmacophore model. They systematically investigate the influence of oxidation states of the metal centers on multi-target antimalarial activity. In vitro experimental results show that the Hg(II) complex exhibits antimalarial efficacy comparable to the clinical drug quinine, with a IC_{50} of 1.98 ± 0.08 μM, versus quinine's IC_{50} of 1.76 μM. Additionally, Zn(II), Cu(II), and Pd(II) complexes demonstrate significant activity, with IC_{50} of 2.39 ± 0.08 μM, 2.67 ± 0.07 μM, and 2.19 ± 0.06 μM, respectively. Molecular docking studies further reveal that binding energies of $Co(PPHmCB)_2$ and $Ni(PPHmCB)_2$ with $PfNDH_2$ are −186.62 and −165.04 kcal/mol, respectively, indicating strong binding affinity. These compounds effectively inhibit enzymatic activity by forming hydrogen bonds and hydrophobic interactions with key amino acid residues of $PfNDH_2$.

3. Applications of Schiff bases in antitumor therapy

Malignant tumors have emerged as a critical global public health challenge, with their high incidence and mortality rates posing persistent threats to modern medicine. Although platinum-based drugs (e.g., cisplatin) remain first-line chemotherapeutic agents for solid tumors, their clinical utility is severely limited by dose-dependent systemic toxicity and acquired resistance mechanisms. Over the past two decades, research focus has shifted toward developing metallo-anticancer agents, among which Schiff base metal complexes stand out due to their structural tunability and multi-mechanistic therapeutic advantages. By integrating the bioactive sites of Schiff base ligands with the redox-regulatory capacity of metal centers, these complexes trigger synergistic antitumor effects, including DNA crosslinking damage, ROS burst, mitochondrial apoptosis pathway activation, and immunomodulation of the tumor microenvironment, thereby offering innovative strategies to overcome limitations of conventional chemotherapy.

3.1 Antitumor activity of Schiff bases metal complexes

Inspired by the clinical efficacy of platinum-based drugs such as cisplatin and oxaliplatin, Schiff base platinum complexes have gained widespread attention [35–37]. Qi et al. [38] design and synthesize two Pt(II) complexes (VB6-Pt1 and VB6-Pt2) (**Figure 9a**) containing copper-coordinating active sites. Studies reveal that these complexes exhibit significant antiproliferative capacity in overcoming tumor cell resistance to cisplatin (DDP). Experimental results show that VB6-Pt1 and VB6-Pt2 demonstrate higher antitumor activity against typical lung cancer cells compared to DDP. Specifically, VB6-Pt1 achieves an IC_{50} of 0.78 μM, effectively reversing DDP resistance in the A549/DDP cell line, with a selectivity index of 26 for mortal MRC-5 fibroblasts, indicating high specificity. In animal studies, VB6-Pt1 inhibits A549 xenograft tumor growth by 81.5%, significantly surpassing DDP's 50.0% inhibition rate. Notably, VB6-Pt1 treatment does not significantly increase P-glycoprotein (P-gp) expression, suggesting its antitumor mechanism may not rely on P-gp upregulation. Mechanistic investigations reveal that VB6-Pt1 overcomes drug resistance by markedly elevating intracellular ROS levels, inducing lysosomal membrane permeabilization, and triggering mitochondrial dysfunction and apoptosis. This discovery provides critical insights for developing novel platinum-based anticancer agents.

In recent years, the development of Ru complexes as second-generation metal-based antitumor drugs has accelerated significantly, demonstrating superior metabolic properties and multi-target mechanisms compared to traditional platinum-based agents. Notably, Ru-Schiff base complexes achieve breakthrough advancements in cancer therapy: these compounds exhibit remarkable antiproliferative activity

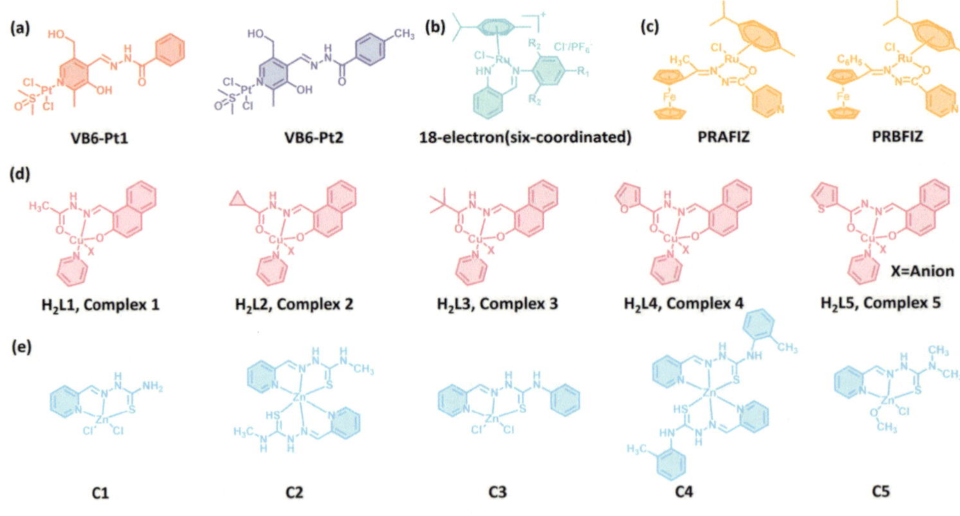

Figure 9.
(a) Molecular structures of VB6-Pt1 and VB6-Pt2, (b) 18-electron(six-coordinated) Schiff base complexes, (c) PRAFIZ and PRBFIZ, (d) Mixed-ligand Cu(II) hydrazone complexes, (e) The crystal structures of Zn(II) complexes.

and potential to overcome drug resistance through mechanisms including regulation of the mitochondrial apoptotic pathway, induction of cell cycle arrest, and inhibition of tumor metastasis. Guo et al. [39] successfully synthesize several novel arene ruthenium-coordinated Schiff base compounds (**Figure 9b**), which demonstrate significant antiproliferative activity against multiple tumor cell lines in vitro. Further studies reveal that these compounds specifically target mitochondria, induce mitochondrial membrane depolarization, and markedly promote intracellular ROS generation, thereby triggering apoptosis. Additionally, they arrest the cell cycle at the G_0/G_1 phase, suppress normal cell division and proliferation, and accelerate apoptotic processes. Collectively, the redox mechanism appears to be the primary pathway through which these compounds exert anticancer effects, providing novel insights and directions for the development of anticancer drugs. Similarly, Sindhu et al. [40] report two novel Ru(II) complexes (PRAFIZ and PRBFIZ) (**Figure 9c**) that demonstrate significant antitumor activity in vitro, with PRAFIZ notably exceeding DDP in efficacy against the A549 cell line. These compounds bind to DNA via an intercalation mode and influence their biological functions through interactions with bovine serum albumin (BSA). Petrova et al. [41] synthesize Ru(III) complexes, particularly RuSalpn, which demonstrate remarkable antitumor activity. These complexes kill cancer cells by inducing apoptosis and autophagy, while also inhibiting cell migration and blocking 3D colony formation, thereby showing potential antimetastatic effects.

Cu(II) is a vital micronutrient that plays a critical role in numerous biological processes within the body. Its unique redox properties and high compatibility with biological systems make Cu(II) one of the most widely used transition metals in the development of anticancer drugs. Additionally, Cu-based compounds are often preferred over platinum-based agents due to their lower toxicity and improved safety profile [42, 43]. Wu et al. [44] synthesize five Cu(II)-based Schiff base complexes, including [Cu(HL1)(pyridine)NO_3] (1), [Cu(HL2)(pyridine)Cl] (2), [Cu(HL3)(pyridine)Br] (3), [Cu(HL4)(pyridine)Br] (4), and [Cu(L5)(pyridine)] (5) (**Figure 9d**). Notably, complex 4 exhibits superior anticancer activity in A549 and A549/DDP cell compared to DDP. Mechanistic studies show that complex 4 induces apoptosis in A549 cancer cells primarily through a ROS-mediated mitochondrial pathway and inhibits cell growth by triggering S phase cell cycle arrest. Additionally, complex 4 demonstrates significant antitumor effects in 3D culture models and shows good hemocompatibility. These findings suggest that complex 4 has the potential to serve as an effective copper-based metal antitumor agent for lung cancer therapy.

Jiang et al. [45] synthesize a series of Zn(II) Schiff base complexes (C1–C5) (**Figure 9e**). C5, with two methyl groups introduced at the N-4 position of the ligand, shows higher lipophilicity and cellular uptake efficiency, thus exhibiting remarkable antitumor activity. C5 exerts its anti-tumor effect through multiple synergistic pathways: *Cell cycle arrest*: C5 up-regulates p27 and p21 while down-regulating CDc25A, CDk2, and cyclin D1, arresting the cell cycle at the S phase to inhibit cell proliferation and division. *Apoptosis induction*: C5 reduces the mitochondrial membrane potential (MMP) via the mitochondrial pathway, up-regulates pro-apoptotic proteins (Bax, Bad, Cyt c, p53), down-regulates anti-apoptotic proteins (Bcl-2, Bcl-xl), activates the caspase pathway, and induces apoptosis. *Lethal autophagy*: C5 increases the expression of autophagy markers LC3-II and Beclin 1 to induce autophagy. The partial reversal of cytotoxicity by the autophagy inhibitor 3-MA indicates that autophagy is lethal. *Anti-angiogenesis*: C5 inhibits VEGF expression, disrupts tumor blood vessel formation, and cuts off the tumor's nutrient supply. Through the synergistic effect of the above multi-pathway, C5 overcomes the drug-resistance issue of single-target agents.

It provides vital experimental evidence for developing non-platinum metal-based antitumor drugs and holds potential for clinical application in multi-target combination therapy.

3.2 Natural product-modified Schiff bases

By directionally coupling natural products with plant-derived bioactive components, we can construct environmentally friendly nanostructured systems. This technology not only significantly reduces toxicity risks and minimizes non-specific accumulation in organs, but also improves biodistribution profiles through precise targeting delivery mechanisms, offering innovative solutions for developing safe, edible biocompatible materials.

Alamri et al. [46] synthesize a series of chitosan Schiff bases (ChCSB1–6) by condensing chitosan (CS) with six aromatic aldehydes (**Figure 10a**) and systematically evaluate their antitumor activities. ChCSB4 (2,6-dichlorinated) and ChCSB5 (2,3-dichlorinated) show moderate antitumor activity against HCT-116 cells, with IC_{50} of 79.31 and 97.05 μg/mL, respectively. The study suggests that these chitosan Schiff bases may exert antitumor effects by inducing cell cycle arrest or causing DNA breakage.

Metal nanoparticles exhibit potential in tumor therapy due to their small size, high penetration, and targeting capabilities. Elsayed et al. [47] extract chitosan from shrimp shells and use it to perform a nucleophilic attack on salicylaldehyde, successfully synthesizing a chitosan Schiff base (**Figure 10b**). Then prepare a novel bimetallic oxide nanomaterial (chitosan Schiff base/$CuFe_2O_4$) by dispersing $CuFe_2O_4$ on the surface of the chitosan Schiff base and assess its antitumor activity against PC3 cells. The chitosan Schiff base/$CuFe_2O_4$ significantly inhibits the proliferation of PC3 cells, with an IC_{50} of 35.3 μg/mL, which is more effective than doxorubicin (IC_{50} = 39.19 μg/mL). Mechanistic studies indicate that the chitosan Schiff base/$CuFe_2O_4$ suppresses the proliferation and survival of PC3 cells by inhibiting the PI3K/AKT/mTOR

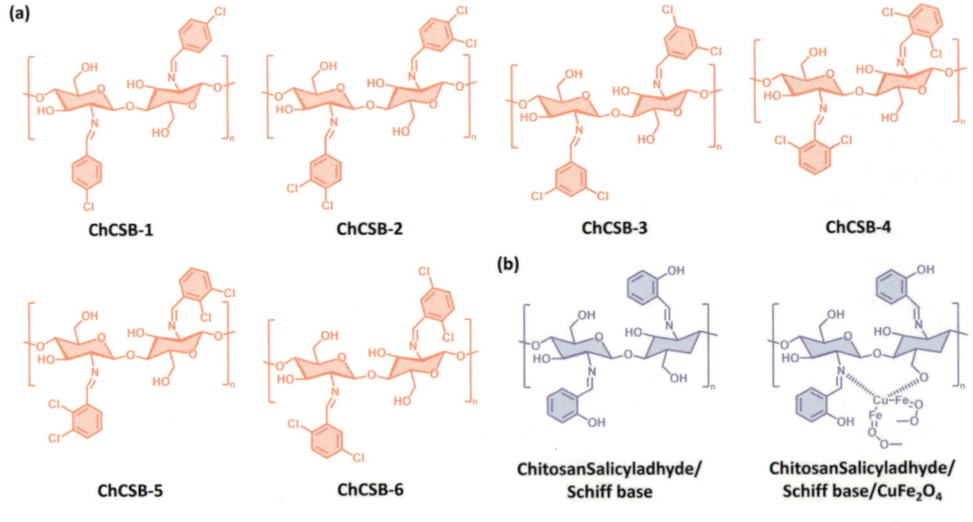

Figure 10.
Molecular structures of chitosan Schiff bases (ChCSB1–6). (b) Schiff base of novel chitosan/salicylaldehyde and the its bimetallic oxide of $CuFe_2O_4$.

signaling pathway and inducing autophagy. Additionally, molecular docking simulations and density functional theory (DFT) calculations support these experimental observations, showing that the chitosan Schiff base/$CuFe_2O_4$ has high chemical activity and biocompatibility.

Zakaria et al. [48] successfully construct six nanocomposite systems, including CS-SB-I-Ag and CS-SB-II-ZnO, through an innovative assembly technique that precisely load Ag, Au, and ZnO nanoparticles onto chitosan Schiff base frameworks. The growth inhibitory effects of these materials on HCT-116 colon cancer cells are systematically evaluated using the MTT colorimetric assay. The results show that the original chitosan Schiff bases CS-SB-I and CS-SB-II exhibit IC_{50} of 112.10 and 98.54 μg/mL, respectively, while the nanocomposites demonstrate significantly enhanced anticancer activity. Among them, the Ag-based nanocomposite systems CS-SB-I-Ag (IC_{50}: 10.99 μg/mL) and CS-SB-II-Ag (12.79 μg/mL) demonstrate the strongest anticancer activity, followed by the gold-based systems CS-SB-I-Au (14.96 μg/mL) and CS-SB-II-Au (26.72 μg/mL), with ZnO composites CS-SB-I-ZnO and CS-SB-II-ZnO showing stable IC_{50} values around 22 μg/mL. Mechanistic studies reveal that the nanoparticles significantly enhance their ability to penetrate cancer cell membranes due to their nanoscale dimensions and high surface area. The anticancer effects are mediated through dual pathways: (1) Nanoparticles induce a burst of ROS, triggering an oxidative stress response that disrupt mitochondrial membrane integrity and inhibit ATP synthesis, ultimately leading to apoptosis or programmed necrosis. (2) Active sites on the nanoparticle surfaces specifically interact with DNA bases and phosphate groups, interfering with genetic material replication and proliferation processes in cancer cells. This research provides crucial theoretical support for developing targeted anticancer strategies based on nanocomposite materials.

3.3 Other Schiff base compounds

Tan et al. [49] synthesize 28 thioether-containing isatin Schiff base derivatives (H1–H28) (**Figure 11**) through molecular design and evaluate their antitumor activity using the MTT assay. Among them, compound H13 exhibits the strongest inhibitory activity against lung cancer A549 cells (IC_{50} = 4.83 μM). A 3D-QSAR CoMFA model is constructed, revealing that the electronic properties of para-substituents on the benzene ring (e.g., 4-methoxy substitution) significantly enhance antitumor activity. Mechanistic studies demonstrate that H13 exerts anticancer effects through dual pathways: (1) inducing autophagic cell death by inhibiting the PI3K/AKT/mTOR signaling pathway, characterized by upregulation of the autophagy marker protein LC3-II; (2) triggering ROS burst leading to mitochondrial membrane potential collapse and activating p53-dependent apoptotic pathways. This compound serves as a lead molecule, providing critical theoretical support for developing autophagy-targeted anticancer drugs, with further exploration of its in vivo efficacy and clinical translation potential required.

Farghadani et al. [50] design a novel indole Schiff base β-diminato compound (LH3), which exhibits significant inhibitory activity against triple-negative breast cancer (MDA-MB-231, IC_{50} = 2.41 ± 0.29 μM) and hormone receptor-positive breast cancer (MCF-7, IC_{50} = 3.51 ± 0.14 μM). Its mechanism of action involves inducing excessive intracellular ROS generation, disrupting mitochondrial membrane potential, and activating the intrinsic apoptotic pathway (e.g., Bax/Bcl-2 imbalance). Additionally, LH3 inhibits cancer cell proliferation by blocking the G2/M phase checkpoint and suppressing the function of CDK/cyclin complexes.

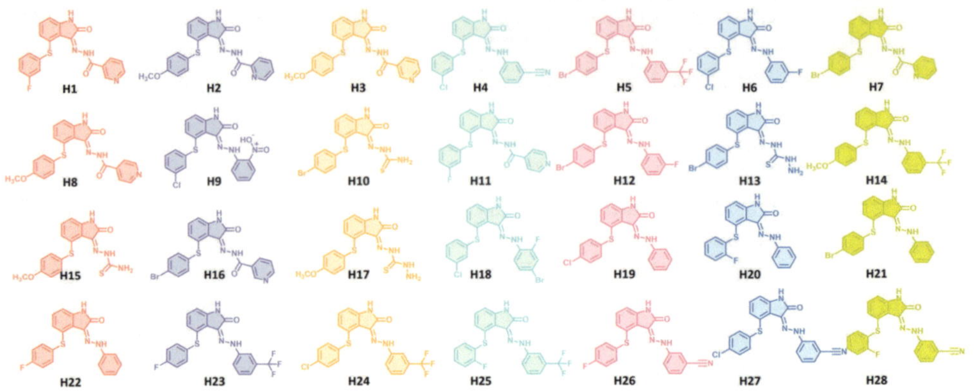

Figure 11.
Design of target compounds H1–H28.

4. Applications of Schiff bases in drug carriers

The pH-responsive hydrogel constructed based on dynamic Schiff base chemistry forms a dynamic network structure through in situ crosslinking of primary amine/aldehyde-functionalized compounds, exhibiting precise pH-responsive properties. The dynamic imine bonds not only endow the material with robust self-healing capability to autonomously repair mechanical damage but also synergize pH-dependent degradation behavior with sustained drug release, establishing an intelligent controlled-release system [51]. Gousia et al. [52] construct a non-toxic, injectable, pH-responsive hydrogel based on Schiff base bonds for targeted drug delivery in tumor microenvironments. Under acidic conditions (pH 5), controlled hydrolysis of imine bonds results in a 70% dry weight loss of the material after 27 days, while the hydrogel maintains structural stability in a physiological neutral environment (pH 7.4) with a 270% increase in wet weight. In an alkaline environment (pH 9), gel shrinkage occurs due to the salting-out effect. The release behavior of the model drug FITC-dextran exhibits environmental dependency: complete release within 13 days under acidic conditions, sustained release over 17 days in a neutral environment, and accelerate drug diffusion under alkaline conditions due to gel contraction. The system also demonstrates self-healing capability, autonomously restoring structural integrity within 30 minutes after cutting damage. Its core advantage lies in achieving precise pH responsiveness and intelligent drug-controlled release through dynamic imine bonds, effectively avoiding the toxicity issues associated with traditional chemical crosslinking agents. Qu et al. [53] develop a Schiff base-crosslinked Tetra-PEG-BSA hydrogel based on a dynamic crosslinking strategy. With its excellent swelling properties, the hydrogel enables efficient drug loading and controlled release. Its low cytotoxicity (with cell viability exceeding 80% at low concentrations) makes it suitable for long-term or repeated administration.

Some natural active components exhibit common limitations—despite possessing significant bioactivity, they suffer from poor solubility, low bioavailability, and inadequate chemical stability. Peng et al. [54] prepare two functionalized nanoparticles via an anti-solvent co-precipitation method. This novel nano-delivery system achieves efficient encapsulation of active ingredients. The system demonstrates unique dual environmental responsiveness: under acidic conditions (pH 5.0), it enables rapid release exceeding 80% within 2 hours, while showing accelerated release behavior in redox environments. Stability tests reveal that the nanoparticles maintain excellent dispersibility after 14 days of storage at 25°C, with particle size increasing only from 118 to 122 nm, and the polydispersity index (PDI) remaining consistently below 0.3. In agricultural applications, the system demonstrates 51.52% preventive efficacy and 48.17% therapeutic efficacy against rice bacterial leaf streak (BLS), representing significant improvement over conventional formulations. Its distinctive environment-responsive release mechanism, long-term storage stability, and biosafety characteristics provide critical technical insights for developing novel intelligent pesticide formulations.

5. Applications of Schiff bases in sensing and imaging

Schiff bases demonstrate significant potential in chemical sensing and bioimaging due to their unique molecular structure (containing $-C{=}N-$) and tunable photoelectric properties. In sensing applications, their selective binding with metal ions or specific biomolecules enables the construction of highly sensitive sensors for detecting environmental pollutants (e.g., heavy metal ions), biomarkers, or pathogens. In imaging, Schiff base derivatives are widely employed as fluorescent probes, leveraging their pH-responsive, redox-responsive, or targeting ligand-modified properties to achieve real-time dynamic imaging at cellular, tissue, and even in vivo levels.

Alnajeebi et al. [55] develop a Schiff base fluorescent chemosensor (S1) that can detect mercury ions (Hg^{2+}) with high sensitivity and selectivity. In a methanol/water (1:9, v/v) solution, S1 exhibits a significant "turn-on" fluorescence response upon binding to Hg^{2+}, with a detection sensitivity of 0.002 ppm. Even in complex systems containing various interfering ions such as Cu^{2+}, Co^{2+}, K^{+}, and Ca^{2+}, S1 maintains its specific recognition of Hg^{2+}. Tests in real water samples, including tap water, river water, and pond water, show that the recovery rate of Hg^{2+} is stable in the range of 90.0% to 104.0%, verifying its reliability in practical detection scenarios. The mechanism of action is as follows: in the free state, S1 undergoes photo-induced electron transfer (PET) due to the lone pair electrons on the nitrogen atom of the imine group, resulting in fluorescence quenching. When Hg^{2+} coordinates with the nitrogen, sulfur, and hydroxyl groups of S1, the PET process is blocked, thereby restoring fluorescence emission.

Sambamoorthy et al. [56] successfully prepare two new Schiff base compounds, PYEB (4-(1-(pyridin-4-ylimino)ethyl)benzene-1,3-diol) and PDEB (4,40-((pyridine-2,3-diylbis(azanylylidene))bis(ethan-1-yl-1-ylidene))bis(benzene-1,3-diol)), via solid-state grinding. Experiments show that PYEB can specifically recognize Fe^{3+} and Cu^{2+} in a methanol/water system, with detection limits of 1.64×10^{-6} M and 2.16×10^{-7} M, respectively. In contrast, PDEB exhibits selectivity only for Cu^{2+}, with a detection limit of 4.54×10^{-4} M. The sensing mechanisms of both compounds are based on coordination chemistry principles. PYEB forms a bidentate coordination mode with Fe^{3+} and Cu^{2+} through imine nitrogen and phenolic oxygen atoms. Interaction with Fe^{3+} leads to a 373% increase in fluorescence intensity due to the inhibition of excited state intramolecular proton transfer (ESIPT) and the chelation-enhanced fluorescence (CHEF) effect. In contrast, interaction with Cu^{2+} results in fluorescence quenching (CHEQ) due to the intersystem crossing (ISC) induced by the paramagnetic nature of Cu^{2+}. PDEB, as a tetradentate ligand, forms a 1:1 complex with Cu^{2+} through two imine nitrogen atoms and two phenolic oxygen atoms. The fluorescence quenching of PDEB has been successfully applied in bioimaging studies of HeLa cells, confirming the potential of this probe for monitoring Cu^{2+} dynamics in living cells. This work has achieved dual functional expansion of Schiff base materials in environmental detection and biomedical imaging through molecular design.

Dua et al. [57] synthesize a dichromene Schiff base probe (COM-CH) via a condensation reaction, achieving selective binding with Th^{4+} through the donor properties of chromene units. Ultraviolet-visible absorption and fluorescence emission tests reveal significant solvatochromic effects in different solvents. In an ACN/H_2O (8:2) system, COM-CH exhibits a "turn-off" fluorescence response to Th^{4+}, with a detection limit as low as 18.7 nM and excellent selectivity against interfering ions such as La^{3+} and Fe^{3+}. Nuclear magnetic resonance (NMR) and Fourier transform infrared (FTIR) spectroscopy combined with DFT calculations confirm the formation of a 2:1 complex between Th^{4+} and the probe's C═N and C═O groups, leading to a reduced energy gap (from 3.50 to 2.29 eV) and spectral redshift. Practical application tests demonstrate Th^{4+} detection in water samples and monazite sand with recovery rates of 94–98%, consistent with ICP-OES standard methods. The study also achieves the first in vivo fluorescence imaging of Th^{4+} in *Caenorhabditis elegans*, validating the probe's low toxicity and applicability in complex biological systems.

6. Conclusions and outlook

Schiff base compounds and their metal complexes, characterized by their distinctive C=N structure and high modifiability, demonstrate broad application prospects in the biomedical sector. They play pivotal roles in the prevention and treatment of diseases, including but not limited to: (1) *Antibacterial activity*—disrupting bacterial cell wall synthesis, damaging membrane integrity, and inhibiting bacterial growth and reproduction; (2) *Antiviral effects*—blocking viral adsorption and invasion into host cells, and inhibiting key enzymes essential for viral replication; (3) *Antifungal action*—impairing ergosterol synthesis in fungal cell membranes and damaging membrane structure and function; (4) *Antiparasitic properties*—disrupting the parasitic survival environments and interfering with metabolic processes; (5) *Antitumor therapy*—inducing tumor cell apoptosis, inhibiting proliferation, and modulating the cell cycle in cancer cells; (6) *Drug delivery*—serving as carriers for targeted and precise drug transport.

In addition, Schiff-based function as probes for biomolecule detection and cellular imaging. Through structural optimization strategies, such as the introduction of functional groups, regulation of metal coordination, and hybridization with natural products, the bioactivity, targeting capability, and stability of Schiff base compounds have been significantly enhanced. These advancements deepen the understanding of their molecular mechanisms and establish a solid theoretical and technical foundation for developing multifunctional drugs and precision medicine applications based on their structural framework.

Future research can further unlock the potential of Schiff base compounds in the following directions:

1. Structural innovation and intelligent design

   Develop Schiff base smart molecules with multiple responsiveness (e.g., triggered release by pH, enzymes, and/or photothermal effects) to enhance their target-delivery efficiency and controlled-release capabilities in complex physiological environments.

2. Multi-technology integration and development

   Leverage AI-assisted molecular design, high-throughput screening, and organoid models to accelerate the discovery and optimization of novel Schiff base drugs while cutting R&D costs.

3. Interdisciplinary application expansion

   Explore Schiff base applications in combined therapies (e.g., chemo-photodynamic therapy), multimodal theranostic integration (closed-loop diagnosis-treatment-monitoring systems), and anti-drug-resistant infections.

4. Green synthesis and sustainability

Develop efficient, low-toxicity, eco-friendly synthesis methods to boost the scaled-up production and ecological compatibility research of Schiff base materials.

Through interdisciplinary integration and technological innovation, Schiff base compounds are expected to become key molecular tools for next-generation precision medicine and smart drug development, offering more competitive solutions to human health challenges.

Acknowledgements

The authors acknowledge the fund help from University of South Africa (UNISA), South Africa National Research Foundation (NRF) (UID 137947, CHN231120165165, UID 141947, and CPRR240414214091).

Conflict of interest

The authors declare no conflict of interest.

Author details

Ruijuan Zhang, Yali Yao and Xinying Liu*
Institute for Catalysis and Energy Solutions, University of South Africa, Florida, South Africa

*Address all correspondence to: liux@unisa.ac.za

References

[1] Golbedaghi R, Tabanez AM, Esmaeili S, Fausto R. Biological applications of macrocyclic Schiff base ligands and their metal complexes: A survey of the literature (2005-2019). Applied Organometallic Chemistry. 2020;**34**:e5884. DOI: 10.1002/aoc.5884

[2] More MS, Joshi PG, Mishra YK, Khanna PK. Metal complexes driven from Schiff bases and semicarbazones for biomedical and allied applications: A review. Materials Today Chemistry. 2019;**14**:100195. DOI: 10.1016/j.mtchem.2019.100195

[3] Rauf A, Shah A, Khan AA, Shah AH, Abbasi R, Qureshi IZ, et al. Synthesis, pH dependent photometric and electrochemical investigation, redox mechanism and biological applications of novel Schiff base and its metallic derivatives. Spectrochimica Acta Part A: Molecular and Biomolecular Spectroscopy. 2017;**176**:155-167. DOI: 10.1016/j.saa.2017.01.018

[4] Boulechfar C, Ferkous H, Delimi A, Djedouani A, Kahlouche A, Boublia A, et al. Schiff bases and their metal complexes: A review on the history, synthesis, and applications. Inorganic Chemistry Communications. 2023;**150**:110451. DOI: 10.1016/j.inoche.2023.110451

[5] Zhang J, Xu L, Wong W-Y. Energy materials based on metal Schiff base complexes. Coordination Chemistry Reviews. 2018;**355**:180-198. DOI: 10.1016/j.ccr.2017.08.007

[6] Antony R, Arun T, Manickam STD. A review on applications of chitosan-based Schiff bases. International Journal of Biological Macromolecules. 2019;**129**:615-633. DOI: 10.1016/j.ijbiomac.2019.02.047

[7] Yadav M, Yadav D, Pal Singh D, Kumar KJ. Pharmaceutical properties of macrocyclic Schiff base transition metal complexes: Urgent need in today's world. Inorganica Chimica Acta. 2023;**546**:121300. DOI: 10.1016/j.ica.2022.121300

[8] Nidhi S, Rao DP, Gautam AK, Verma A, Gautam Y. Schiff bases and their possible therapeutic applications: A review. Results in Chemistry. 2025;**13**:101941. DOI: 10.1016/j.rechem.2024.101941

[9] Santu A, Manoj E. Coumarin based Schiff base proligands as promising antioxidant, antibacterial and anticancer agents: DFT and molecular docking correlations. Journal of Molecular Structure. 2025;**1333**:141650. DOI: 10.1016/j.molstruc.2025.141650

[10] Sundari ACM, Jha A. Sonochemical synthesis of Schiff base of substituted benzoxazolone: A quantum chemical, antibacterial evaluation and molecular docking study. Chemistry Select. 2025;**10**(5):e202404442. DOI: 10.1002/slct.202404442

[11] Aldholmi M, Marchand P, Ourliac-Garnier I, Le Pape P, Ganesan A. A decade of antifungal leads from natural products: 2010-2019. Pharmaceuticals (Basel). 2019;**12**(4):182. DOI: 10.3390/ph12040182

[12] Umer SM, Solangi M, Khan KM, Saleem RS. Indole-containing natural products 2019-2022: Isolations, reappraisals, syntheses, and biological activities. Molecules. 2022;**27**(21):7586. DOI: 10.3390/molecules27217586

[13] Ekiert HM, Szopa A. Biological activities of natural products

II. Molecules. 2022;**27**(5):1519. DOI: 10.3390/molecules27051519

[14] Damena T, Zeleke D, Desalegn T, Demissie TB, Eswaramoorthy R. Synthesis, characterization, and biological activities of novel vanadium(IV) and cobalt(II) complexes. ACS Omega. 2022;**7**:4389-4404. DOI: 10.1021/acsomega.1c06205

[15] El-Gammal OA, Mohamed FS, Rezk GN, El-Bindary AA. Structural characterization and biological activity of a new metal complexes based of Schiff base. Journal of Molecular Liquids. 2021;**330**:115522. DOI: 10.1016/j.molliq.2021.115522

[16] Hou W, Xu H. Incorporating selenium into heterocycles and natural products—from chemical properties to pharmacological activities. Journal of Medicinal Chemistry. 2022;**65**:4436-4456. DOI: 10.1021/acs.jmedchem.1c01859

[17] de Oliveira PS, Lima BJ, DSWM B, DNGH M, Almeida MR, Ribeiro AD, et al. Zn (II) complex with vanillin derived Schiff base: Antifungal, antibacterial, antioxidant and anticholinesterase activities. Future Medicinal Chemistry. 2025;**17**:767-778. DOI: 10.1080/17568919.2025.2479421

[18] Juyal VK, Thakuri SC, Panwar M, Rashmi PO, Perveen K, Bukhari NA, et al. Manganese(II) and zinc(II) metal complexes of novel bidentate formamide-based Schiff base ligand: Synthesis, structural characterization, antioxidant, antibacterial, and in-silico molecular docking study. Frontiers in Chemistry. 2024;**12**:1414646. DOI: 10.3389/fchem.2024.1414646

[19] Abu-Dief AM, El-Sagher HM, Shehata MR. Fabrication, spectroscopic characterization, calf thymus DNA binding investigation, antioxidant and anticancer activities of some antibiotic azomethine Cu(II), Pd(II), Zn(II) and Cr(III) complexes. Applied Organometallic Chemistry. 2019;**33**:e4943. DOI: 10.1002/aoc.4943

[20] Malanovic N, Lohner K. Gram-positive bacterial cell envelopes: The impact on the activity of antimicrobial peptides. Biochimica et Biophysica Acta (BBA)-Biomembranes. 2016;**1858**:936-946. DOI: 10.1016/j.bbamem.2015.11.004

[21] Wolohan P, Reichert DE. Use of binding energy in comparative molecular field analysis of isoform selective estrogen receptor ligands. Journal of Molecular Graphics and Modelling. 2004;**23**:23-38. DOI: 10.1016/j.jmgm.2004.03.002

[22] Berhanu AL, Gaurav MI, Malik AK, Aulakh JS, Kumar V, Kim K-H. A review of the applications of Schiff bases as optical chemical sensors. TrAC Trends in Analytical Chemistry. 2019;**116**:74-91. DOI: 10.1016/j.trac.2019.04.025

[23] Uddin MN, Samina AS, Alam SMR. Biomedical applications of Schiff base metal complexes. Journal of Coordination Chemistry. 2020;**73**:3109-3149. DOI: 10.1080/00958972.2020.1854745

[24] Verma C, Quraishi MA. Recent progresses in Schiff bases as aqueous phase corrosion inhibitors: Design and applications. Coordination Chemistry Reviews. 2021;**446**:214105. DOI: 10.1016/j.ccr.2021.214105

[25] Azzouzi M, Ouchaoui AA, Azougagh O, El Hadad SE, Abou-salama M, Oussaid A, et al. Synthesis, crystal structure, and antiviral evaluation of new imidazopyridine-Schiff base derivatives: in vitro and in silico anti-HIV studies. RSC Advances.

2024;**14**:36902-36918. DOI: 10.1039/d4ra07561g

[26] Abbas G, Irfan A, Ahmed I, Al-Zeidaneen FK, Muthu S, Fuhr O, et al. Synthesis and investigation of anti-COVID19 ability of ferrocene Schiff base derivatives by quantum chemical and molecular docking. Journal of Molecular Structure. 2022;**1253**:132242. DOI: 10.1016/j.molstruc.2021.132242

[27] Martins DOS, Ruiz UEA, Santos IA, Oliveira IS, Guevara-Vega M, de Paiva REF, et al. Exploring the antiviral activities of the FDA-approved drug sulfadoxine and its derivatives against chikungunya virus. Pharmacological Reports. 2024;**76**:1147-1159. DOI: 10.1007/s43440-024-00635-z

[28] Hao Y, Feng Y, Dong Y, Ren Y, Huang J, Ma H, et al. Synthesis and antifungal properties of 1,2,4-Triazole Schiff base agents based on a 3D-QSAR model. Chemistry & Biodiversity. 2024;**21**:e202302064. DOI: 10.1002/cbdv.202302064

[29] Buran K. Metal complexes of sulfamethazine/benzoin-based Schiff base ligand: Synthesis, characterization, DFT calculations, and antimicrobial activities. Turkish Journal of Biology. 2025;**49**:118-126. DOI: 10.55730/1300-0152.2729

[30] Liu W, Qin Y, Liu S, Xing R, Yu H, Li P. Synthesis and characterization of slow-release chitosan oligosaccharide pyridine Schiff base copper complexes with antifungal activity. Journal of Agricultural and Food Chemistry. 2024;**72**:3872-3883. DOI: 10.1021/acs.jafc.3c04601

[31] Dziduch K, Greniuk D, Wujec M. The current directions of searching for antiparasitic drugs. Molecules. 2022;**27**(5):1534. DOI: 10.3390/molecules27051534

[32] Aggarwal S, Paliwal D, Kaushik D, Gupta GK, Kumar A. Pyrazole Schiff base hybrids as anti-malarial agents: Synthesis, in vitro screening and computational study. Combinatorial Chemistry & High Throughput Screening. 2018;**21**(3):194-203. DOI: 10.2174/1386207321666180213092911

[33] Rani M, Devi J, Kumar B, Rathi M. Unveiling anti-malarial, antimicrobial, antioxidant efficiency and molecular docking study of synthesized transition metal complexes derived from heterocyclic Schiff base ligands. Chemistry–An Asian Journal. 2024;**19**:e202400676. DOI: 10.1002/asia.202400676

[34] Bagul AD, Kumar M, Alanazi AM, Tufail A, Tufail N, Gaikwad DD, et al. Experimental and computational evaluation of anti-malarial and antioxidant potential of transition metal (II) complexes with tridentate Schiff base derived from pyrrolopyrimidine. Biometals. 2024;**37**:1713-1737. DOI: 10.1007/s10534-024-00636-8

[35] Kurz T, Eaton JW, Brunk UT. Redox activity within the lysosomal compartment: Implications for aging and apoptosis. Antioxidants & Redox Signaling. 2009;**13**:511-523. DOI: 10.1089/ars.2009.3005

[36] Kaminaga K, Hamada R, Usami N, Suzuki K, Yokoya A. Targeted nuclear irradiation with an X-ray microbeam enhances total JC-1 fluorescence from mitochondria. Radiation Research. 2020;**194**:511-518. DOI: 10.1667/rr15110.1

[37] Perelman A, Wachtel C, Cohen M, Haupt S, Shapiro H, Tzur A. JC-1: Alternative excitation wavelengths

facilitate mitochondrial membrane potential cytometry. Cell Death & Disease. 2012;**3**:e430. DOI: 10.1038/cddis.2012.171

[38] Qi J, Zheng Y, Li B, Wei L, Li J, Xu X, et al. Mechanism of vitamin B6 benzoyl hydrazone platinum(II) complexes overcomes multidrug resistance in lung cancer. European Journal of Medicinal Chemistry. 2022;**237**:114415. DOI: 10.1016/j.ejmech.2022.114415

[39] Guo L, Li P, Li J, Gong Y, Li X, Wen T, et al. Potent half-sandwich 16−/18-electron iridium(III) and ruthenium(II) anticancer complexes with readily available amine–imine ligands. Inorganic Chemistry. 2023;**62**:21379-21395. DOI: 10.1021/acs.inorgchem.3c03471

[40] Sindhu M, Kalaivani P, Prabusankar G, Sivasamy R, Prabhakaran R. Preparation of new organo-ruthenium(II) complexes and their nucleic acid/albumin binding efficiency and in vitro cytotoxicity studies. Dalton Transactions. 2024;**53**:3075-3096. DOI: 10.1039/d3dt04017h

[41] Petrova Z, Mocanu T, Spasov R, Hanganu A, Marinescu G, Culita DC, et al. Antitumor activity of ruthenium(III) complexes with [N_2O_2]-tetradentate Schiff base ligands. Journal of Inorganic Biochemistry. 2025;**266**:112853. DOI: 10.1016/j.jinorgbio.2025.112853

[42] Han TY, Guan TS, Iqbal MA, Haque RA, Sharmila Rajeswari K, Khadeer Ahamed MB, et al. Synthesis of water soluble copper(II) complexes: Crystal structures, DNA binding, oxidative DNA cleavage, and in vitro anticancer studies. Medicinal Chemistry Research. 2014;**23**:2347-2359. DOI: 10.1007/s00044-013-0824-9

[43] da Silva DA, De Luca A, Squitti R, Rongioletti M, Rossi L, Machado CML, et al. Copper in tumors and the use of copper-based compounds in cancer treatment. Journal of Inorganic Biochemistry. 2022;**226**:111634. DOI: 10.1016/j.jinorgbio.2021.111634

[44] Wu Y, Hou L, Lan J, Yang F, Huang G, Liu W, et al. Mixed-ligand copper(II) hydrazone complexes: Synthesis, structure, and anti-lung cancer properties. Journal of Molecular Structure. 2023;**1279**:134986. DOI: 10.1016/j.molstruc.2023.134986

[45] Jiang M, Pang J, Jia X, Chu Y, Li W, Liang H, et al. Development of a zinc(ii) 2-pyridinecarboxaldehyde thiosemicarbazone complex with remarkable antitumor and antiangiogenic activities. Dalton Transactions. 2023;**52**:6029-6040. DOI: 10.1039/d3dt00148b

[46] Alamri AA, Borik RMA, El-Wahab AHFA, Mohamed HM, Ismail KS, El-Aassar MR, et al. Synthesis of Schiff bases based on chitosan, thermal stability and evaluation of antimicrobial and antitumor activities. Scientific Reports. 2025;**15**:892. DOI: 10.1038/s41598-024-73610-6

[47] Elsayed GH, Fahim AM. Studying the impact of chitosan salicylaldehyde/Schiff base/$CuFe_2O_4$ in PC3 cells via theoretical studies and inhibition of PI3K/AKT/mTOR signalling. Scientific Reports. 2025;**15**:4129. DOI: 10.1038/s41598-025-86096-7

[48] Zakaria N, Kandile NG, Mohamed MI, Zaky HT, Mohamed HM. Superior remedy colon cancer HCT-116 cells via new chitosan Schiff base nanocomposites: Synthesis and characterization. International Journal of Biological Macromolecules. 2024;**281**:135916. DOI: 10.1016/j.ijbiomac.2024.135916

[49] Tan H, Zhang G, Xu C, Lei X, Chen J, Long H, et al. Synthesis of novel 4-substituted isatin Schiff base derivatives as potential autophagy inducers and evaluation of their antitumour activity. Molecular Diversity. 2025;**29**:1983-2000. DOI: 10.1007/s11030-024-10954-1

[50] Farghadani R, Lim HY, Abdulla MA, Rajarajeswaran J. Novel indole Schiff base β-diiminato compound as an anti-cancer agent against triple-negative breast cancer: In vitro anticancer activity evaluation and in vivo acute toxicity study. Bioorganic Chemistry. 2024;**152**:107730. DOI: 10.1016/j.bioorg.2024.107730

[51] Pappalardo R, Boffito M, Cassino C, Caccamo V, Chiono V, Ciardelli G. Schiff-base cross-linked hydrogels based on properly synthesized poly(ether urethane)s as potential drug delivery vehicles in the biomedical field: Design and characterization. ACS Omega. 2024;**9**:45774-45788. DOI: 10.1021/acsomega.4c03157

[52] Gousia SS, Ahanger FA, Nazir N, Shaheen A, Tak UN, Manhas AA, et al. Cyclodextrin modified biocompatible chitosan-cinnamaldehyde Schiff base hydrogels: Their antibacterial, antioxidant, and drug delivery potential. International Journal of Biological Macromolecules. 2025;**306**:141523. DOI: 10.1016/j.ijbiomac.2025.141523

[53] Qu Y, Li J, Jia X, Yin L. Schiff base-crosslinked tetra-PEG-BSA hydrogel: Design, properties, and multifunctional functions. Journal of Functional Biomaterials. 2025;**16**(2):69. DOI: 10.3390/jfb16020069

[54] Peng J, Zhou L, Chen J, Hu D, Gan X. Zein and resveratrol Schiff base nanocomplexes: An efficient delivery system to enhance the antibacterial efficacy of berberine. International Journal of Biological Macromolecules. 2025;**306**:141496. DOI: 10.1016/j.ijbiomac.2025.141496

[55] Alnajeebi AM, Yahya R, Shafie A, Ashour AA, Felemban MF, Tayeb FJ. Highly sensitive and selective detection of Hg^{2+} ions and antibacterial activity using a Schiff base derivative. Journal of Fluorescence. 2024. DOI: 10.1007/s10895-024-04070-4 [Online ahead of print]

[56] Sambamoorthy S, Thamaraichelvan G, Karikalan A, Kumar SS. Heterocyclic fluorescent Schiff base chemosensors for the detection of Fe(III) and Cu(II) ions. Luminescence. 2024;**39**:e4739. DOI: 10.1002/bio.4739

[57] Dua A, Saini P, Goyal S, Selvam P, Ashok Kumar SK, Thiruppathi G, et al. Chromene-chromene Schiff base as a fluorescent chemosensor for Th^{4+} and its application in bioimaging of Caenorhabditis elegans. Methods. 2024;**225**:28-37. DOI: 10.1016/j.ymeth.2024.03.002

Section 4

Sensors and Analytical Applications

Chapter 8

Recent Advancements in Schiff Bases as Chemosensors

Isha Murtaza, Sohail Anjum Shahzad and Zulfiqar Ali Khan

Abstract

Chemosensor is a molecule or device that detects or responds to chemical species known as analytes. Chemosensors bind to specific analytes, triggering a measurable response. Schiff bases are widely explored as chemosensors due to unique properties such as chelating ability and binding to analytes including metal ions. The Schiff base can form a conjugated system, leading to a change in optical properties upon binding with analytes. Schiff bases exhibit fluorescence properties, making them suitable for optical sensing applications. Beside medicinal and biological applications, several Schiff bases are reported for the detection of analytes such as metal ions, anions, and nitroaromatic compounds. Schiff bases are inexpensive, easy to synthesize, and readily available laboratories.

Keywords: Schiff bases, chemosensors, fluorescent chemical sensors, nitroaromatic compounds, metal cations, anions

1. Introduction

Chemosensors are molecular or supramolecular devices that are able to detect specific analytes *via* fluorescence, electrochemical signals, or visible color changes. In fields like industrial monitoring, environmental safety, and biomedical diagnostics, where precise/accurate, quick, and careful detection of hazardous substances is crucial, the chemosensors are becoming more and more important [1–6].

Schiff bases appear as highly effective among various classes of chemosensors because of their ease of synthesis, structural simplicity, strong coordination ability, and tunable photophysical characteristics. Their structure contains an azomethine or imine bond (>C=N-). Schiff bases can detect a wide variety of noxious species by different mechanisms like colorimetric shifts, enhancements, and fluorescence quenching owing to their π-conjugation and excellent chelating properties [7–10].

The detection of nitroaromatic compounds (NACs), particularly those used in explosives such as 2,4,6-trinitrophenol (TNP), 2,4,6-trinitrotoluene (TNT), 2,4-dinitrotoluene (DNT), and nitrobenzene (NB), is of significant importance due to their widespread use in military, industrial, and agricultural applications and their serious environmental and health hazards. NACs are chemically stable, highly persistent in soil and groundwater, and resistant to natural biodegradation processes, leading to their accumulation in ecosystems over long periods. Their presence in contaminated sites poses risks such as carcinogenicity, mutagenicity, liver dysfunction, reproductive

 IntechOpen

toxicity, and groundwater pollution. Furthermore, NACs are commonly found in unexploded landmines, improvised explosive devices, and waste from munitions manufacturing, presenting ongoing threats to both civilian and military populations. Traditional detection methods—such as ion mobility spectrometry, gas chromatography, and trained canines—often suffer from limitations including high operational cost, complexity, delayed response times, and limited portability. In contrast, fluorescence-based sensors, including Schiff base derivatives, offer rapid, highly sensitive, cost-effective, and potentially field-deployable solutions. These optical chemosensors enable real-time detection of NACs with minimal instrumentation, making them especially promising for applications in anti-terrorism efforts, forensic investigations, environmental monitoring, and remediation of explosive residues [11, 12]. **Table 1** listed high-energy nitroaromatic compounds' (NACs) names, abbreviations, and vapor pressure in torr, giving the top priority to being detected [11].

Metal cations such as Fe^{3+} and Cu^{2+} are essential trace elements required for various physiological processes in humans, including oxygen transport, enzymatic redox reactions, connective tissue formation, and neural function. Iron, primarily present in hemoglobin, myoglobin, and several enzymes, regulates cellular respiration and immune function. Copper supports iron metabolism, red blood cell production, and enzymatic antioxidant defense. However, an imbalance of these metal ions—either through deficiency or excess—can lead to severe disorders. Iron deficiency results in anemia, impaired cognitive development, and increased maternal-fetal risks, while copper deficiency or toxicity can contribute to osteoporosis, neurological damage, cardiovascular risk, and inherited diseases such as Wilson's and Menkes disease. Schiff base-based fluorescent chemosensors offer distinct advantages for detecting such metal ions due to their strong coordination capabilities and ability to undergo fluorescence or color changes upon complexation. These changes, driven by mechanisms like ligand-to-metal charge transfer (LMCT), and intramolecular charge transfer (ICT), have enabled the development of ON/OFF sensing systems with practical applications in biological, environmental, and industrial contexts [13–15].

Additionally, the sensing of toxic anions such as fluoride (F^-), cyanide (CN^-), acetate (AcO^-), and arsenite (AsO_2^-) is crucial due to their widespread use and hazardous nature. Many of these anions are biologically active or environmentally persistent, leading to conditions such as water contamination, metabolic disorders, and organ dysfunction. Schiff base chemosensors are particularly attractive for anion sensing because they enable naked-eye detection through distinct color changes, driven by interactions such as hydrogen bonding, electron transfer, intramolecular charge transfer (ICT), and deprotonation. These mechanisms facilitate real-time, low-cost, and highly selective sensing—often without the need for advanced instrumentation [16, 17].

Nitro aromatic Compound(s) (NACs)	Abbreviation	Vapor pressure (torr)
2,4,6-Trinitrophenol (Picric acid)	TNP (PA)	5.8×10^{-9}
Nitrobenzene; 1,3,5-trinitrobenezene & *o-,m-,p-*dinitrobenzene	NB	9.00×10^{-4}
2,4,6-Trinitrotoluene	TNT	5.50×10^{-6}
2,4-Dinitrotoluene; 2,6-Dinitrotoluene	DNT	2.63×10^{-4}
4-nitrotoluene, 3-nitrotoluene & 2-nitrotoluene	NT	4.89×10^{-3}

Table 1.
Common nitroaromatic compound(s) (NACs) as explosive(s): name, abbreviation, and vapor pressure data [11].

The strong binding of Schiff bases to a variety of analytes due to their multiple donor atoms (e.g., N, O) and conjugated electronic systems renders them highly favorable for the construction of chemosensors. Their capacity to show differential optical responses to analyte interaction, for instance, turn-on/turn-off fluorescence and visible color change, enables both qualitative and quantitative sensing of analytes frequently in real time and without the need for advanced instrumentation [18–20].

Based on these features, this chapter presents a comprehensive overview of recent advancements in Schiff base-based chemosensors, focusing on their applications in the detection of nitroaromatic compounds, metal cations, and anions. Their structural design, optical behavior, sensing mechanism, and practical applications in biological and environmental systems are also elaborated on in detail.

1.1 Chemistry of Schiff bases

Schiff bases are typically synthesized through the condensation of primary amines with aldehydes or ketones under mild conditions. The first steps toward the synthesis and investigation of these substances were taken by Hugo Schiff's groundbreaking work in the nineteenth century with the evolution of organic chemistry. Their unique physicochemical properties are due to the azomethine or imine bond (>C=N-), and the R^1, R^2, and R^3 groups in the Schiff base are organic substituents (**Figure 1**). Schiff bases occur in various structural forms, involving mononuclear, binuclear, and polymeric frameworks that affect their electronic behavior and chelating ability [21, 22].

1.1.1 Structural and electronic features

The azomethine or imine bond (>C=N-) is the core structural element of Schiff bases that provides both planarity and rigidity and contributes to the development of stable coordination complexes with metal ions. The presence of additional donor atoms like carboxyl, sulfur, or hydroxyl groups increases their binding affinity and allows for selective interactions with specific analytes. Depending on their molecular design, Schiff bases can be mono , di , or polydentate, which could influence their sensing behavior and chelating strength [23–25].

The electronic structure of Schiff bases, particularly those with extended conjugation, is highly tunable *via* synthetic modifications. Electron density, solubility, and fluorescence properties can be modulated by the substituent groups on the aromatic ring or heteroatoms in the backbone, thus modifying the sensor's sensitivity and selectivity [26–29].

1.2 Schiff bases as chemosensors for detection of nitroaromatic compounds (NACs)

Schiff bases are being increasingly investigated for the detection of nitroaromatic compounds (NACs) like TNT, DNT, nitrophenols, and picric acid. NACs are

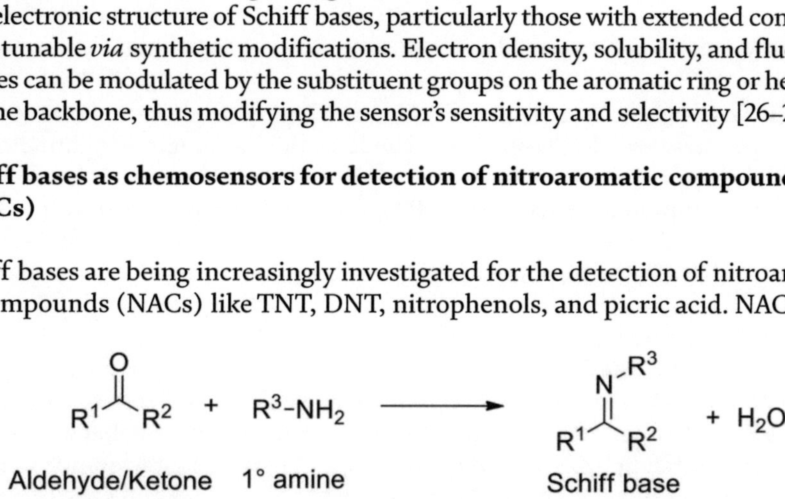

Figure 1.
General reaction of Schiff base synthesis [20].

electron-deficient in nature because of the presence of multiple nitro groups that enable various sensing mechanisms when interacting with Schiff base derivatives [12].

Nitroaromatic compounds (NACs) sensing by Schiff base-based chemosensors could be done by multiple interconnected mechanisms driven by the strong electron-withdrawing nature of nitro groups. A key pathway involves electron transfer from excited-state electron-rich Schiff base to electron-deficient NACs that results in significant fluorescent quenching and enables "turn-off" detection mode and the process is called photoinduced electron transfer (PET). In systems with donor-acceptor architectures, NAC binding can also disrupt intramolecular charge transfer (ICT) processes, which leads to characteristic shifts in emission wavelength (hypsochromic or bathochromic) and changes in fluorescence intensity. Planar NAC molecules further stabilize ground-state complexes *via* π-π stacking interactions with aromatic Schiff base frameworks that contribute to static quenching or observable colorimetric shifts. Moreover, hydrogen bonding and dipole-dipole interactions with nitro groups can enhance binding specificity, while selective functionalization with electron-donating groups (e.g., -OH, $-NH_2$) tailors sensor sensitivity and affinity, which often enables naked-eye detection under ambient or UV light. All these mechanisms establish Schiff base fluorescent sensors as highly selective, sensitive, and versatile platforms for the rapid detection of NACs in environmental, security, and forensic applications [30–35].

Sensing of nitroaromatic compounds (NACs) is really imperative, as they are linked with biological and environmental hazards because of their toxicity and use in explosives, dyes, and pesticides, and Schiff base-based chemosensors have garnered great attention for their detection. Fluorescence quenching through fluorescence resonance energy transfer (FRET) or photoinduced electron transfer (PET) are the mechanisms typically involved in the detection, as well as charge transfer mechanisms where the binding of NACs induces electronic changes. Because of the strong electron-withdrawing nature of NACs, its interactions with the electron-rich Schiff base moiety are enhanced, resulting in detectable optical shifts. High sensitivity with detection limits in the nanomolar to micromolar range is obtained by functionalizing with donor-acceptor groups. Hydrogen bonding and π-π interactions with NACs help in achieving selectivity. Applications include security and defense for explosive residue detection, environmental monitoring for pollutants like TNT and DNT, and industrial safety for monitoring hazardous nitroaromatic intermediates [30].

Ayşegül Köse and Mehmet Tümer developed a highly selective and sensitive Schiff base chemosensor **1** (**Figure 1**) *via* condensation of 5-aminoisophthalic acid with 2,3-dihydroxybenzaldehyde and evaluated it as a fluorometric sensor for nitroaromatic compounds (NACs) in DMSO. Chemosensor 1 exhibited blue fluorescence (400-450 nm) when excited in the 250–274 nm range. Various NACs, that is, nitrobenzene (NB), 2,4-nitrophenol (DNP), 4-nitrophenol, and 1,3,5-trinitrophenol (TNP), were tested, and chemosensor 1 showed the highest selectivity and sensitivity for DNP, with a quenching constant (K_{SV}) of 2.4 × 104 M^{-1} and a limit of detection (LOD) of 2.77 μM (**Table 2**). The quenching mechanism involves π-π stacking and hydrogen bonding interactions between the chemosensor **1** and electron-deficient NACs. The Stern-Volmer plot for sensor **1** showed linear quenching behavior toward all NACs. No significant background quenching and external interference were observed. These findings establish chemosensor **1** as a selective and sensitive fluorescent turn-off chemosensor for environmental detection of DNP and related nitroaromatic explosives (**Figure 2**) [31].

Sensor	Signal type	Sensing Mechanism	Analyte	LOD	Wavelength λem/λex/λabs	Medium	Selectivity	pH range	Ref.
1	Turn-off fluorescence	π-π stacking and hydrogen bonding	2,4-dinitrophenol (DNP)	2.77 μM	Em: (400-450 nm) Ex: (250–274 nm)	DMSO	Highly selective for DNP	–	[31]
2	Turn-off fluorescence	Fluorescence quenching	Picric acid (PA)	0.77 μM	–	–	Selective over phenol	–	[32]
3	AIEE	Aggregation induced emission (AIE)	2,4-dinitrophenol (DNP). Nitrobenzene (NB)	—	—	—	Selective for NACs	—	[32]
4	AIEE	AIE, Excimer formation	DNP, NB	—	Excimer emission at 690 nm	—	Selective for NACs	—	[32]
5	Turn-off fluorescence	Static quenching, FRET, Coulombic Interaction, IFE	Picric acid (TNP)	10 nM	469 nm/378 nm	THF:H_2O 90:10	highly selective for phenolic NACs	—	[33]
6	Turn-off fluorescence	Static quenching, donor-acceptor interaction	2,4,6-trinitrophenol (TNP)	0.253 μM	437 nm/393 nm	Aqueous	Highly selective for TNP over others	—	[30]
7	Turn-off fluorescence	FRET, PET	NB, DNP,TNP,NP	—	410–492 nm	—	Highly selective for NB	—	[34]
8	Turn-off fluorescence	FRET, PET	NB, DNP,TNP,NP	—	410–492 nm	—	Highly selective for NB	—	[34]
9	Turn-off fluorescence	FRET, PET	Nitrobenzene (NB)	0.013 μM	410–492 nm	—	Highly selective for NB	—	[34]
10	Colorimetric (UV-Vis shift)	H-bonding, charge transfer	Picric acid	0.692 μM	238 nm, 322 nm → 337 nm	THF:H_2O	Selective for picric acid	(2–11)	[35]

Table 2.
Various parameters for nitroaromatic compounds (NACs) analysis with chemosensors (**1–10**).

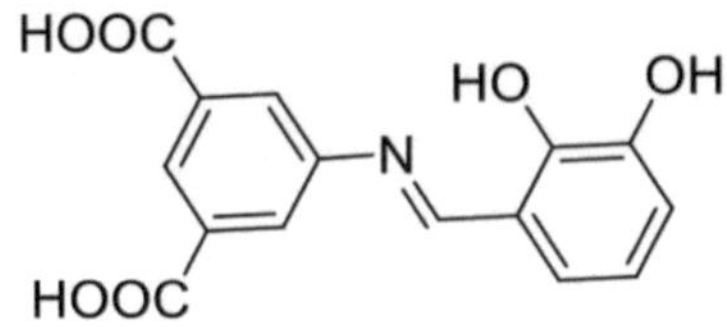

1

Figure 2.
Fluorometric sensor **1** *for NACs showed highest selectivity for 2,4-dintrophenol (2,4-DNP) [31].*

Mustafa Bal synthesized three Schiff base compounds, **2**, **3**, and **4**, as chemosensors incorporating pyrene, anthracene, and naphthalene units, respectively, through condensation reactions with butane-1,4-diamine and evaluated their fluorimetric response to nitroaromatic explosives (**Figure 3**). These fluorescent probes demonstrated diverse sensing mechanisms, including conventional fluorescence quenching and AIEE. Chemosensor **2** (pyrene-based) exhibited the highest fluorescence quenching upon interaction with picric acid (PA), showing a Ksv value to 16,700 M^{-1} and a LOD of 0.77 μM, particularly in the first and second emission zones. On the contrary, the chemosensors **3** and **4** showed emission enhancement because of aggregation upon exposure to nitroaromatic compounds like 2,4-dinitrophenol and nitrobenzene, indicating the potential for AIE-based sensing. All three chemosensors showed distinct photophysical profiles with strong Stokes shifts, and chemosensor **4** exhibited excimer emission at 690 nm due to ICT (**Table 2**). Selectivity studies showed clear differentiation between nitroaromatics and structurally similar phenols. These findings suggest that structurally tuned Schiff bases with polycyclic aromatic fluorophores can serve as effective platforms for the detection of explosive-related nitroaromatic compounds *via* both quenching and aggregation-induced mechanisms [32].

A novel Schiff base chemosensor **5** synthesized through condensation between 1-pyrene carboxaldehyde and dimethyl 5-aminoisophthalate was reported by Panigrahi et al. for highly selective detection of phenolic nitroaromatic compounds (**Figure 4**). The chemosensor **5** formed blue-emissive nanoaggregates in THF/water mixtures (**Table 2**), displaying aggregation-induced emission (AIE) properties. The fluorescence was significantly quenched in the presence of picric acid (2,4,6-trinitrophenol, TNP) with a remarkably low detection limit of 10 nM. The quenching mechanism involves a combination of static quenching, FRET, Coulombic interactions, and the inner filter effect (IFE). Stern-Volmer analysis confirmed high quenching efficiency with a K_{SV} of 2.03×10^4 M^{-1} for TNP. The nanoaggregates were stable, photostable, and biocompatible and could detect TNP in both aqueous and solid phases. Encapsulation in Pluronic P123 reduced cytotoxicity and allowed for successful live-cell imaging that confirmed chemosensor **5**'s role as a fluorescent sensor and bioimaging agent [33].

Altun et al. synthesized a novel Schiff base chemosensor **6** through condensation of thienopyridine-based amine with 3-nitrobenzaldehyde and evaluated it for the selective sensing of 2,4,6-trinitrophenol (TNP) in aqueous media (**Figure 5**). The chemosensor **6** exhibited strong blue fluorescence with an emission maximum at 437 nm upon excitation at 393 nm and a substantial turn-off response (64.5% quenching) in the presence of TNP. Stern-Volmer analysis revealed a quenching constant of 4.16×10^4 M^{-1} with a LOD of 0.253 μM, and a 1:1 binding stoichiometry was confirmed *via* Job's plot. The fluorescence quenching mechanism was attributed to static quenching facilitated by donor-acceptor interactions between the Schiff

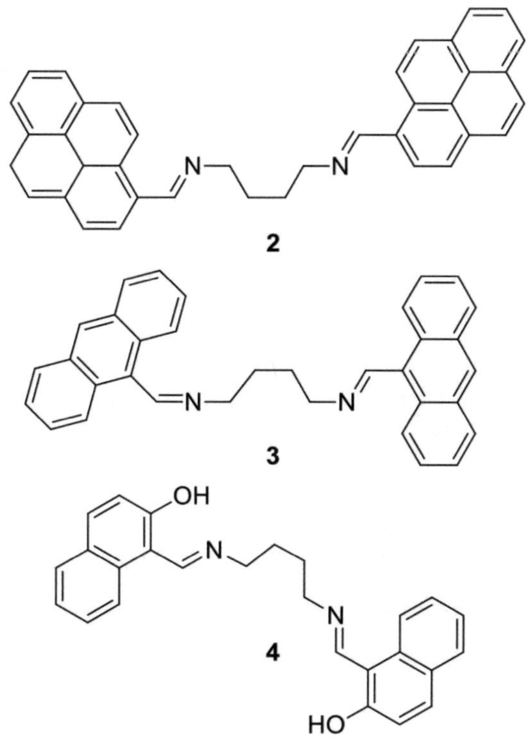

Figure 3.
Schiff base chemosensor **2, 3, 4** *structures for the detection of NACs [32].*

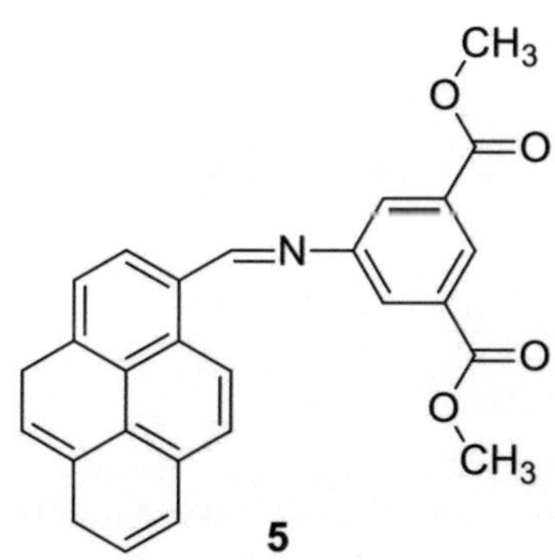

Figure 4.
Chemosensor **5** *for the selective detection of 2,4,6-trinitrophenol (TNP) [33].*

base sensor **6** and the electron-deficient nitro groups of TNP (**Table 2**). The sensor **6** also demonstrated excellent selectivity over other NACs (e.g., TNT, DNP, DNB, and DNT), with no observable quenching from competing analytes or metal ions. Its high photostability, aqueous solubility, and visible fluorescence color change render it a promising candidate for environmental sensing and explosive residue detection [30].

Kose and coworkers synthesized three Schiff base hydrazide-imine conjugates (chemosensors **7, 8,** and **9**) incorporating pyrene, anthracene, and phenanthrene moieties, respectively, for the selective detection of nitroaromatic explosives (**Figure 6**).

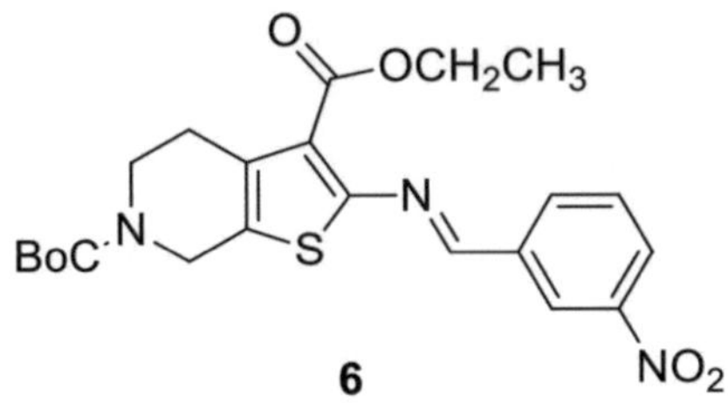

6

Figure 5.
Chemosensor **6** *for detection of NACs (TNT, DNP, DNB, and DNT) [30].*

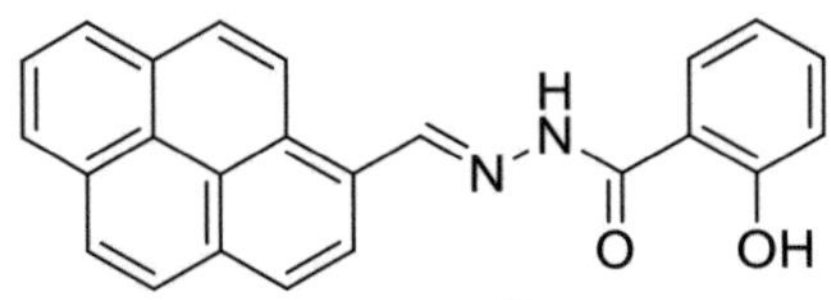

7

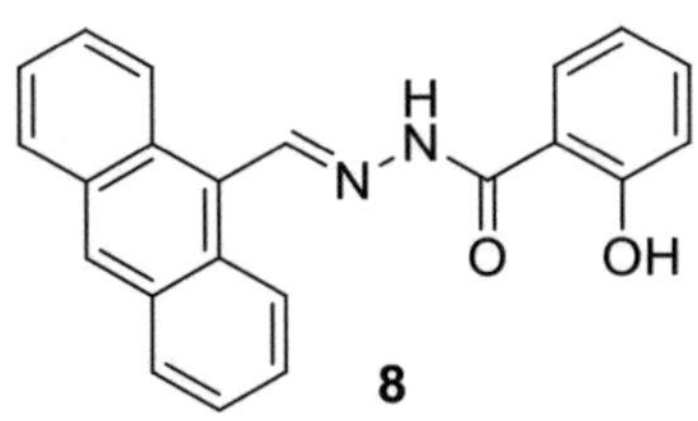

8

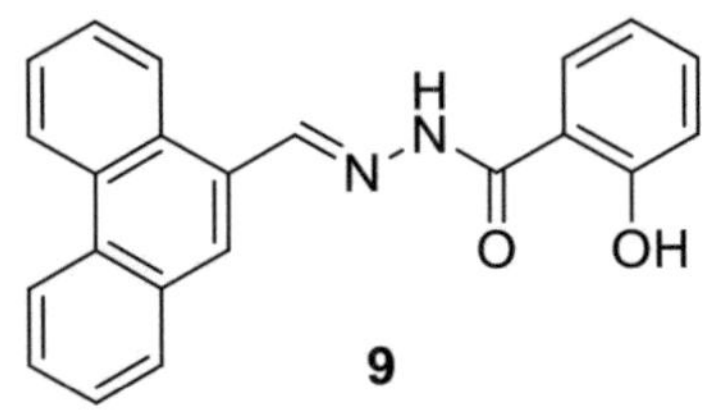

9

Figure 6.
Chemosensor 7, 8, *and* 9 *for selective detection of NACs [34].*

These chemosensors exhibited strong aggregation-induced emission (AIE) in solution with large Stokes shifts because of π-π stacking and intramolecular hydrogen bonding. Fluorescence studies revealed selective quenching in the presence of NACs like nitrobenzene (NB), 2,4-dinitrophenol (DNP), 2,4,6-trinitrophenol (TNP), and 4-nitrophenol (NP). Chemosensor **9** demonstrated the highest quenching constant toward NB ($K_{SV} = 7.86 \times 10^5\ M^{-1}$) with a detection limit of 0.013 μM, followed by chemosensor **8** ($K_{SV} = 5.49 \times 10^5\ M^{-1}$) and chemosensor 7 ($K_{SV} = 2.20 \times 10^5\ M^{-1}$). The mechanism involved PET and FRET facilitated by electron-deficient NACs. Emission/excitation wavelengths varied per sensor, ranging from 410 to 492 nm (**Table 2**). Stern-Volmer analysis confirmed excellent selectivity and sensitivity, particularly for NB. These Schiff base chemosensors are promising candidates for fluorescence-based detection of explosive-related nitroaromatics in environmental and forensic applications [34].

Apurva Goel and Rajesh Malhotra synthesized a pyranone-functionalized Schiff base chemosensor **10**, DFA (3-[-1-(4-fluorophenylimino)

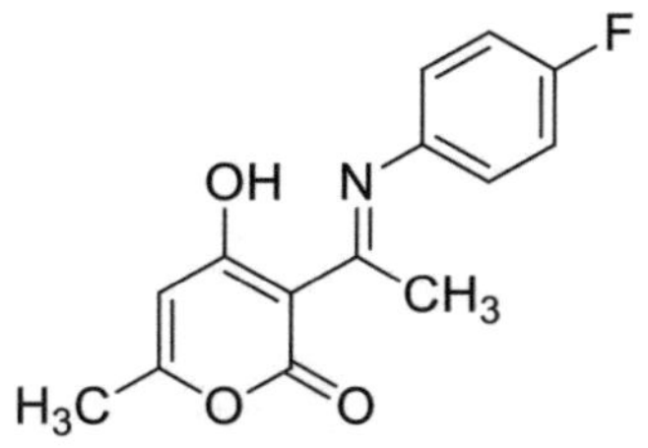

10

Figure 7.
Chemosensor **10** *for selective detection of picric acid (PA) [35].*

ethyl]-4-hydroxy-6-methyl-pyran-2-one), *via* condensation of dehydroacetic acid with 4-fluoroaniline and demonstrated excellent sensitivity for picric acid (2,4,6-trinitrophenol) detection (**Figure 7**). Chemosensor **10** exhibited two distinct UV-Vis absorption bands at 238 nm and 322 nm when characterized by NMR, HR-MS, and FT-IR, which shifted upon picric acid binding (red shift to 337 nm), indicating 1:1 binding stoichiometry confirmed by NMR titration and Job's plot. The detection mechanism is based on hydrogen bonding and charge-transfer interactions, further supported by DFT and NBO analyses. Chemosensor **10** showed a strong association constant (Ka = $1.3 \times 10^5\ M^{-1}$) and a low limit of detection (LOD = 6.92×10^{-7} M) (**Table 2**). The chemosensor **10** displayed high selectivity for picric acid over various nitroaromatics and non-nitroaromatic explosives, functional across a wide pH range (2–11) with excellent photostability and response time (~40 s). Aggregation studies and FESEM images confirmed aggregation-induced spectral changes in THF/water mixtures. These properties highlight chemosensor **10**'s potential as a cost-effective and robust sensor for environmental and security-related picric acid monitoring [35].

1.3 Schiff bases as chemosensors for detection of metal cations

Schiff bases are well-known ligands for metal ion detection because of the presence of donor atoms like nitrogen (from the azomethine group) and oxygen (from phenolic or alcoholic hydroxyl groups). Their ability to form stable complexes with transition and heavy metal cations forms the basis for their use as optical and fluorescent chemosensors [36–39].

The detection of metal cations by Schiff base chemosensors involves multiple interconnected structural and photophysical mechanisms. Upon complexation with metal cations like Fe^{3+}, Cu^{2+}, Al^{3+}, or Hg^{2+}, Schiff bases can undergo ligand-to-metal charge transfer (LMCT), wherein electron density shifts from the ligand to the vacant orbitals of the metal, which typically results in fluorescence quenching or colorimetric transition. Metal binding can induce chelation-enhanced quenching (CHEQ) through the creation of new non-radiative pathways or chelation-enhanced fluorescence (CHEF) by suppressing non-radiative decay and rigidifying the Schiff base framework. Coordination often leads to the structural planarization and extension of conjugation within the Schiff base backbone, producing hypsochromic (blue) or bathochromic (red) shifts in absorption or emission spectra. Moreover, metal cations can promote or disrupt intramolecular charge transfer (ICT) processes, which influence fluorescence behavior depending on the polarity and electron flow within the complex. The selective detection of metal cations is also governed by the geometry and stoichiometry of the binding sites; for illustration, tridentate Schiff bases accommodate ferric (Fe^{3+}) ion, while bidentate designs favor binding to cupric (Cu^{2+}) ion more efficiently. These mechanisms establish Schiff base derivatives as sensitive and versatile platforms for the real-time detection of toxic and biologically significant metal cations in environmental, industrial, and clinical applications [40–53].

Copper is an essential trace element involved in various biological processes, but its excess can cause toxic effects like organ damage and oxidative stress. Furthermore, due to its extensive use in agriculture and industry, cupric (Cu^{2+}) ion pollution is a growing concern for both human health and the environment. Consequently, the sensitive and selective detection of Cu^{2+} ion in aqueous environments is crucial [40].

Khan et al. [41] designed Schiff base chemosensor **11** and evaluated it as a fluorimetric and colorimetric chemosensor for the detection of Cu^{2+} ions in aqueous methanol (MeOH:H_2O, 8:2 v/v) (**Figure 8**). When chemosensor **11** formed a complex with Cu^{2+}, the color changed from yellowish-brown to light yellow, along with significant changes in the UV-Vis spectrum, and new absorption bands appeared at 470, 320, and 250 nm. Fluorescence studies showed pronounced quenching of the chemosensor **11** emission at 438 nm (λex = 330 nm) with a quenching constant of $5.24 \times 10^4\ M^{-1}$ and an LOD of 0.268 μM (**Table 3**). The optimal sensing occurred at pH 7.5, and the system demonstrated excellent selectivity over other metal cations, including Na^+, K^+, Mg^{2+}, Zn^{2+}, and Fe^{3+}, which caused negligible fluorescence changes. Job's plot revealed a 2:1 ligand-to-metal binding stoichiometry supported by reversible binding behavior confirmed through three EDTA-mediated regeneration cycles. Sensor **11** maintained high photostability and gave a rapid response within 2 minutes. The sensing mechanism, involved ligand-to-metal charge transfer (LMCT) and coordination of Cu^{2+} to imine nitrogen and phenolic oxygen donor atoms. The method was successfully validated in real samples (soil, pond water, and tap water) and achieved recovery values between 95.00 ± 0.5% and 103.33 ± 0.18%. These results position chemosensor **11** as a selective, reusable, sensitive, and fast fluorescent sensor for Cu^{2+} detection in environmental applications [41].

Kalyani Rout and colleagues designed a triazole-appended bis-Schiff base chemosensor **12** synthesized *via* condensation of 3,5-diamino-1,2,4-triazole and syringaldehyde (**Figure 9**). Chemosensor **12** functions as a colorimetric and fluorescent "turn-on" dual chemosensor for Cu^{2+} and lead (Pb^{2+}) ions in CH_3OH-tris buffer (1:1 v/v, pH 7.2) (**Table 3**). The chemosensor **12** exhibited strong fluorescence enhancement upon Cu^{2+} and Pb^{2+} binding due to ICT processes and PET inhibition. Upon metal ion complexation, distinct color changes were observed: colorless to yellow for Cu^{2+} and colorless to light yellow for Pb^{2+}. Detection limits were 1.24 μM

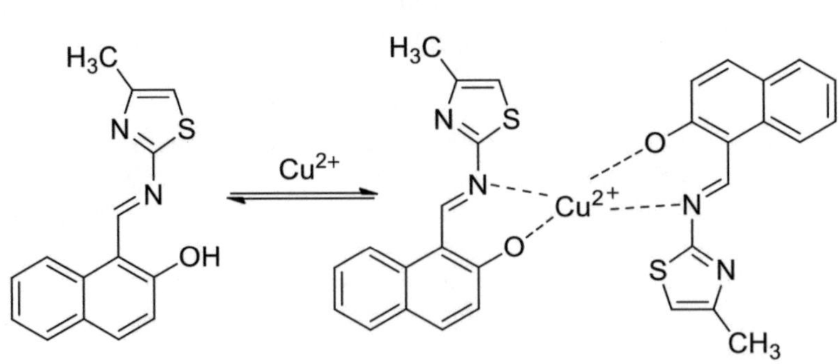

11

Figure 8.
Fluorimetric and colorimetric sensor **11** *for the detection of* Cu^{2+} *and its metal complex [41].*

DOI: http://dx.doi.org/10.5772/intechopen.1011342

Sensor	Signal Type	Sensing Mechanism	Analyte	LOD	Wavelength λem/λex/λabs	Medium	Selectivity	pH range	Ref.
11	Turn-off fluorescence and colorimetric	LMCT/coordination to N/O	Cu^{2+}	0.268 μM	Em: 438 nm, Abs: 470/320/250 nm	$MeOH:H_2O$ 8:2	Excellent over Na^+, K^+, Mg^{2+}, Zn^{2+}, Fe^{3+}	7.5	[41]
12	Turn-off fluorescence and colorimetric	PET, inhibition, ICT	Cu^{2+}, Pb^{2+}	Cu^{2+}: 1.24 μM, Pb^{2+}: 0.99 μM	Em: 412 nm (Cu^{2+}), 440 nm (Pb^{2+})	CH_3OH–tris buffer	Dual selectivity	4–12	[42]
13	Colorimetric	Cu^{2+} coordination to N/O	Cu^{2+}	42.09 nm	Abs: 462 nm	CH_3CN	Highly selective, reversible	~7.5	[43]
14	Turn-on fluorescence	CHEF, inhibition of C=N isomerization	Al^{3+}	2.59×10^{-7} M	Em: 533 nm, Ex: 320 nm	$EtOH:H_2O$ (1:1)	High selectivity over ions	7.5–8.0	[44]
15	Turn-on fluorescence	ICT, CHEF, C=N isomerization inhibition	Al^{3+}	24.8 nM	Em: 410 nm, Ex: 330 nm	DMSO–HEPES buffer	High selectivity	7.4	[45]
16	Turn-on fluorescence + colorimetric	Hydrolysis of C=N, PET inhibition	Cr^{3+}	0.5 μM	Em: 568 nm	Aqueous	High selectivity	6–10	[46]
17	Turn-off fluorescence + colorimetric	Binding-induced non-radiative decay	Hg^{2+}	1.2 μM	—	$CH_3CN:H_2O$	High selectivity	—	[47]
18	Turn-on fluorescence	CHEF via N, O donors	Zn^{2+}, Cd^{2-}	Zn^{2+}: 2.7 nM, Cd^{2+}: 6.6 nM	Em: 545 nm (Zn^{2+}), 560 nm (Cd^{2+})	$DMSO:H_2O$ (9:1)	High selectivity	3–12	[48]
19	Turn-on fluorescence	CHEF, ICT	Zn^{2+}	150 nM	Em: 490 nm	$MeCN:H_2O$ (1:1)	Highly selective	—	[49]
20	Turn-on fluorescence	ESIPT suppression	Zn^{2+}	11 nM	Em: 530 nm, Ex: 340 nm	$DMSO:H_2O$	High selectivity	—	[50]

Sensor	Signal Type	Sensing Mechanism	Analyte	LOD	Wavelength λem/λex/λabs	Medium	Selectivity	pH range	Ref.
21	Turn-on fluorescence	PET inhibition, CHEF	Cd^{2+}	14.8 nM	Em: 510 nm, Ex: 380 nm	ACN:H_2O (8:2)	High selectivity over Zn^{2+}	6–10	[51]
22	Turn-on fluorescence + colorimetric	CHEF, inhibition of PET	Sn^{2+}	314 nM	Em: 460 nm, Ex: 320 nm	DMSO:H_2O	Highly selective over 16 ions	—	[52]
23	Turn-off fluorescence, ratiometric	Coordination via C=N groups	Cu^{2+}, Cd^{2+}, Ni^{2+}, Co^{2+}	Cu^{2+}: 2.36 μM, Cd^{2+}: 41.0 μM, Ni^{2+}: 2.04 μM, Co^{2+}: 1.65 μM	Em: 410 nm, Ex: 352 nm	DMSO:HEPES (5:1)	Highly selective for Cu^{2+}	7.5	[53]

Table 3.
*Various parameters for metal cation analysis with chemosensors (**11–23**).*

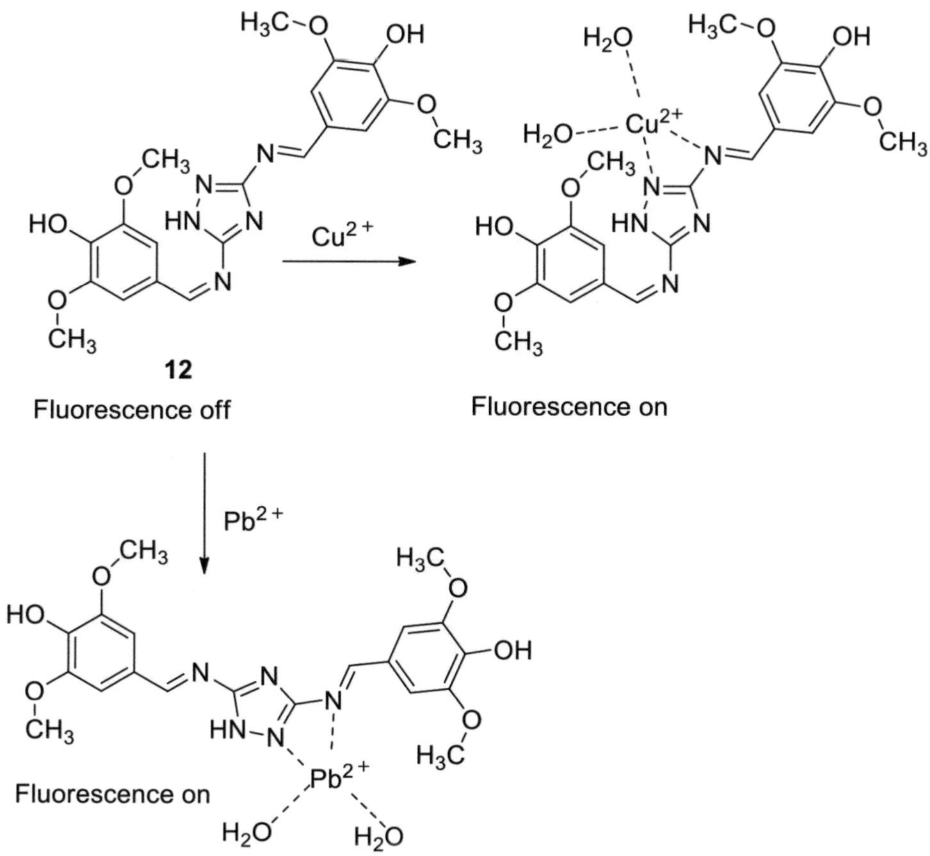

Figure 9.
Fluorescent "turn-on" dual sensor **12** *for detection of* Cu^{2+} *and* Pb^{2+} *ions [42].*

(Cu^{2+}) and 0.99 μM (Pb^{2+}) fluorometrically and much lower colorimetrically. Binding stoichiometries were 1:1 for Cu^{2+} and 1:2 for Pb^{2+}. The chemosensor **12** maintained performances across pH 4-12 (**Table 3**), exhibited reversibility with sulfide (S^{2-}) ion and EDTA, and was successfully applied for real water sample analysis and live cell imaging and demonstrated strong selectivity against competing metal ions [42].

Ram Kumar and coworkers synthesized a Schiff base chemosensor **13** *via* the condensation of dihydroxybenzene-based hydrazone and 2,4-dihydroxybenzaldehyde (**Figure 10**). Chemosensor **13** acts as a highly selective and sensitive colorimetric chemosensor for Cu^{2+} ions. The presence of Cu^{2+} induces a distinct color change from colorless to yellow and a new UV-visible absorption band at 462 nm (in CH_3CN, pH 7.5). Job's plot confirmed 1:1 binding stoichiometry, and the Ka was determined as 3.02×10^4 M^{-1}. The LOD was calculated to be 42.09 nM (**Table 3**). DFT calculations supported the Cu^{2+} coordination through the Schiff base's phenolic oxygen atoms and imine nitrogen. Furthermore, reversibility with EDTA and practical applications like strip-based detection and real water sample analysis further emphasized the chemosensor **13**'s utility. The chemosensor **13** also mimicked an "INHIBIT" molecular logic gate through optical signal modulation [43].

Feyza Kolcu and Ismet Kaya synthesized a carbazole-based Schiff base chemosensor **14** *via* a one-step condensation of 9-ethyl-9H-carbazol-3-amine with

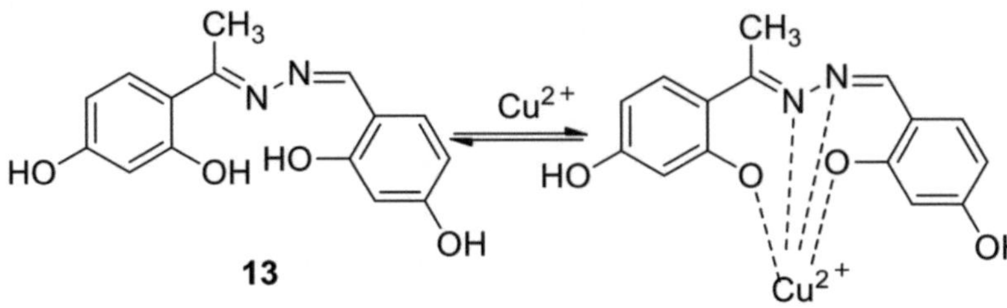

Figure 10.
Colorimetric sensor **13** *for* Cu^{+2} *ion detection [43].*

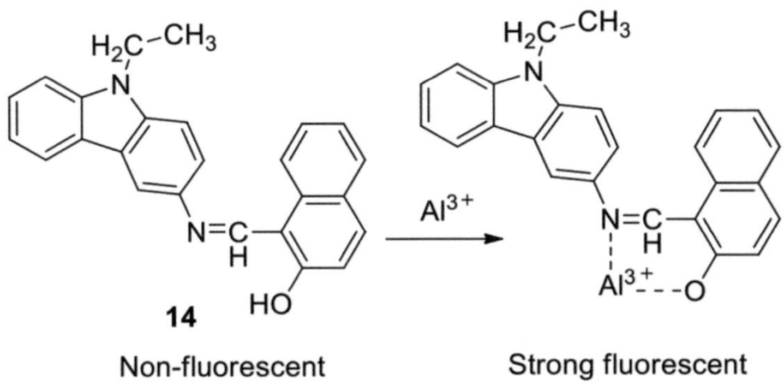

Figure 11.
Chemosensor **14** *for detection of* Al^{+3} *ion [44].*

2-hydroxy-1-naphthaldehyde and developed it as a turn-on fluorescent chemosensor for the selective and sensitive detection of Al^{3+} ions in EtOH:H_2O (1:1, v/v) medium (**Figure 11**). Chemosensor **14** displayed negligible fluorescence due to excited state intramolecular proton transfer (ESIPT) and (>C=N-) isomerization when characterized by FT-IR, ^{1}H NMR, ^{13}C NMR, LC-MC/MS, and UV-Vis spectroscopy. Upon Al3+ binding, a significant fluorescence enhancement at 533 nm (excitation at 320 nm) and a visible fluorescence color change (yellow to white under 366 nm UV light) were observed, credited to the inhibition of C=N isomerization and the CHEF effect. The chemosensor **14** exhibited a 1:1 stoichiometry with Al^{3+}, confirmed by Job's plot and LC-MS/MS (m/z 410 for [L + Al(III) + H_2O + H]$^+$). It demonstrated high selectivity over other tested metal cations and anions, and the L-Al^{3+} complex formation was reversible using EDTA. The detection limit was 2.59×10^{-7} M, and the association constant was 5×10^4 M^{-1} (**Table 3**). Optimal sensing occurred at pH 7.5–8.0 with strong photostability over 3600 s and a fluorescence quantum yield of 21.3%. These findings confirmed the potential of the Schiff base chemosensor **14** as a practical fluorescent sensor for Al^{3+} ions in environmental and biological systems [44].

Dathees et al. [45] synthesized a Schiff base chemosensor **15** *via* condensation of naphthol-based aldehyde and substituted aromatic diamine (**Figure 12**). Chemosensor **15** functioned as a highly selective fluorogenic "turn-on" chemosensor for Al^{3+} ions in a DMSO-H_2O buffer medium (HEPES, pH 7.4) (**Table 3**). Upon Al^{3+} binding, a substantial fluorescence enhancement at 410 nm (λex = 330 nm) occurred because of chelation-enhanced fluorescence (CHEF), intramolecular charge transfer (ICT), and restricted C=N isomerization. The chemosensor **15** showed excellent selectivity for Al^{3+} over 20 metal cations with a 1:1 binding stoichiometry and an association constant of 6.19×10^3 M^{-1}. The LOD was remarkably low at 24.8 nM. DFT

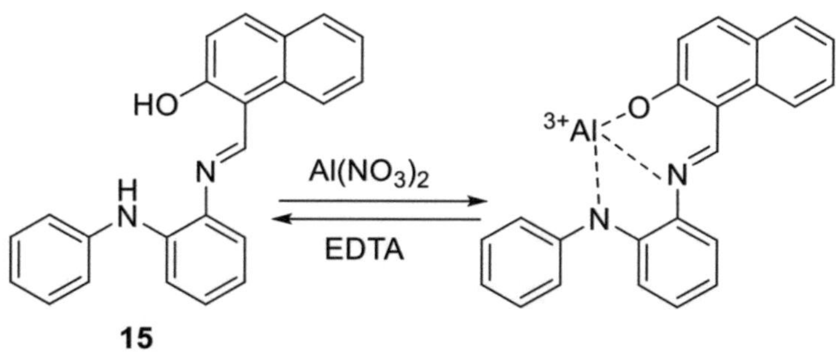

Figure 12.
Fluorogenic "turn-on" sensor **15** *for Al^{+3} ion detection [45].*

studies confirmed binding through N- and O-donor sites and revealed significant HOMO-LUMO energy gap reduction upon complexation. Chemosensor **15** also demonstrated reversibility with EDTA, and successful application in real water and soil samples confirmed its practical utility in environmental monitoring [45].

Khan et al. [46] synthesized a novel water-soluble Schiff base chemosensor **16** *via* the condensation of thiazole-based amine and naphthol-derived aldehyde and developed it for the sensitive and selective detection of Cr^{3+} ions in aqueous solution (**Figure 13**). Chemosensor **16** exhibited a distinct fluorescence "turn-on" response and a colorimetric shift (yellow to colorless) specifically upon interaction with Cr^{3+} because of the selective hydrolysis of the azomethine (>C=N-) group. This hydrolysis released 2-hydroxy-1-naphthaldehyde, enhanced fluorescence at 568 nm by inhibiting PET (photoinduced electron transfer) from the imine group to the fluorophore. The chemosensor **16** showed excellent selectivity and displayed negligible response to a wide range of interfering metal cations (e.g., Na^+, K^+, Zn^{2+}, Cd^{2+}, Pb^{2+}, Cu^{2+}, Hg^{2+}, Al^{3+}). Spectroscopic investigations (UV-Vis, fluorescence, FT-IR, ^1H NMR, ^{13}C NMR) confirmed the structural changes upon Cr^{3+} binding. The LOD was calculated to be 0.5 μM (**Table 3**), significantly below the environmental protection agency's permissible limit of 10 μM for Cr^{3+} in drinking water. Fluorescence intensity remained stable over a pH range of 6–10, which confirmed the sensor's suitability for environmental and biological applications. Real sample analysis in soil demonstrated high recovery rates (95–99%), emphasizing chemosensor **16**'s practicality as a low-cost, dual-mode (colorimetric/fluorometric) sensor for Cr^{3+} monitoring [46].

Ahran Kim et al. [47] synthesized a novel Schiff base-based chemosensor **17** *via* the condensation of 3-nitrobenzohydrazide and phenyl-derived acrylaldehyde to selectively sense Hg^{2+} ions (**Figure 14**). The chemosensor **17** exhibited excellent selectivity and sensitivity for Hg^{2+} among various metal cations through colorimetric and fluorescence responses (**Table 3**). Spectroscopic techniques (^1H NMR, FT-IR, ESI-MS) and single-crystal X-ray diffraction confirmed the Schiff base structure and its binding mode with Hg^{2+}. Upon interaction with Hg^{2+}, chemosensor **17** displayed a significant fluorescence "turn-off" effect because of a strong binding interaction that induced non-radiative decay, supported by Job's plot indicating 1:2 stoichiometry. The detection limit for Hg^{2+} was 1.2 μM, which demonstrated its efficiency even in aqueous media (CH_3CN/H_2O). Density Functional Theory (DFT) studies further validated the structural and electronic characteristics of the chemosensor **17** and its

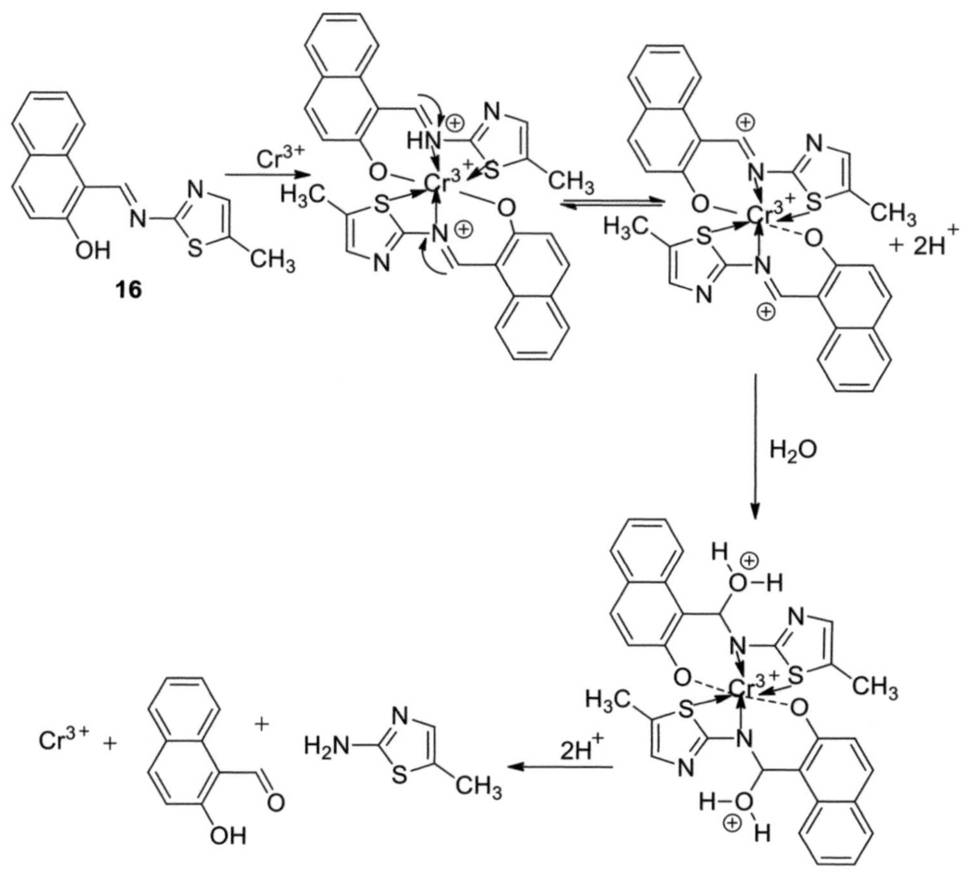

Figure 13.
Water soluble Schiff base sensor **16** *for detection of* Cr^{+3} *ion [46].*

metal complex. The chemosensor **17** also showed practical applicability in detecting mercury in real water samples, which highlights its potential for environmental monitoring due to high selectivity, ease of synthesis, and strong binding affinity toward toxic Hg^{2+} ions [47].

A novel yanilinyl Schiff base chemosensor **18** was synthesized through the condensation of 2,6-diformyl-4-methylphenol and 2-hydroxy-4-methoxybenzaldehyde with hydrazine by Sinha et al. [48] and developed as a multi-responsive chemosensor for the fluorogenic detection of Zn^{2+} and Cd^{2+} ions in a DMSO-water (9:1, v/v) buffered medium (pH 7.2) (**Figure 15**). Upon excitation at 482 nm, chemosensor **18** displayed strong fluorescence enhancement at 545 nm for Zn^{2+} and 560 nm for Cd^{2+} with LOD of 2.7 nM and 6.6 nM, respectively. The chemosensor **18** demonstrated excellent selectivity with minimal interference from other metal cations and functioned effectively across a wide pH range (3–12) (**Table 3**). UV-Vis and fluorescence titrations along with Job's plot, Benesi-Hildebrand analysis, ESI-MS, and ^1H NMR confirmed 1:1 stoichiometry for both Zn^{2+} and Cd^{2+} complexes. DFT calculations supported the binding interactions and electronic transitions. The fluorescence turn-on effect is attributed to chelation-enhanced fluorescence (CHEF) *via* N-, O-donor centers. Additionally, reversibility was demonstrated using Na_2EDTA, enabling logic gate applications. A practical TLC-based test kit was also developed that successfully

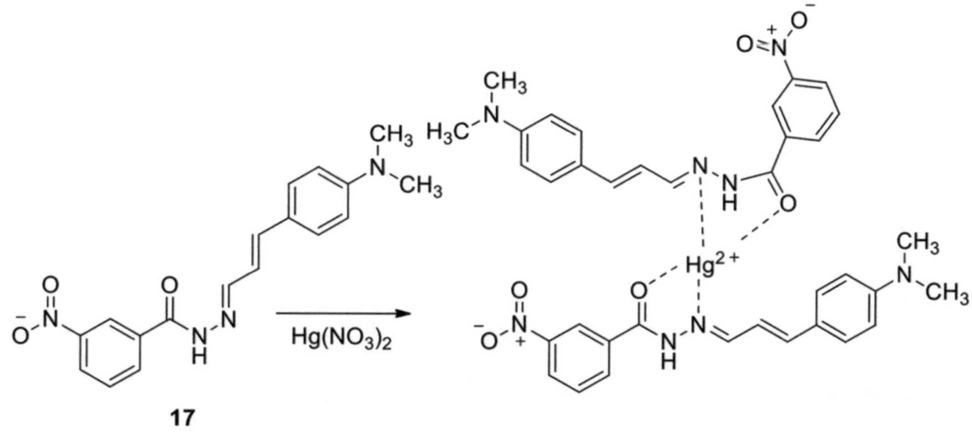

Figure 14.
Sensor **17** *for detection of Hg^{2+} ion [47].*

detected Zn^{2+} and Cd^{2+} in drinking water with recovery rates above 90%, which highlighted the potential of chemosensor **18** as a selective, low-cost, and portable fluorescent chemosensor [48].

A Schiff base ligand, N,N′-phenylenebis(salicylideaminato) chemosensor **19**, was synthesized by Hosseini et al. [49] *via* a condensation reaction and developed as a selective fluorescence "turn-on" chemosensor for Zn^{2+} detection in MeCN-H_2O (1:1, v/v) medium (**Figure 16**). Upon coordination with Zn^{2+}, a significant fluorescence enhancement was observed due to chelation-enhanced fluorescence (CHEF) because of the suppression of non-radiative decay through complex-induced rigidity. The emission maximum shifted from 515 nm (free ligand) to 490 nm (Zn^{2+} complex), and a new ICT band was identified in UV-Vis spectra at 411 nm. Fluorescence titration confirmed a 1:1 binding stoichiometry with a high stability constant (log K_0 = 6.96 M^{-1}) and a low detection limit of 1.5 × 10^{-7} M (**Table 3**). Importantly, the chemosensor **19** demonstrated excellent selectivity for Zn^{2+} over biologically and environmentally relevant metal cations, including Cd^{2+}, Hg^{2+}, Cu^{2+}, Ni^{2+}, and Co^{2+}, which either showed negligible response or quenched fluorescence. Competitive assays revealed minimal interference even in the presence of excess background ions. The chemosensor **19** was further validated through real sample analysis of tap, river,

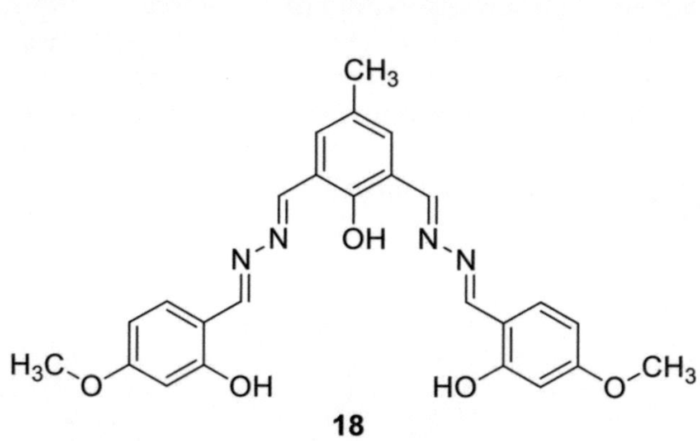

Figure 15.
Fluorescent chemosensor **18** *for detection of Zn^{+2} and Cd^{+2} ion [48].*

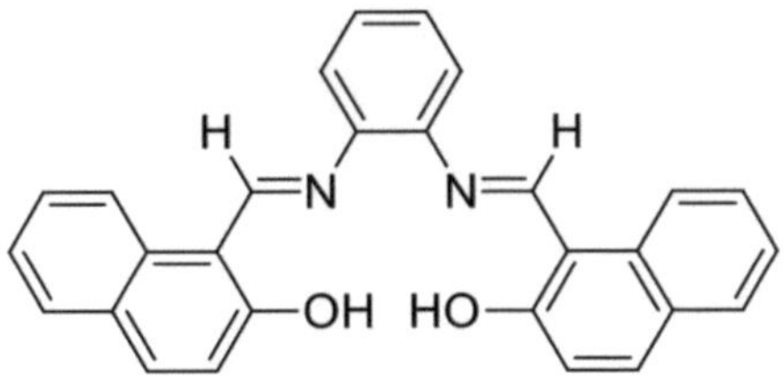

19

Figure 16.
Sensor **19** *for environmental monitoring of* Zn^{+2} *ion in tap, river, and industrial waste [49].*

and industrial wastewater with recovery rates close to 100% that confirmed its utility for practical Zn^{2+} detection in environmental monitoring applications [49].

Behura et al. [50] synthesized a dinitrophenylhydrazine-derived Schiff base chemosensor **20**, which functions as a highly selective fluorescent turn-on chemosensor for Zn^{2+} in aqueous DMSO medium (**Figure 17**). The chemosensor **20** exhibited weak fluorescence due to ESIPT, but upon binding Zn^{2+}, the ESIPT was suppressed, which resulted in a ~ 200-fold fluorescence enhancement at 530 nm (λex = 340 nm). The L:Zn^{2+} complex formed in a 1:1 stoichiometry with a high association constant (6.35×10^7 M^{-1}) and a low LOD (1.1×10^{-8} M) (**Table 3**). The chemosensor **20** demonstrated excellent selectivity over competing metal cations and is effective for Zn^{2+} detection in living HeLa cells as well as in solid phase. Its biocompatibility and reversibility with EDTA were also confirmed. These qualities underscore the suitability of chemosensor **20** for real-time, aqueous-phase Zn^{2+} detection in environmental and biological contexts [50].

Mohanasundram et al. [51] synthesized a quinoline-functionalized Schiff base ligand chemosensor **21** from 2-hydroxy-1-naphthaldehyde and 2-hydrazinoquinoline and evaluated it as a highly selective turn-on fluorescent chemosensor for Cd^{2+} ions. In a CAN:H_2O (8,2) medium, chemosensor **21** (**Figure 18**) exhibited negligible fluorescence due to C=N isomerization and photoinduced electron transfer (PET). Upon binding with Cd^{2+}, the fluorescence enhanced significantly (CHEF effect) with a 37-fold increase at 510 nm (λex = 380 nm) (**Table 3**). The binding stoichiometry was 2:1 (ligand:metal) with a high association constant (1.77×10^5 M^{-1}) and a detection limit of 14.8 nM. The chemosensor **21** demonstrated strong selectivity for Cd^{2+} over Zn^{2+} and other metal cations and maintained fluorescence within pH 6–10. Interference from Zn^{2+} could be masked using CN^-. Mechanistic insight was supported by NMR and DFT analysis, which confirmed the involvement of binding sites such as -OH, imine-N, and quinoline-N in Cd^{2+} coordination. The chemosensor **21** proved effective for detecting Cd^{2+} in real water samples with recoveries above 95% and highlighted its potential in environmental monitoring [51].

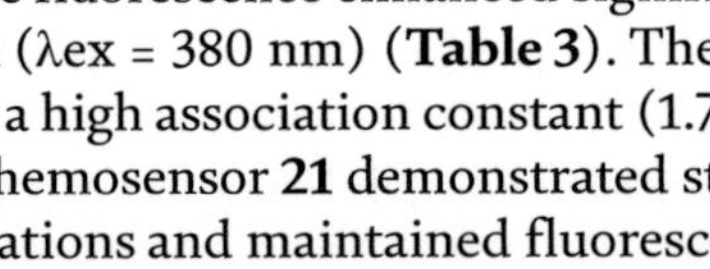

20

Figure 17.
Sensor **20** *for* Zn^{+2} *ion detection in living Hela cell line [50].*

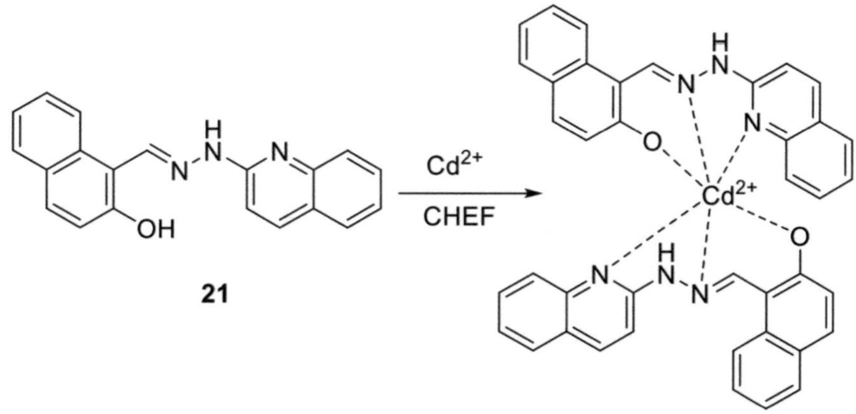

Figure 18.
*Turn-on fluorescent sensor **21** for Cd^{+2} ion sensing in water samples [51].*

Kolcu et al. [52] developed a Schiff base chemosensor **22** synthesized *via* a one-step condensation of bithiophene-based aldehyde and triphenylamine derived diamine (**Figure 19**). The chemosensor **22** was applied as a highly selective "turn-on" fluorescent chemosensor for Sn^{2+} ions in aqueous DMSO medium. In the absence of Sn^{2+}, the chemosensor **22** exhibited negligible fluorescence due PET and C=N isomerization. Upon addition of Sn^{2+}, a 33-fold increase in emission at 460 nm (λex = 380 nm) was observed due to CHEF and inhibition of PET and isomerization (**Table 3**). The L:Sn^{2+} complex followed a 1:2 (ligand:metal) stoichiometry with a high binding constant (Ka = 6.8 × 10^9 M^{-2}) and a low detection limit

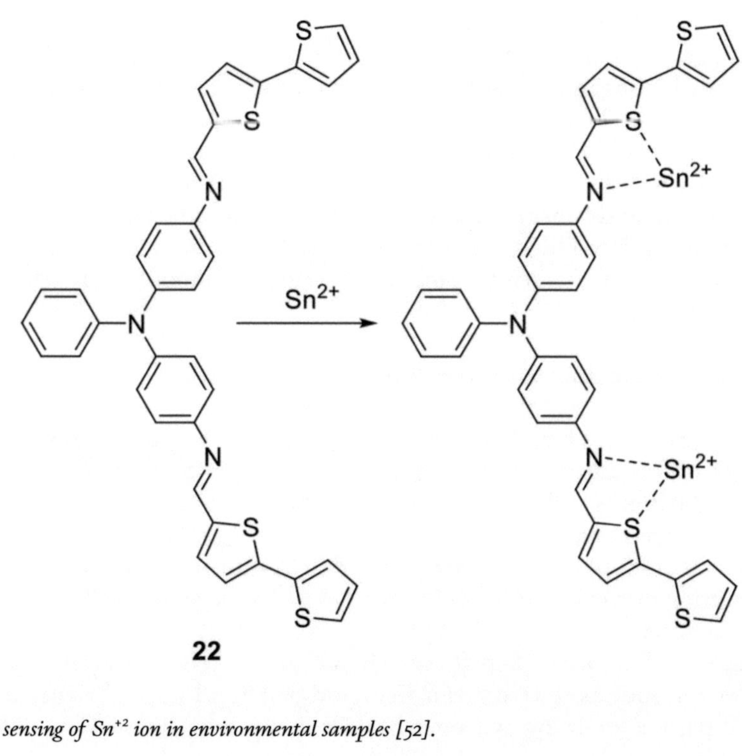

Figure 19.
*Sensor **22** for sensing of Sn^{+2} ion in environmental samples [52].*

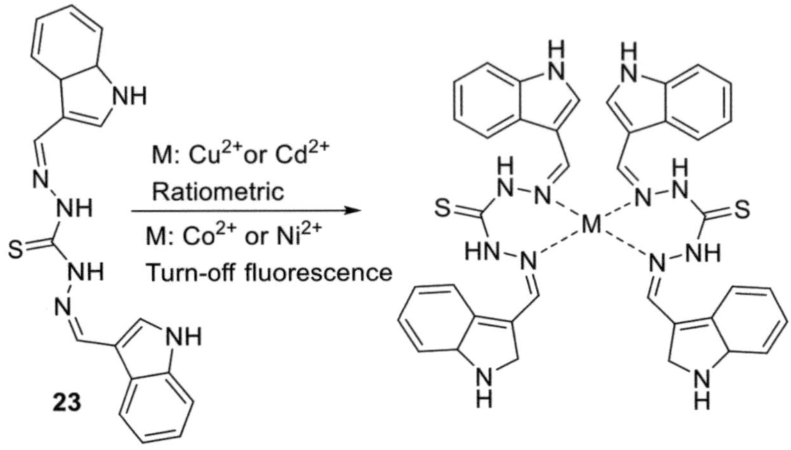

Figure 20.
Chemosensor **23** *for sensing metal cations (Cu^{2+}, Cd^{2+}, Ni^{2+} and Co^{2+}) [53].*

(LOD = 3.14 × 10^{-7} M). The chemosensor **22** exhibited excellent selectivity for Sn^{2+} over sixteen competing metal cations and showed strong photostability and reversibility as confirmed by EDTA-based decomplexation. A distinct colorimetric change (yellow to deep blue under UV) supported practical applicability for real-time detection of Sn^{2+} in environmental samples [52].

Revanna et al. [53] synthesized a novel Schiff base chemosensor **23** as a versatile fluorescent chemosensor by reacting indole-3-carbaldehyde with indole-based carbothiohydrazide derivative for the detection of Cu^{2+}, Cd^{2+}, Ni^{2+}, and Co^{2+} ions (**Figure 20**). Chemosensor **23** exhibited ratiometric sensing and fluorescence "turn-off" behavior in a DMSO:HEPES buffer medium (5:1 v/v, pH 7.5) with significant emission changes at 410 nm (λex = 352 nm) upon metal cation binding. The chemosensor **23** formed 2:1 ligand-to-metal complexes, which was confirmed by Job's plot and ESI-MS analysis. The LOD for Cu^{2+}, Cd^{2+}, Ni^{2+}, and Co^{2+} were 2.36, 41.0, 2.04, and 1.65 μM, respectively. Binding constants revealed strong affinity, especially for Cu^{2+} (Ka = 4.11 × 10^5 M^{-1}). DFT and TD-DFT calculations supported the charge transfer interaction and coordination through the azomethine (>C=N-) groups. The chemosensor **23** exhibited high selectivity toward Cu^{2+} even in the presence of other metal cations, suggesting its strong potential for real-world multi-ion detection applications [53].

1.4 Schiff bases as chemosensors for detection of anions

Although Schiff bases have been extensively explored for cation sensing, their application in anion detection has also gained increasing attention due to their ability to form hydrogen bonds and electrostatic interactions with negatively charged species. Anions like cyanide (CN^-), fluoride (F^-), arsenite (AsO_2^-), and acetate (AcO^-) are of interest because of their industrial, biological, and environmental significance [9]. The detection mechanisms for anions by Schiff bases generally involve the following key interactions:

The detection of anions by Schiff base chemosensors relies on several key interaction mechanisms, such as electron transfer, hydrogen bonding, and structural modulation. Schiff bases containing hydrogen bond donor groups like -OH or -NH readily interact with basic anions like fluoride (F^-), cyanide (CN^-), and acetate (AcO^-), leading to perturbations in the electronic environment and resulting in distinct colorimetric or fluorescence changes. Anions can also engage with electron-deficient sites at the azomethine carbon (>C=N-) of the Schiff base, particularly when electron-withdrawing groups are present, which facilitates nucleophile interactions that could alter the optical properties of the sensor. Intramolecular charge transfer (ICT) processes within Schiff base frameworks are often modulated upon anion binding and produce characteristic hypsochromic or bathochromic shifts in absorption or emission spectra. Deprotonation of labile protons (e.g., -OH or -NH) can occur in cases involving highly basic anions that lead to the formation of anionic species with extended conjugation and pronounced color or fluorescence shifts. The overall sensing response is also influenced by the geometry, size, and hydration energy of the target anion, with smaller, highly charged anions typically exhibiting stronger interactions [55–64]. These mechanisms collectively enable Schiff base chemosensors to achieve selective, sensitive, and often naked-eye-detectable recognition of a broad range of anionic species in biological, environmental, and industrial settings.

The tendency of cyanide to bind with Fe^{2+} ions in cytochromes and significantly change the electron transport chain makes it one of the most hazardous species, despite its widespread use in the polymer, fiber, gold, and electroplating industries [54]. Numerous chemosensors based on Schiff bases have been created that use a visual color change to identify cyanide (CN^-) in a variety of environmental and other samples.

Parchegani et al. [55] synthesized a novel eco-friendly chitosan-based Schiff base chemosensor **24** *via* condensation of chitosan with an azo-functionalized salicylaldehyde derivative for the selective detection of biologically and environmentally significant anions, that is, cyanide (CN^-), acetate (AcO^-), and bicarbonate (HCO^{3-}), in semi-aqueous media. The chemosensor **24** exhibited distinct colorimetric responses (colorless to pale yellow/orange/red) and bathochromic shifts in UV-Vis spectra upon interaction with these anions in DMSO/H_2O (90:10, v/v) (**Figure 21**). The detection limits were exceptionally low at 6.28×10^{-7} M for CN^-, 6.4×10^{-7} M for AcO^-, and 1.06×10^{-5} M for HCO^{3-} ions, which confirmed the chemosensor's high sensitivity (**Table 4**). FT-IR analysis confirmed that the binding mechanism is governed primarily by hydrogen bonding interactions rather than coordination with the imine (>C=N-) group. The chemosensor **24** also demonstrated excellent selectivity, minimal interference from other anions, and reversibility that makes it suitable for repeated use. Its

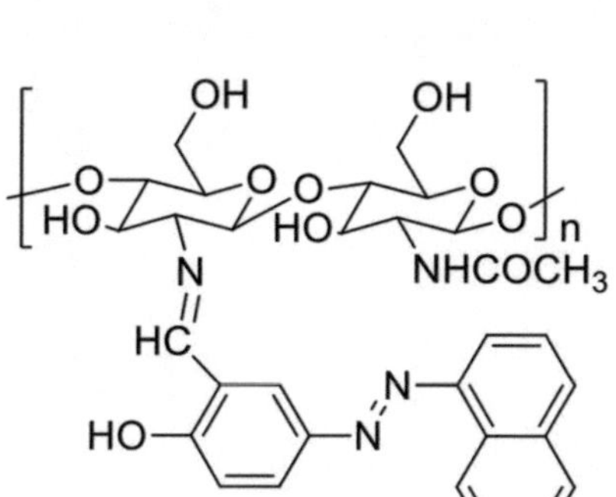

24

Figure 21.
Chitosan based sensor **24** *for detection of cyanide (CN^-), acetate (AcO^-), and bicarbonate (HCO_3^-) in semi-aqueous media [55].*

Sensor	Signal Type	Sensing Mechanism	Analyte	LOD	Wavelength	Medium	Selectivity	pH range	Ref.
24	Colorimetric	H-bonding	CN^-, AcO^-, HCO_3^-	CN^-: 0.628 μM, AcO^-: 0.64 μM, HCO_3^-: 10.6 μM	UV shift	DMSO:H_2O (90:10)	Highly selective	6–12	[55]
25	Colorimetric + Turn-off fluorescence	Nucleophilic addition to C=N	CN^-	1.3 μM	—	Aqueous DMSO	Highly selective	—	[56]
26	Colorimetric + Fluorescence	H-bonding, deprotonation	F^-, OH^-	—	Abs: 530 nm, Em: 400–450 nm	CH_3CN	Selective for F^-, OH^-	—	[58]
27	Colorimetric + Fluorescence	H-bonding, deprotonation, HF_2^- formation	F^-, OH^-	F^-: 48 μM	Abs: 530 nm, Em: 400–450 nm	CH_3CN	Highly selective	—	[58]
28	Colorimetric	H-bonding, ICT modulation	F^-	—	Abs: 375 → 475 nm	DMSO	Highly selective	—	[59]
29	Turn-on fluorescence	PET suppression, ICT	PO_4^{3-}	1.02 μM	Abs: 389 → 426 nm, Em: 540 nm	DMSO:H_2O (1:1)	Highly selective	—	[61]
30	Turn-on fluorescence	ICT inhibition, C=N isomerization	I^-	8 nM	—	DMSO:H_2O (1:1)	Highly selective	7–10	[62]
31	Turn-on fluorescence	ICT inhibition, C=N isomerization	I^-	11 nM	—	DMSO:H_2O (1:1)	Highly selective	7–10	[62]
32	Colorimetric	H-bonding, deprotonation	CN^-, AcO^-	CN^-: 1.4 μM, AcO^-: 26 μM	Abs: 522 nm (AcO^-), 497 nm (CN^-)	DMF:H_2O (9:1)	Highly selective	—	[64]

Table 4.
*Various parameters for anions analysis with chemosensors (**24–32**).*

performance remained stable over a broad pH range (6–12), and successful application in real food samples (biscuits, apple seeds, vinegar chips) confirmed its practical utility in trace anion detection under environmentally friendly conditions [55].

Chemchem et al. [56] developed a new thiophene-based Schiff base chemosensor, **25**, for the selective detection of (CN^-) anions through both fluorescence and colorimetric modes. The chemosensor **25** (**Figure 22**) exhibited a distinct ratiometric color change from yellow to colorless and a fluorescence "turn-off" response upon interaction with CN^- in aqueous DMSO medium. The detection mechanism involves nucleophilic addition of CN^- to the imine (>CH=N-) group, disrupting conjugation and thus altering both absorbance and emission properties. Chemosensor **25** demonstrated excellent selectivity for CN^- over other common anions (F^-, Cl^-, Br^-, I^-, NO_3^-, AcO^-, $H_2PO_4^-$, SO_4^{2-}) with an LOD of 1.3 μM (**Table 4**), below the WHO guideline for cyanide in drinking water. The 1:1 binding stoichiometry was confirmed through ^1H NMR titration and Job's plot. Spectral analyses supported the covalent interaction of CN^- with the electrophilic imine carbon. These qualities emphasize the utility of chemosensor **25** for sensitive and selective detection of cyanide with potential applications in environmental monitoring [56].

Although fluoride is crucial for healthy teeth, excessive levels of it in drinking water can lead to unaesthetic dental fluorosis [57]. Schiff bases are also a candidate for chemosensing of F^- ions. Two pyrrole-based Schiff base chemosensors (**26** and **27**) were synthesized by Velmathi et al. [58] through condensation of pyrrole-2-carboxaldehyde with 1,4-phenylenediamine and 2-nitro-1,4-phenylenediamine, respectively, and evaluated for selective detection of fluoride (F^-) and hydroxide (OH^-) anions through both colorimetric and fluorescence modes (**Figure 23**). Chemosensor **26** turned from colorless to yellow on the addition of F^- or OH^-. Chemosensor **27**, which bears a nitro substituent, exhibited more pronounced color changes and enhanced sensitivity than chemosensor **26** because of stronger electron-withdrawing effects that facilitate hydrogen bonding and deprotonation of the pyrrole-NH proton. Upon addition of F^- or OH^- in CH_3CN, the chemosensor **27** changed from yellow to permanganate in color and showed a red-shifted absorption band at 530 nm together with fluorescence enhancement from 400 to 450 nm (**Table 4**). NMR and FTIR confirmed hydrogen bonding and the formation of HF_2^- species at higher fluoride concentrations. Binding constants (Ka) were determined to be $2.08 \times 10^4\ M^{-1}$ for

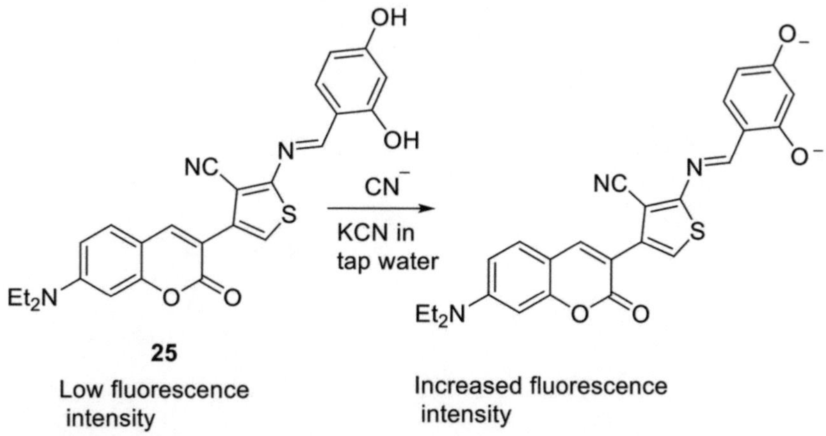

Figure 22.
Sensor 25 *for selective detection of cyanide ion in drinking water samples [56].*

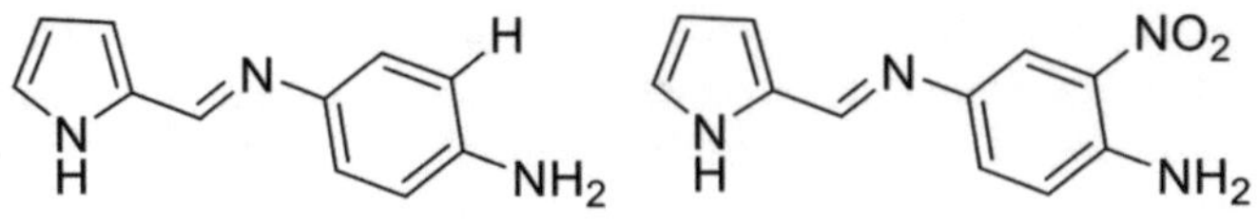

Figure 23.
*Sensor **26** and **27** for chemosensing of fluoride ion [58].*

F^- and $3.47 \times 10^4\ M^{-1}$ for OH^- with chemosensor **27**. Selectivity studies revealed negligible response to Cl^-, Br^-, and I^-, which establishes these Schiff bases as efficient, sensitive, and selective dual-mode chemosensors for anion detection in aprotic solvents [58].

Yang et al. [59] stated a Schiff base-based ratiometric and colorimetric chemosensor **28** for the selective detection of fluoride (F^-) anion that exhibited a naked-eye-visible color change from light yellow to dark orange in DMSO medium (**Figure 24**). The detection mechanism involves strong hydrogen bonding interactions between the phenolic -OH of the Schiff base and the F^- ion, which resulted in deprotonation that led to intramolecular charge transfer (ICT) modulation. This interaction caused a significant red shift in the UV-Vis absorption spectrum (from 375 nm to 475 nm) that indicated enhanced ICT because of the increased electron density upon F^- binding. The chemosensor **28** displayed high selectivity for F^- (**Table 4**) over other common anions (Cl^-, Br^-, I^-, NO_3^-, HSO_4^-, $H_2PO_4^-$, CH_3COO^-) with minimal interference. The colorimetric shift was easily observable, which makes the chemosensor **28** suitable for low-cost and real-time fluoride detection without instrumentation [59].

Phosphates are essential for several biochemical reactions and play a crucial role in numerous biological processes like DNA and RNA synthesis, intercellular signal transduction, and ATP energy supply. Nevertheless, excessive phosphate deposition can lead to environmental issues like oxygen depletion in water and algae eutrophication [60]. This underscores the imperative need for efficient and sensitive detection

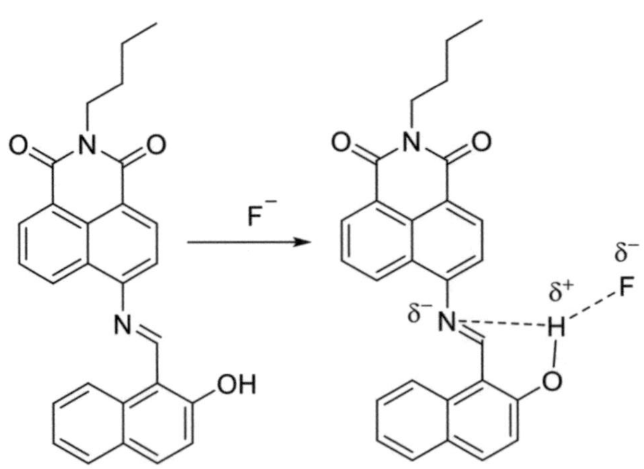

Figure 24.
*Sensor **28** for selective detection of fluoride ion [59].*

methods for phosphate anions to monitor both environmental and biological systems. A new fluorescent chemosensor **29** containing a 1,8-naphthalimide fluorophore was synthesized by Radujevic et al. [61] for the selective detection of phosphate (PO_4^{3-}) anions in aqueous DMSO (1:1, v/v) (**Figure 25**). The chemosensor **29** exhibited a significant fluorescence enhancement at 540 nm upon PO_4^{3-} addition due to hydrogen bonding-induced suppression of PET and enhancement of ICT within the conjugated system (**Table 4**). UV-Vis titration showed a notable red shift in absorption maxima from 389 to 426 nm with clear isosbestic points that confirmed stable complex formation. The binding stoichiometry was determined to be 1:1 with a high association constant (Ka) of $3.8 \times 10^4\ M^{-1}$. The detection limit was found to be 1.02 μM, which indicated excellent sensitivity. The chemosensor **29** demonstrated strong selectivity for PO_4^{3-} over structurally and electronically similar anions like $H_2PO_4^-$, HPO_4^-, NO_3^-, and SO_4^{2-}. Its low detection threshold, water compatibility, and fluorimetric readout make it promising for practical applications in biological phosphate detection and environmental phosphate monitoring [61].

Two novel fluorene-based Schiff base chemosensors, **30** and **31,** were developed by Junaid et al. [62] for highly sensitive and the selective detection of iodide ions (I^-) through a fluorescence "turn-on" mechanism. These donor-acceptor-type chemosensors, **30** and **31**, were synthesized *via* condensation with hydroxyl-substituted aromatic aldehydes that exhibited fluorescence enhancement upon binding I^- with LOD of 8 and 11 nM, respectively (**Table 4**). The sensing response is due to the inhibition of intramolecular charge transfer (ICT) and C=N isomerization, which induces conformational rigidity and restores fluorescence. The sensors demonstrated excellent selectivity for I^- over other anions (e.g., F^-, Cl^-, AcO^-, CN^-, PO_4^{3-}) even under competitive conditions and maintained photostability across a wide pH range (7–10). Spectroscopic studies, Job's plots, ^1H NMR titration, and DFT analysis confirmed

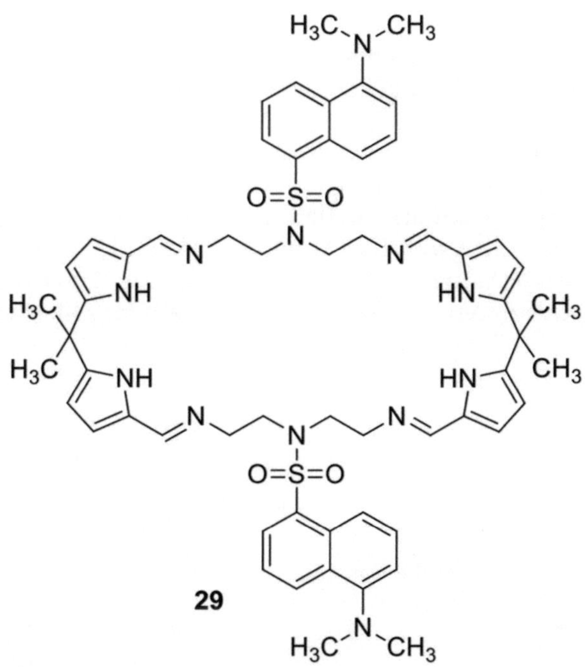

29

Figure 25.
*Sensor **29** for selective detection of phosphate ion [61].*

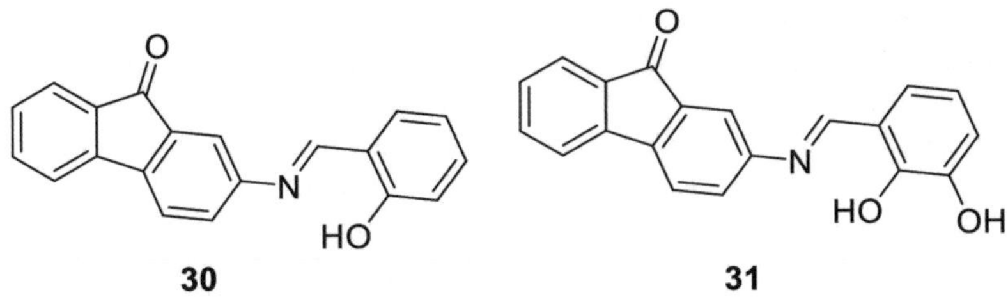

Figure 26.
Sensor **30** *and* **31** *for detection of iodide ion [62].*

a 1:2 binding stoichiometry and strong host-guest interaction (Ka = 4.2×10^6 M^{-1}). Practical applications included paper-strip sensors (LOD ~ 5 μM), blood serum analysis, food and water sample quantification (recovery >97%), and live cell imaging in HeLa cells with no observed cytotoxicity. Additionally, reversible fluorescence switching *via* alternate I^-/Cu^{2+} addition enabled the design of an INHIBIT logic gate that highlighted the multifunctionality of these Schiff base chemosensors **30** and **31** (**Figure 26**) [62].

Numerous enzymes, antibodies, and metabolic processes all utilize acetate [63]. By condensing 4-aminoantipyrine with 1-(3-formyl-4-hydroxyphenylazo)-3-nitrobenzene, Arabahmadi [64] created a novel azo-Schiff base chemosensor **32** (**Figure 27**), which was then tested for the colorimetric detection of cyanide (CN^-) and acetate (AcO^-) anions. The chemosensor **32** exhibited distinct color changes from yellow to purple upon interaction with CN^- and AcO^- in DMF/H_2O (9:1, v/v), allowing for naked eye detection without instrumentation. UV-Vis titrations revealed the appearance of new absorption bands at 522 nm for AcO^- and 497 nm for CN^-, along with well-defined isosbestic points, which confirmed the formation of stable complexes. The sensing mechanism was due to partial deprotonation and hydrogen bonding of the phenolic -OH group by the anions that resulted in intramolecular charge transfer and bathochromic shifts (**Table 4**). The detection limits were determined as 1.4×10^{-6} M for CN^- and 2.6×10^{-5} M for AcO^- with corresponding binding constants (Ka) of 3.07×10^4 M^{-1} and 4.2×10^3 M^{-1}, respectively. The chemosensor **32** showed high selectivity with no significant colorimetric response to other common anions. Furthermore, the chemosensor **32** demonstrated reversibility upon treatment with EDTA that enabled multiple detection cycles. These features position the azo-Schiff base chemosensor **32** as a promising, reusable, and low-cost colorimetric sensor for hazardous anions like cyanide and acetate [64].

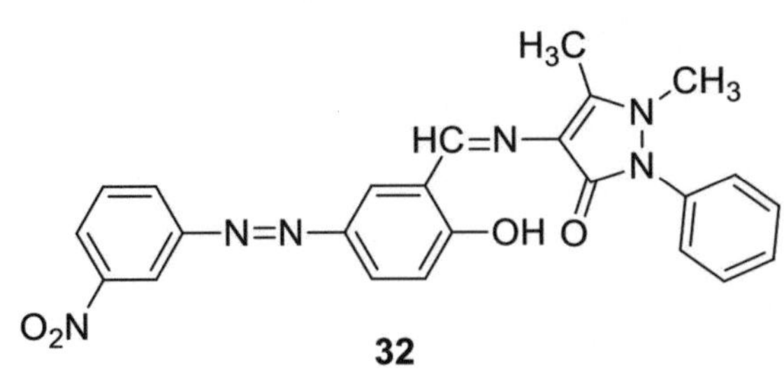

Figure 27.
Sensor **32** *for selective colorimetric detection of cyanide (CN^-) and acetate (AcO^-) [64].*

2. Conclusion

The use of Schiff bases as powerful chemosensors for the detection of metal cations, anion, and nitroaromatic chemicals is covered in length in this chapter. Recent developments in Schiff bases as chemosensors have demonstrated their adaptability and effectiveness in detecting a variety of analytes, including metal cations, nitroaromatic explosives, and anions that are important to the environmental monitoring, disease diagnosis, and cell imaging. The interaction between chemosensors and analytes has been explained by a number of different mechanisms, mechanisms including metal-ligand charge transfer (MLCT), excited-state intramolecular proton transfer (ESIPT), fluorescence resonance energy transfer (FRET), photoinduced electron transfer (PET), intramolecular charge transfer (ICT), and chelation enhanced fluorescence (CHEF) mechanism. In both aqueous and solid-phase environments, Schiff base sensors have demonstrated low LOD, high selectivity, and quick response times. Sensors are widely used in biological imaging, explosive residue detection, and environmental monitoring, as demonstrated by the many documented systems operating in actual samples and living cells. All of these findings show that Schiff bases are useful, affordable, and practical platforms for detecting analytes that need to be detected in real time, selectively, and sensitively. In conclusion, we anticipate that research on novel and enhanced chemosensors will keep growing and evolving to detect as-yet-unidentified analytes.

Glossary

AIE (aggregation induced emission): A photophysical phenomenon in which the molecules that are weakly fluorescent in solution show strong fluorescence upon aggregation.

AIEE (aggregation induced emission enhancement): A phenomenon where photoluminescence of molecules increases significantly upon aggregation as compared to their emission in solution.

CHEF (chelation enhanced fluorescence): A phenomenon where the fluorescence intensity of a molecule increases significantly when it forms a chelate with a metal ion because of the suppression of non-radiative decay.

CHEQ (chelation enhanced quenching): A phenomenon where the fluorescence intensity of a molecule decreases upon metal ion binding due to new non-radiative pathways created *via* coordination.

DFT (density functional theory): The electronic structure of many-body systems, such as atoms, molecules, and solids, can be studied computationally using this method.

ESIPT (excited-state intramolecular proton transfer): A mechanism where a molecule, upon excitation undergoes a rapid internal proton transfer within its structure, typically results in fluorescence suppression.

FESEM (field emission scanning electron microscopy): A microscopy technique that provides high resolution morphological analysis of a sample by using a narrow, focused beam of electrons.

FRET (fluorescence resonance energy transfer): A process in which, when two molecules are near together, energy is transferred non-radiatively from an excited donor molecule to an acceptor molecule.

ICT (intramolecular charge transfer): The transfer of electrons from a donor region to an acceptor region within a single molecule that changes the overall electronic distribution.

IFE (inner filter effect): A non-radiative mechanism where the loss of fluorescence intensity happens due to the absorption of excitation or emission light by the molecule itself.

Job's plot (metod of continuous variation): A technique used to determine the stoichiometry of a sensor-analyte complex by varying molar ratios.

LMCT (ligand-to-metal charge transfer): It describes the transfer of an electron within a complex from a ligand-based molecular orbital to the metal-based molecular orbital.

NBO analysis (natural bond orbital analysis): A computational chemistry method that analyze the electron density distribution in molecules to provide a more intuitive picture of chemical bonding.

PET (photoinduced electron transfer): The phenomenon of transferring an electron from a photoexcited donor molecule to a ground-state acceptor molecule.

Stern-volmer analysis: A graphical method used to study luminescence quenching, especially in the context of fluorescence.

Turn-on fluorescence: A phenomenon where a nonfluorescent or mildly fluorescent molecule exhibits a notable increase in fluorescence intensity upon interacting with a specific analyte or stimulus.

Turn-off fluorescence: It involves reducing or eliminating fluorescence intensity that occurs when analyte interaction activates non-radiative decay processes.

Fluorescence quenching: A phenomenon where the fluorescence intensity of a fluorophore (a molecule that emits light when excited) is reduced or eliminated due to interactions with other molecules or substances.

Author details

Isha Murtaza[1], Sohail Anjum Shahzad[2] and Zulfiqar Ali Khan[1]*

1 Department of Chemistry, Government College University Faisalabad, Faisalabad, Pakistan

2 Department of Chemistry, COMSATS University Islamabad, Abbottabad, Pakistan

*Address all correspondence to: zulfiqarchem@gmail.com

References

[1] Wu D, Sedgwick AC, Gunnlaugsson T, Akkaya EU, Yoon J, James TD. Fluorescent chemosensors: The past, present and future. Chemical Society Reviews. 2017;**46**(23):7105-7123. DOI: 10.1039/c7cs00240h

[2] Zhou X, Lee S, Xu Z, Yoon J. Recent progress on the development of chemosensors for gases. Chemical Reviews. 2015;**115**(15):7944-8000. Available from: https://pubs.acs.org/doi/10.1021/cr500567r

[3] Al-Saidi HM. Recent advancements in organic chemosensors for the detection of Pb^{2+}: A review. Chemical Papers. 2023;**77**(9):4807-4822. DOI: 10.1007/s11696-023-02836-x

[4] Tümay SO, Yeşilot S. Highly selective turn-on fluorescence determination of mercury ion in food and environmental samples through novel anthracene and pyrene appended Schiff bases. Journal of Photochemistry Photobiology A: Chemistry. 2021;**407**:113093. DOI: 10.1016/j.jphotochem.2020.113093

[5] Al-Saidi HM, Khan S. Recent advances in thiourea based colorimetric and fluorescent chemosensors for detection of anions and neutral analytes: A review. Critical Reviews in Analytical Chemistry. 2024;**54**(1):93-109. DOI: 10.1080/10408347.2022.2063017

[6] Alhamami MAM, Mohammed AAY, Algethami JS, Al-Saidi HM, Khan S, Alharthi SS. Highly sensitive and selective Schiff base chemosensor for Cu^{2+} and 2, 4-D detection: A promising analytical approach. Microchemical Journal. 2024;**197**:109817. DOI: 10.1016/j.microc.2023.109817

[7] Aulakh JS, Kumar V, Kim KH. A review of the applications of Schiff bases as optical chemical sensors. Trends in Analytical Chemistry. 2019;**116**:74e91. DOI: 10.1016/j.trac.2019.04.025

[8] Kumar A, Virender, Saini M, Mohan B, Shayoraj, Kamboj M. Colorimetric and fluorescent Schiff base sensors for trace detection of pollutants and biologically significant cations: A review (2010-2021). Microchemical Journal. 2022;**181**:107798. DOI: 10.1016/j.microc.2022.107798

[9] Junaid HM, Batool M, Harun FW, Akhter MS, Shabbir N. Naked eye chemosensing of anions by Schiff bases. Critical Reviews in Analytical Chemistry. 2022;**52**(3):463-480. DOI: 10.1080/10408347.2020.1806703

[10] Kumari N, Singh S, Baral M, Kanungo BK. Schiff bases: A versatile fluorescence probe in sensing cations. Journal of Fluorescence. 2023;**33**(3):859-893. DOI: 10.1007/s10895-022-03135-6

[11] Sun X, Wang Y, Lei Y. Fluorescence based explosive detection: From mechanisms to sensory materials. Chemical Society Reviews. 2015;**44**(22):8019-8061. DOI: 10.1039/C5CS00496A

[12] Ju K-S, Parales RE. Nitroaromatic compounds, from synthesis to biodegradation. Microbiology and Molecular Biology Reviews. 2010;**74**(2):250-272. DOI: 10.1128/MMBR.00006-10

[13] Afrin A, Swamy PC. Novel Schiff base derivatives for the detection of one-to-multi metal ions and tracking the live cell imaging. Coordination Chemistry Reviews. 2023;**494**:215327. DOI: 10.1016/j.ccr.2023.215327

[14] Sharma V, Chandra R, Dutta S, Sahu D, Patra GK. Schiff base and organic ligand stabilized metal nanoparticles as potential chemosensors for hazardous metal ions: Design, principle, optical signaling mechanism and application. Inorganica Chimica Acta. 2024;**573**(1):122321. DOI: 10.1016/j.ica.2024.122321

[15] Kumar R, Singh B, Gahlyan P, Kumar R, Pani B. Recent developments on the colorimetric and fluorometric detection of 3d block metal ions using Schiff base probes. Journal of Molecular Structure. 2023;**1289**:135859. DOI: 10.1016/j.molstruc.2023.135859

[16] Yapar G, Demir N, Kiraz A, Özkat GY, Yıldız M. Synthesis, biological activities, antioxidant properties, and molecular docking studies of novel bis-schiff base podands as responsive chemosensors for anions. Journal of Molecular Structure. 2022;**1266**:133530. DOI: 10.1016/j.molstruc.2022.133530

[17] Bhalla P, Malhotra K, Tomer N, Malhotra R. Binding interactions and sensing applications of chromone derived Schiff base chemosensors via absorption and emission studies: A comprehensive review. Inorganic Chemistry Communications. 2022;**146**:110026. DOI: 10.1016/j.inoche.2022.110026

[18] Udhayakumari D, Inbaraj V. A review on Schiff base fluorescent chemosensors for cell imaging applications. Journal of Fluorescence. 2020;**30**:1203-1223. DOI: 10.1007/s10895-020-02570-7

[19] Musikavanhu B, Liang Y, Xue Z, Feng L, Zhao L. Strategies for improving selectivity and sensitivity of Schiff base fluorescent chemosensors for toxic and heavy metals. Molecules. 2023;**28**(19):6960. DOI: 10.3390/molecules28196960

[20] Khan S, Chen X, Almahri A, Allehyani ES, Alhumaydhi FA, Ibrahim MM, et al. Recent developments in fluorescent and colorimetric chemosensors based on schiff bases for metallic cations detection: A review. Journal of Environmental Chemical Engineering. 2021;**9**(6):106381. DOI: 10.1016/j.jece.2021.106381

[21] Hameed A, Al-Rashida M, Uroos M, Ali SA, Khan KM. Schiff bases in medicinal chemistry: A patent review (2010-2015). Expert Opinion on Therapeutic Patents. 2017;**27**(1):63-79. DOI: 10.1080/13543776.2017.1252752

[22] Berhanu AL, Gaurav MI, Malik AK, Aulakh JS, Kumar V, Kim K-H. A review of the applications of Schiff bases as optical chemical sensors. TrAC Trends in Analytical Chemistry. 2019;**116**:74-91. DOI: 10.1016/j.trac.2019.04.025

[23] Da-Silva CM, Da-Silva DL, Modolo LV, Alves RB, De-Resende MA, Martins CVB, et al. Schiff bases: A short review of their antimicrobial activities. Journal of Advanced Research. 2011;**2**(1):1-8. DOI: 10.1016/j.jare.2010.05.004

[24] Rajimon KJ, Elangovan N, Khairbek AA, Thomas R. Schiff bases from chlorine substituted anilines and salicylaldehyde: Synthesis, characterization, fluorescence, thermal features, biological studies and electronic structure investigations. Journal of Molecular Liquids. 2023;**370**:121055. DOI: 10.1016/j.molliq.2022.121055

[25] Abd-El-Aziz A, Li Z, Zhang X, Elnagdy S, Mansour MS, ElSherif A, et al. Advances in coordination chemistry of Schiff base complexes: A journey from nanoarchitectonic design to biomedical applications. Topics in Current Chemistry. 2025;**383**:8. DOI: 10.1007/s41061-025-00489-w

[26] Goshisht MK, Patra GK, Tripathi N. Fluorescent Schiff base sensors as a versatile tool for metal ion detection: Strategies, mechanistic insights, and applications. Materials Advances. 2022;**3**(6):2612-2669. DOI: 10.1039/D1MA01175H

[27] Soroceanu A, Bargan A. Advanced and biomedical applications of Schiff-base ligands and their metal complexes: A review. Crystals. 2022;**12**(10):1436. DOI: 10.3390/cryst12101436

[28] Gurusamy S, Sankarganesh M, Sathish V, Thanasekaran P, Mathavan A. A novel colorimetric, selective fluorescent "turn-off" chemosensor and biomolecules binding studies based on iodosalicylimine Schiff-base derivative. Journal of Photochemistry and Photobiology A: Chemistry. 2022;**425**:113674. DOI: 10.1016/j.jphotochem.2021.113674

[29] Jia Y, Li J. Molecular assembly of Schiff base interactions: Construction and application. Chemical Reviews. 2015;**115**(3):1597-1621. Available from: https://pubs.acs.org/doi/10.1021/cr400559g

[30] Altun A, Buldurun K, Turan N. Enhanced luminescent probe utilizing Schiff base ligand for 2, 4, 6-trinitrophenol detection in aqueous media. Journal of the Institute of Science and Technology. 2025;**15**(1):228-240. DOI: 10.21597/jist.1540673

[31] Köse A, Tümer M. Two Schiff base compounds derived from 5-aminoisophthalic acid: Chemsensors properties for sensing of metal ions and nitroaromatic compounds. Hacettepe Journal of Biology and Chemistry. 2023;**51**(1):1-12. DOI: 10.15671/hjbc.1078505

[32] Bal M. Fluorescent materials containing polycyclic aromatic compounds: Synthesis, fluorimetric detection of nitroaromatic compounds and color properties. Polycyclic Aromatic Compounds. 2024;**44**(6):4044-4064. DOI: 10.1080/10406638.2023.2244631

[33] Panigrahi A, Sahu BP, Mandani S, Nayak D, Giri S, Sarma TK. AIE active fluorescent organic nanoaggregates for selective detection of phenolic-nitroaromatic explosives and cell imaging. Journal of Photochemistry and Photobiology A: Chemistry. 2019;**374**:194-205. DOI: 10.1016/j.jphotochem.2019.01.029

[34] Kose M, Kırpık H, Kose A. Fluorimetric detections of nitroaromatic explosives by polyaromatic imine conjugates. Journal of Molecular Structure. 2019;**1185**:369-378. DOI: 10.1016/j.molstruc.2019.03.003

[35] Goel A, Malhotra R. Efficient detection of picric acid by pyranone based Schiff base as a chemosensor. Journal of Molecular Structure. 2022;**1249**:131619. DOI: 10.1016/j.molstruc.2021.131619

[36] Taha A, Farooq N, Singh N, Hashmi AA. Recent developments in Schiff base centered optical and chemical sensors for metal ion recognition. Journal of Molecular Liquids. 2024;**401**:124678. DOI: 10.1016/j.molliq.2024.124678

[37] Alam MZ, Alimuddin, Khan SA. A review on Schiff base as a versatile fluorescent chemo-sensors tool for detection of Cu^{2+} and Fe^{3+} metal ion. Journal of Fluorescence. 2023;**33**(4):1241-1272. DOI: 10.1007/s10895-022-03102-1

[38] Afrin A, Jayaraj A, Gayathri MS, Swamy CAP. An overview of Schiff base-based fluorescent turn-on probes: A potential candidate for tracking live cell imaging of biologically active metal ions.

Sensors and Diagnostics. 2023;**2**(5):988-1076. DOI: 10.1039/D3SD00110E

[39] Kwon N, Hu Y, Yoon J. Fluorescent chemosensors for various analytes including reactive oxygen species, biothiol, metal ions, and toxic gases. ACS Omega. 2018;**3**(10):13731-13751. DOI: 10.1021/acsomega.8b01717

[40] Oliveri IP, Attinà A, Bella SD. A zinc (II) schiff base complex as fluorescent chemosensor for the selective and sensitive detection of copper (II) in aqueous solution. Sensors. 2023;**23**(8):3925. DOI: 10.3390/s23083925

[41] Khan S, Muhammad M, Al-Saidi HM, Hassanian AA, Alharbi W, Alharbi KH. Synthesis, characterization and applications of Schiff base chemosensor for determination of Cu^{2+} ions. Journal of Saudi Chemical Society. 2022;**26**(4):101503. DOI: 10.1016/j.jscs.2022.101503

[42] Rout K, Manna AK, Sahu M, Mondal J, Singh SK, Patra GK. Triazole-based novel bis Schiff base colorimetric and fluorescent turn-on dual chemosensor for Cu^{2+} and Pb^{2+}: Application to living cell imaging and molecular logic gates. RSC Advances. 2019;**9**(44):25919-25931. DOI: 10.1039/C9RA03341F

[43] Kumar R, Singh B, Gahlyan P, Verma A, Bhandari M, Kakkar R, et al. An innovative Schiff-base colorimetric chemosensor for the selective detection of Cu^{2+} ions and its applications. RSC Advances. 2024;**14**(32):23083-23094. DOI: 10.1039/D4RA03097D

[44] Kolcu F, Kaya I. Carbazole-based Schiff base: A sensitive fluorescent 'turn-on' chemosensor for recognition of Al (III) ions in aqueous-alcohol media. Arabian Journal of Chemistry. 2022;**15**(7):103935. DOI: 10.1016/j.arabjc.2022.103935

[45] Dathees TJ, Paul SPM, Sanmugam A, Abiram A, Murugan S, Kumar RS, et al. Naphthalene derived Schiff base as a reversible fluorogenic chemosensor for aluminium ions detection. Spectrochimica Acta Part A: Molecular and Biomolecular Spectroscopy. 2024;**308**:123732. DOI: 10.1016/j.saa.2023.123732

[46] Khan S, Muhammad M, Algethami JS, Al-Saidi HM, Almahri A, Hassanian AA. Synthesis, characterization and applications of Schiff base chemosensor for determination of Cr (III) ions. Journal of Fluorescence. 2022;**32**(5):1889-1898. Available from: https://link.springer.com/article/10.1007/s10895-022-02990-7

[47] Kim A, Kim S, Kim C. A conjugated Schiff base-based chemosensor for selectively detecting mercury ion. Journal of Chemical Sciences. 2020;**82**(132):1-7. Available from: https://link.springer.com/article/10.1007/s12039-020-01789-y

[48] Purkait R, Dey S, Sinhaa C. A multi-analyte responsive chemosensor vanilinyl Schiff base: Fluorogenic sensing of Zn(ii) , Cd (ii) and I^-. New Journal of Chemistry. 2018;**42**(20):16653-16665. DOI: 10.1039/C8NJ03165G

[49] Hosseini M, Vaezi Z, Ganjali MR, Faridbod F, Abkenar SD, Alizadeh K, et al. Fluorescence "turn-on" chemosensor for the selective detection of zinc ion based on Schiff-base derivative. Spectrochimica Acta Part A: Molecular and Biomolecular Spectroscopy. 2010;**75**(3):978-982. DOI: 10.1016/j.saa.2009.12.016

[50] Behura R, Dash PP, Mohanty P, Behera S, Mohanty M, Dinda R, et al. A Schiff base luminescent chemosensor for selective detection of Zn^{2+} in aqueous medium. Journal of Molecular Structure.

2022;**1264**:133310. DOI: 10.1016/j.molstruc.2022.133310

[51] Mohanasundaram D, Bhaskar R, Kumar GGV, Rajesh J, Rajagopal G. A quinoline based Schiff base as a turn-on fluorescence chemosensor for selective and robust detection of Cd^{2+} ion in semi-aqueous medium. Microchemical Journal. 2021;**164**:106030. DOI: 10.1016/j.microc.2021.106030

[52] Kolcu F, Erdener D, Kaya İ. Synthesis and characterization of a highly selective turn-on fluorescent chemosensor for Sn^{2+} derived from diimine Schiff base. Synthetic Metals. 2021;**272**:116668. DOI: 10.1016/j.synthmet.2020.116668

[53] Revanna BN, Madegowda M, Neelufur SA, Rangaswamy J, Naik N. A multi-ion detection Schiff base chemosensor: Turn-off and ratiometric fluorescence sensing of Ni^{2+}, Co^{2+}, Cd^{2+}, and Cu^{2+} ions. Journal of Molecular Structure. 2025;**1337**:142183. DOI: 10.1016/j.molstruc.2025.142183

[54] Chandra R, Ghorai A, Patra GK. A simple benzildihydrazone derived colorimetric and fluorescent 'on–off-on' sensor for sequential detection of copper (II) and cyanide ions in aqueous solution. Sensors and Actuators B: Chemical. 2018;**255**:701-711. DOI: 10.1016/j.snb.2017.08.067

[55] Parchegani F, Amani S, Zendehdel M. Eco-friendly chitosan Schiff base as an efficient sensor for trace anion detection. Spectrochimica Acta Part A: Molecular and Biomolecular Spectroscopy. 2021;**255**:119714. DOI: 10.1016/j.saa.2021.119714

[56] Chemchem M, Yahaya I, Aydıner B, Seferoğlu N, Doluca O, Merabet N, et al. A novel and synthetically facile coumarin-thiophene-derived Schiff base for selective fluorescent detection of cyanide anions in aqueous solution: Synthesis, anion interactions, theoretical study and DNA-binding properties. Tetrahedron. 2018;**74**(48):6897-6906. DOI: 10.1016/j.tet.2018.10.008

[57] Petersen PE, Lennon MA. Effective use of fluorides for the prevention of dental caries in the 21st century: The WHO approach. Community Dentistry and Oral Epidemiology. 2004;**32**(5):319-321. DOI: 10.1111/j.1600-0528.2004.00175.x

[58] Velmathi S, Reena V, Suganya S, Anandan S. Pyrrole based Schiff bases as colorimetric and fluorescent chemosensors for fluoride and hydroxide anions. Journal of Fluorescence. 2012;**22**:155-162. DOI: 10.1007/s10895-011-0942-z

[59] Yang L, Liu Y-L, Liu C-G, Fu Y, Ye F. A naked-eye visible colorimetric and ratiometric chemosensor based on Schiff base for fluoride anion detection. Journal of Molecular Structure. 2021;**1236**:130343. DOI: 10.1016/j.molstruc.2021.130343

[60] Yang S, Huang Y, Lu A, Wang Z, Li H. A highly selective and sensitive sequential recognition probe Zn^{2+} and $H_2PO_4^-$ based on chiral Thiourea Schiff base. Molecules. 2023;**28**(10):4166. DOI: 10.3390/molecules28104166

[61] Radujević A, Penavic A, Pavlović RZ, Badjić JD, Anzenbacher-Jr P. Cross-reactive binding versus selective phosphate sensing in an imine macrocycle sensor. Chem. 2022;**8**(8):2228-2244. DOI: 10.1016/j.chempr.2022.05.010

[62] Junaid HM, Waseem MT, Khan ZA, Munir F, Sohail S, Farooq U, et al. Fluorenone-based fluorescent and colorimetric sensors for selective

detection of I–ions: Applications in hela cell imaging and logic gate. ACS Omega. 2022;**7**(11):9730-9742. Available from: https://pubs.acs.org/doi/10.1021/acsomega.1c07279

[63] Jin R, Sun W. Theoretical study of thiourea derivatives as chemosensors for fluoride and acetate anions. Science China Chemistry. 2012;**55**:1428-1434. DOI: 10.1007/s11426-012-4660-4

[64] Arabahmadi R. Synthesis of a new azo Schiff base colorimetric chemosensor for detection of cyanide and acetate anions. Applied Chemistry Today. 2024;**19**(70):75-86. DOI: 10.22075/CHEM.2023.29355.2132

Chapter 9

Carbazole-Based Schiff Bases: Structural Insights and Applications toward Metal Ion Detection

Syeda Aaliya Shehzadi and Mustaghees Ur Rehman

Abstract

Carbazole-based Schiff bases have emerged as versatile chemosensors for detecting toxic and essential metal ions, addressing critical environmental and health concerns arising from metal pollution. This chapter explores their structural design, emphasizing the integration of the electron-rich carbazole core with imine (–C=N–) functionalities, which enhance π-conjugation, thermal stability, and selective metal coordination. Detection of transition metal ions such as Fe^{3+}, Cr^{3+}, Al^{3+}, and Cu^{2+} has been successfully achieved using carbazole-based Schiff bases with a detection limit up to nanomolar range. Heavy metals, such as Hg and Pb, have also been detected using such materials. These sensors operate *via* mechanisms such as photoinduced electron transfer (PET), excited-state intramolecular proton transfer (ESIPT), chelation-enhanced fluorescence (CHEF), and aggregation-induced emission enhancement (AIEE), enabling "turn-on" or "turn-off" optical responses. This chapter highlights advancements in logic gate integration, smartphone-based detection, and AIEE-active probes for on-site monitoring. Challenges such as hydrolytic instability and aqueous solubility are addressed through structural modifications, including sulfonate groups or nanomaterial integration. Future directions emphasize push-pull architectures, near-infrared emission, and user-friendly formats like hydrogel strips. By bridging synthetic versatility with functional adaptability, carbazole-Schiff bases offer scalable solutions for environmental, industrial, and biomedical metal sensing, underscoring their potential in next-generation detection technologies.

Keywords: chemosensor, carbazole, Schiff base, metal ion detection, fluorescent probes

1. Introduction

The increasing industrialization and anthropogenic activities have led to the widespread distribution of heavy and transition metal ions in various ecosystems, posing serious threats to both environmental and human health. Thus, detection of metal ions in environmental and biological systems has emerged as a critical area of research, driven by the dual imperatives of safeguarding human health and preserving ecological balance.

Heavy metals such as mercury (Hg^{2+}), lead (Pb^{2+}), and chromium (Cr^{3+}/Cr^{6+}) are notorious for their toxicity even at trace concentrations, contaminating water supplies, soil, and food chains. Conversely, essential ions like iron (Fe^{3+}), zinc (Zn^{2+}), copper (Cu^{2+}), and cobalt (Co^{2+}) play vital roles in biological processes but become harmful when dysregulated. Traditional analytical methods for metal ion detection, including atomic absorption spectroscopy (AAS) and inductively coupled plasma mass spectrometry (ICP-MS), offer high sensitivity but suffer from limitations such as costly instrumentation, complex sample preparation, and lack of real-time monitoring capabilities. This has spurred interest in developing chemosensors—molecular probes that combine selectivity, sensitivity, and practicality for on-site detection. Among the plethora of sensing platforms, Schiff bases, being dynamic ligands formed *via* condensation of amines (NH_2) and carbonyl (C=O) groups, have gained prominence due to their synthetic versatility, strong metal-binding affinities, and tunable optical responses. Schiff bases are typically formed *via* condensation reactions between primary amines and carbonyl compounds, resulting in imine (–C=N–) functionalities that serve as key binding sites for metal ions.

In recent years, Schiff base sensors and their metal ion complexes have gathered significant attention as a prominent research focus, driven by their diverse practical applications across scientific disciplines. These include environmental monitoring and biological cell imaging, where they enable precise detection of target ions within cellular structures, tissues, and organelles [1–5]. Additionally, their customizable design and multi-stimuli-responsive characteristics make them valuable in optoelectronic systems [6–8]. The inherent molecular switching behavior of Schiff bases has further been exploited in advanced technologies such as molecular keypads and logic gates [9, 10].

The carbazole moiety, a nitrogen-containing heterocycle with an electron-rich aromatic core, provides a rigid, planar scaffold that enhances π-conjugation, thermal stability, and luminescent properties. When integrated with the imine (–C=N–) group, a hallmark of Schiff bases, these hybrid systems exhibit synergistic effects: the carbazole unit amplifies fluorescence or colorimetric signals, while the imine group acts as a selective coordination site for metal ions. This dual functionality enables carbazole-Schiff bases to act as "turn-on" or "turn-off" sensors, with detectable changes in absorption or emission spectra upon metal binding. Moreover, their modular synthesis allows for precise structural modifications, such as introducing electron-donating/withdrawing substituents or extending conjugation, to optimize sensitivity toward specific ions.

Over the past decade, researchers have harnessed these attributes to design carbazole-Schiff base sensors for detecting a wide array of metal ions. For instance, derivatives functionalized with hydroxyl (-OH) or thiol (-SH) groups demonstrate a high affinity for Hg^{2+}, a potent neurotoxin, *via* soft-soft interactions with sulfur or nitrogen donors. Similarly, carboxylate-appended ($-COO^-$) carbazole-Schiff bases selectively bind Fe^{3+} through chelation-enhanced quenching (CHEQ) mechanisms, enabling quantification in biological fluids [11]. The selectivity of these sensors often arises from the geometric and electronic complementarity between the ligand's binding pocket and the target ion. Beyond selectivity, the sensitivity of carbazole-Schiff bases, often achieving detection limits in the nanomolar range, is attributed to their amplified signal transduction. The extended π-system of carbazole facilitates intramolecular charge transfer (ICT) or photoinduced electron transfer (PET) processes, which are perturbed upon metal coordination, leading to pronounced optical changes.

Recent innovations, such as incorporating aggregation-induced emission (AIE) motifs or graphene oxide nanocomposites, have further improved detection thresholds, enabling trace-level analysis in complex matrices like wastewater or cellular environments [12, 13]. Although some extensive review articles have addressed the potential of Schiff bases and their derivatives toward metal ion detections, [14–18] a comprehensive summary specifically addressing the potential of carbazole-based Schiff bases and their use in this context remains absent in the literature.

The current chapter provides a comprehensive exploration of carbazole-based Schiff bases as next-generation metal ion sensors. It begins by elucidating the structural design principles that govern their sensing behavior, emphasizing the interplay between the carbazole core, Schiff base linker, and auxiliary functional groups, with detailed discussion on the mechanisms of ion recognition, mechanisms underlying metal ion recognition, spectroscopic responses including chelation dynamics, electronic interactions, and steric effects, followed by structure-activity relationships.

Through comprehensive literature examples, the emerging trends and opportunities in this rapidly evolving field are identified and highlighted. By reviewing recent literature and critically analyzing structure-property relationships, this chapter aims to inspire innovative approaches to sensor design while bridging the gap between academic research and industrial implementation.

2. Detection of transition metal ions (Fe, Cr, Al, Cu)

Transition metal ions are among the most widely invested metals worldwide. Many industries, such as mining, electronics, and battery manufacturing, release metal ions as waste. They can contaminate water, soil, and food chains if not treated and regulated, causing serious diseases in humans and animals. Strict regulatory limits on metal concentrations by government bodies demand sensitive detection methods, essential for compliance and advancements in sensors, catalysis, and materials science.

2.1 Detection of Fe^{3+} and Cr^{3+}

Fe^{3+} is vital for biological processes like oxygen transport and cellular metabolism, but its imbalance (deficiency or excess) leads to disorders such as anemia, organ damage, and diabetes, highlighting the need for precise detection methods to monitor its levels for human health. Li *et al.* [19] have synthesized a carbazole-based Schiff base fluorescent probe by reacting f 3-(2-formyl)thienyl-9H-hexylcarbazole (**1**) with 4-hydroxy-3-nitroaniline (**2a**) to give probe 1a (**P1a**) and 3-nitroaniline (**2b**) to give probe 1b (**P1b**). The thiophene group was introduced into the 3-position of carbazole to extend the conjugated π-bridge and also to provide a new coordination binding site from the sulfur atom of thiophene, as shown in **Figure 1**. The conjugation extended from carbazole moiety to thiophene to benzene ring *via* **–C=N–** group. The structures of synthesized probes were confirmed *via* IR and NMR spectroscopies. Probe **1a** showed dramatic enhancement of the fluorescent intensity upon addition of Fe^{3+} or Cr^{3+} in the presence of other metal ions, such as Na^{+}, Ni^{2+}, K^{+}, Ag^{+}, Ca^{2+}, Cd^{2+},

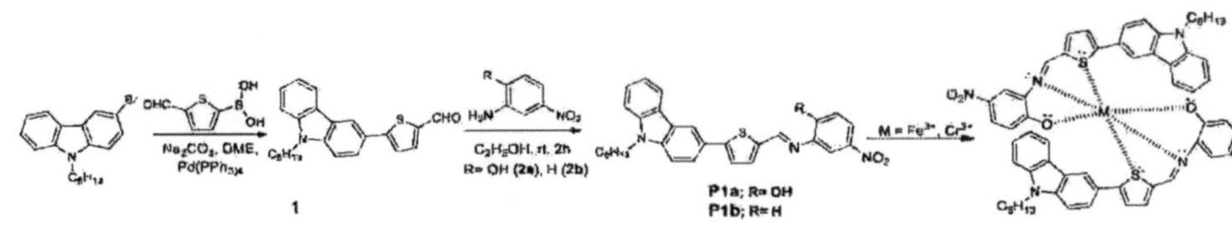

Figure 1.
*The synthetic route of probe 1 (**P1a-b**) and binding mode of **P1a** with M (M = $Fe3^{+}$ and $Cr3^{+}$).*

Hg^{2+}, Zn^{2+}, Co^{2+}, Mn^{2+}, Mg^{2+}, and Fe^{2+}, which had no distinct influence on fluoresce intensity. The emission peak was red-shifted from 495 to 502 nm. While probe **1b** showed an increase in fluorescent intensity in almost all other ions, suggesting that **1b** had no selective recognition for the tested metal ions. This suggested that chemosensors without a phenolic –OH group in the ortho position exhibited weak selective recognition for tested metal ions, indicating that the O atom of –OH is the significant recognition site for metal ions.

The fluorescence enhancement of **P1a** upon binding with Fe^{3+} or Cr^{3+} was perhaps due to inhibition of -C=N isomerization and ESIPT processes that typically quench fluorescence in the unbound state; the selective coordination with these metal ions stabilized the molecular structure, suppressing non-radiative decay pathways and resulting in increased fluorescence intensity. The authors also investigated the binding stoichiometry and binding mode of **P1a** for both Fe^{3+} and Cr^{3+}. The binding stoichiometry was 2:1 for both ions ($P1/M^{3+}$) as determined by Job's plot. Binding mode was confirmed by ^{1}H-NMR, upon coordination with Cr^{3+}, the proton signals of **P1a** broadened and exhibited downfield shift for the hydroxyl, imino, and thienyl protons, indicating that the oxygen atom of the phenolic –OH, the nitrogen atom of the imine group (–N=CH), and the sulfur atom of the thienyl ring served as key binding sites, as presented in **Figure 1**. The absence of the ortho –OH group in **P1b** correlates with its diminished metal ion recognition ability.

By increasing the number of nitrogen coordination sites, a much stronger effect can be expected. He *et al.* [20] conceived this idea and synthesized a dimeric Schiff-base probe 2 (**P2**) by reacting 9-hexylcarbazole-3-carbaldehyde (3) with 2,3-diaminomaleonitrile in ethanol under reflux conditions as presented in **Figure 2**. The hexyl chains enhanced the solubility in organic solvents like DMF, while the Schiff base (–CH=N–) and nitrile (–CN) groups provided coordination sites for metal ions. Both absorbance and fluorescence spectra were studied to find the selective response of probe **P2** to a variety of 18 metal ions.

P2 exhibited high selectivity for Fe^{3+} over 17 other metal ions, as shown in **Figure 3**, during fluorescence studies in DMF with 60% quenching at 372 nm exclusively with Fe^{3+}, with no significant response to competing ions. Competitive

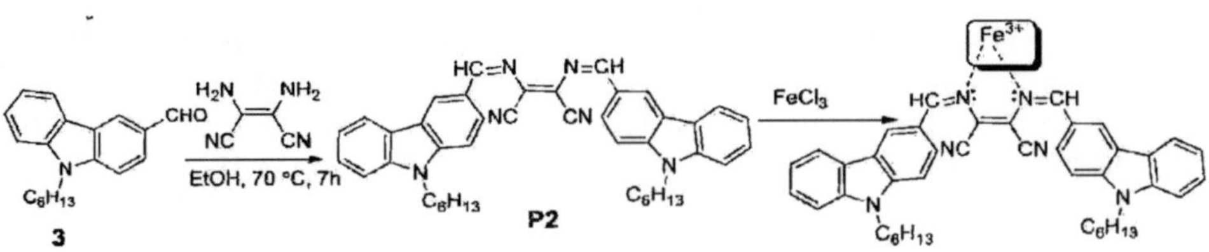

Figure 2.
Synthesis and binding mode of probe P2 with Fe^3+.

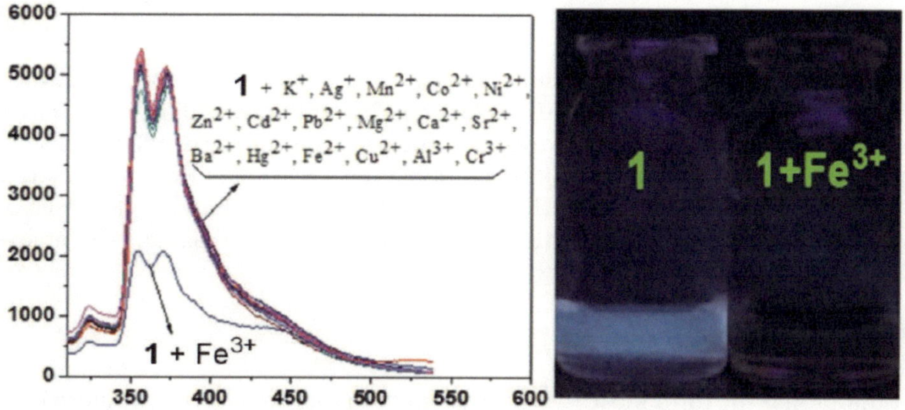

Figure 3.
UV response of **P2** *with* Fe^{3+} *and other metal ions in DMF solvent, adapted from Ref. [20].*

experiments confirmed minimal interference. The detection limit obtained was 3.75×10^{-8} M, which is well below the EPA's permissible limit (5.35 μM) for drinking water. Job's plot and Benesi-Hildebrand analyses established a 1:1 binding stoichiometry between **P2** and Fe^{3+}. The association constant ($K_a = 7.98 \times 10^6\ M^{-1}$) indicated strong affinity. Binding occurred *via* coordination of Fe^{3+} to both nitrogen atoms of the imino group (–C=N–), forming a rigid 5-membered chelate (**Figure 2**). This interaction disrupted the ligand's excited-state processes, leading to fluorescence quenching. In the unbound state, **P2** emitted strongly due to restricted C=N isomerization and ICT. Fe^{3+} coordination induced ligand-to-metal charge transfer (LMCT) by dissipating energy non-radiatively and thus quenching fluorescence. Reversibility with EDTA confirmed that the process is dynamic and practical for reuse.

Iron detection by imino nitrogen and phenolic oxygen is highly pragmatic, as in another report by Nandhakumar *et al.* [21], a carbazole-derived Schiff base fluorescent chemosensor (**P3**) was synthesized *via* condensation of 9-ethyl-9H-carbazol-3-amine (4) and salicylaldehyde (5) in ethanol, yielding a 73% of **P3** as summarized in **Figure 4**. The structure was characterized by NMR, IR, and mass spectrometry.

The authors tested **P3** for the detection of Fe^{3+} and arginine amino acids. The synthesized probe exhibited a sequential "on-off-on" fluorescence response, where Fe^{3+} induced a *turn-off via* ICT from the carbazole nitrogen to Fe^{3+}, while arginine (Arg) restored fluorescence (*turn-on*). Job's plot and Benesi-Hildebrand analysis revealed a

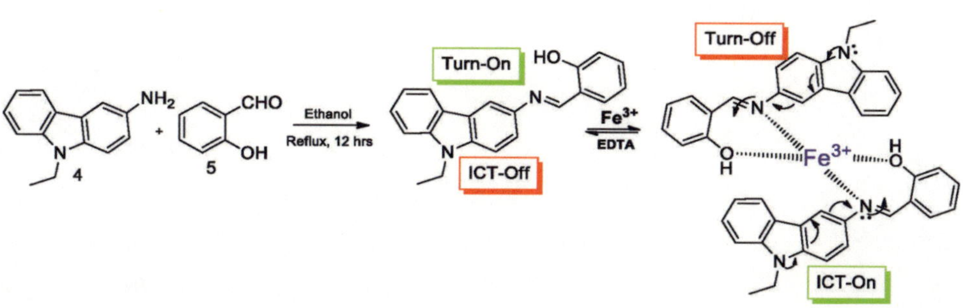

Figure 4.
Synthesis of **P3** *and binding mechanism toward* Fe^{3+}.

2:1 (P3:Fe^{3+}) binding stoichiometry with a binding constant $K_a = 2.09 \times 10^4 M^{-1}$ and a low detection limit (LOD) of 12.22 nM for Fe^{3+}. Coordination involved the carbazole nitrogen, imine nitrogen, and phenolic oxygen, confirmed by FT-IR and ^{1}H-NMR shifts. The sensor operated in DMF/H_2O (1,1, pH 7.4) with reversibility *via* EDTA and pH stability (6–8). Authors also investigated the detection ability of **P3** in molecular logic gates smartphone-based on-site detection of Fe^{3+} & Arg *via* RGB colorimetry. The detection of Fe^{3+} ion in A549 cells using live-cell imaging showed quenching of fluorescence upon treating cells with $FeCl_3$ solution, while a real-sample Fe^{3+} analysis in water with 99–101% recovery was achieved.

2.2 Detection of Al^{3+}

Aluminum is the third most abundant element in the Earth's crust, comprising roughly 8% by weight, and is extensively used across industries such as transportation (aircraft, cars), construction (window frames, roofing), packaging (beverage cans, foil), and electrical applications (power lines, wiring) and in many consumer goods like cookware, appliances, and furniture. Exposure to aluminum is usually not harmful, but excessive contact with Al^{3+} can bioaccumulate in the body and is implicated in neurological disorders (e.g., Alzheimer's and Parkinson's disease), kidney damage, chronic renal failure, and bone softening due to its interference with physiological processes [22, 23]. Consequently, developing highly selective and sensitive Al^{3+} detection methods is essential for monitoring environmental and biological systems and mitigating their adverse health and ecological impacts. A challenge in developing chemosensors for Al^{3+} is its lower coordination ability than other transition metals, and it is difficult to identify spectroscopically due to its lack of photophysical characteristics. However, O- and N- being strong hetero donor centers could capture Al^{3+} in soluble form.

Another design similar to **P3** was reported by Babu *et al.* [24] based on Schiff base, derived from **4** and 3,5-diiodo salicylaldehyde (**6**) in ethanol, as shown in **Figure 5**. The strong electron donor effect of two iodo groups in **P4** could enhance the donor ability of the =OH group.

In 90% aqueous CH_3CN, the probe **P4** exhibited a weak emission at 540 nm (λ_{ex} = 385 nm) due to non-radiative decay *via* ESIPT and -C=N isomerization. Among various transition metal cations added, such as Sn^{2+}, Cd^{2+}, Mn^{2+}, Al^{3+}, Cr^{3+}, Co^{2+}, Cu^{2+}, Hg^{2+}, Pb^{2+}, Zn^{2+}, and Fe^{3+}, only Al^{3+} induced a ~ 6-fold "turn-on" fluorescence with a red shift to 559 nm and the appearance of new absorption bands at 428 and 496 nm, which was attributed to ground-state complex formation and Metal to Ligand Charge

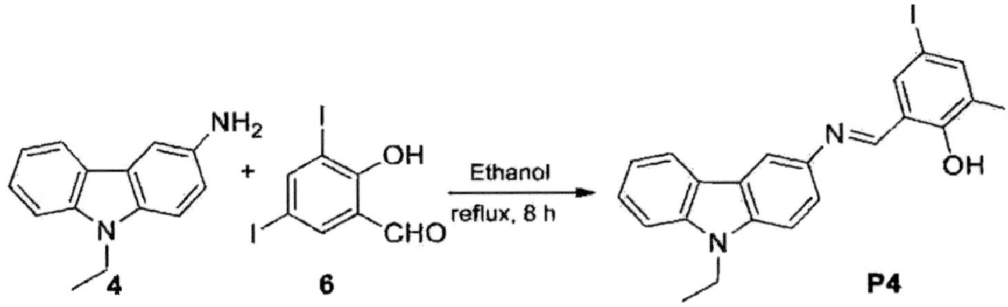

Figure 5.
*Synthetic route to probe 4 (**P4**).*

Transfer (MLCT) transitions. Benesi-Hildebrand and Job's plot analyses revealed a 1:1 of P4-Al^{3+} stoichiometry with K_a = 5.44 × 10^4 L mol^{-1}. The LOD was determined as 0.403 mg μL^{-1}. The mechanism of binding involved the coordination of phenolic O- and imine N-donors, conferring high selectivity for Al^{3+} over a panel of competing cations. The authors did not propose a binding mode or bounded structure of complex.

Replacing diiodo salicylaldehyde with naphthaldehyde led to enhanced fluorescence due to extended π-conjugation, as witnessed by a carbazole Schiff-based Al^{3+} detector, reported by Kaya *et al.* [25] where they condensed **4** and 2-hydroxy-1-naphthaldehyde (7) in ethanol under refluxed conditions to get probe **P5** as shown in **Figure 6**. They tested the probe **P5** for both cations (K^+, Ag^+, Ba^{2+}, Mn^{2+}, Mg^{2+}, Sn^{2+}, Hg^{2+}, Ca^{2+}, Co^{2+}, Zn^{2+}, Cu^{2+}, Ni^{2+}, Pb^{2+}, Al^{3+}, Fe^{3+}, Cr^{3+}, and Cr^{6+}) and anions (F^-, Cl^-, Br^-, I^-, CO_3^{2-}, HCO_3^-, and HPO_4^-) in aqueous media. The probe demonstrated high selectivity toward Al^{3+} over 16 competing metal cations and seven anions.

Fluorescence studies revealed an 11-fold emission enhancement at 533 nm upon Al^{3+} addition, with no significant response to other ions. Competitive experiments confirmed that Al^{3+} detection remains unaffected in the presence of interfering species, revealing its specificity. The binding stoichiometry between **P5** and Al^{3+} was recognized as 1:1 using Job's plot and Benesi-Hildebrand analyses. The K_a value was calculated as 5 × 10^4 M^{-1}, indicating strong affinity. Binding occurred *via* coordination of Al^{3+} to the imine nitrogen (–C=N–) and the deprotonated hydroxyl oxygen (O-), forming a rigid, planar complex as presented in **Figure 5**. This interaction suppresses excited-state intramolecular proton transfer (ESIPT) and C=N isomerization, which is a key non-radiative decay pathway in the free ligand. The mechanism of detection was the "turn-on" fluorescence response based on chelation-enhanced fluorescence (CHEF) effect. Authors believed that a bathochromic shift in UV-Vis spectra from 462 to 479 nm on binding showed the involvement $n \rightarrow \pi^*$ electronic transition due to lone pairs of nitrogen and oxygen atoms resulting in a six-membered ring after chelation. The LOD calculated was 2.59 × 10^{-7} M, below the WHO-recommended threshold for Al^{3+} in drinking water (2.41 μM), suggesting its utility in practical applications.

In another report by Kaya *et al.* [26] the selective Al^{3+} detection was achieved by synthesizing carbazole Schiff-base probe 6 (**P6**) having benzene ring as π-linker between carbazole and imino group while introducing –OH and –OCH_3 groups on aldehyde partner. The synthesis pathway is shown in **Figure 7**. The electron-donating ability of methoxy group enhanced the coordination power of the designed probe for low-coordinating metals. The **P6** was tested for various metal ions (Co^{2+}, Cu^{2+}, Ni^{2+}, Fe^{3+}, Cr^{3+}, K^+, Zn^{2+}, Mn^{2+}, Al^{3+} and Ag^+) by fluorescence study, it exhibited weak fluorescence intensity (λ_{ex} = 320 nm, λ_{em} = 465 nm) in free form due to –C=N isomerization and ESIPT, but the fluorescence intensity increased 108 times with Al^{3+}, with no prominent change for other metal ions.

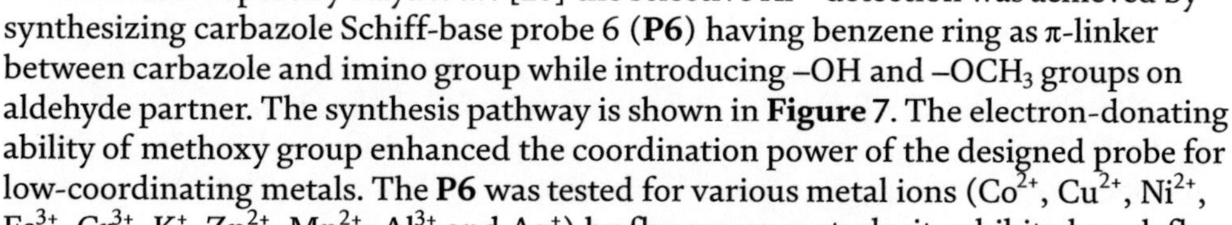

Figure 6.
*Synthesis and binding mode of probe 5 (**P5**) with Al^{3+}.*

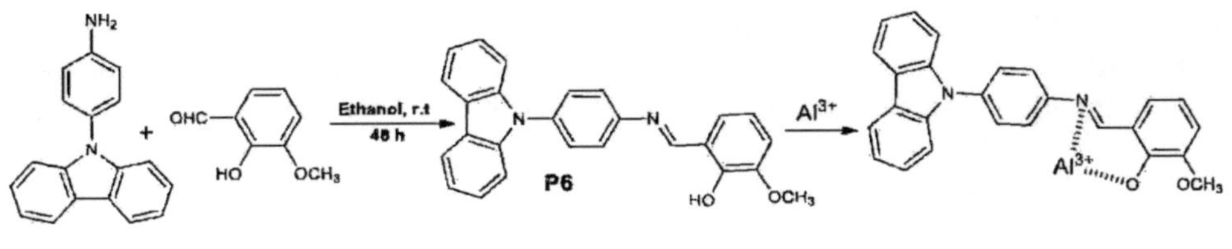

Figure 7.
Synthetic pathway and binding mechanism of **P6** *for* Al^{3+}.

The LOD was 9.29×10^{-7} M, and K_a value using the modified Benesi-Hildebrand equation was $1.64 \times 10^4\ M^{-1}$. The binding stoichiometry of the complex between **P6** and Al^{3+} was stated at a ratio of 2:1, but in binding mechanism, it showed 1:1. Although this sensor worked well in DMSO/H_2O (1:1) environment, most of such chemosensors suffer from water solubility issues. This limitation makes them impractical for real-world problems. Water is a strong solvent, and its interactions with Al^{3+} (through hydration) can outcompete the interactions with potential ligands in a detector. This usually results in weak, unstable, or even insoluble Al^{3+} complexes, making it difficult to create a reliable detector. Thus, the development of a stable, water-soluble Al^{3+} ion detector remains a challenge due to its lower coordination tendency and leading to weak interactions in water. Chemli *et al.* [27] has devised a symmetric optical probe (**P7**) by condensing 2,7-bisformyl-N-pentylcarbazole (8) and 2-aminophenol (9) moieties in methanol at 80°C for 4 h to give **P7** as shown in **Figure 8**. Structure was confirmed by IR, Mass, and NMR spectroscopic methods.

This design led to AIEE utilizing ESIPT in THF/H_2O mixtures, with enhanced fluorescence observed at 480 nm when the water fraction reached 70%. Upon addition of Al^{3+}, a distinct color change from yellow to colorless was observed under daylight, and an intense blue emission under UV light, indicating a highly selective "turn-on" fluorescence response. Among a panel of metal ions, only Al^{3+} caused significant spectral changes. Fluorescence and UV-Vis titrations confirmed a 1:1 binding stoichiometry (Job's plot) and an association constant (K_a) of $1.15 \times 10^4\ M^{-1}$ (Benesi-Hildebrand method). The LOD was determined as 112 nM (UV-Vis) and 9.4 nM (fluorescence), well below the WHO guideline of 7.4 μM for Al^{3+} in drinking water.

DFT calculations revealed that the regions with high electronegativity, having the strongest attraction to cations binding sites, were observed around the nitrogen atom of the imine groups (–C=N–) and the oxygen of the phenol groups (–OH) (**Figure 9**). The binding mechanism was attributed to Al^{3+}-induced hydrolysis of the imine bond, regenerating 2,7-bisformylcarbazole, as confirmed by ^{1}H-NMR spectral

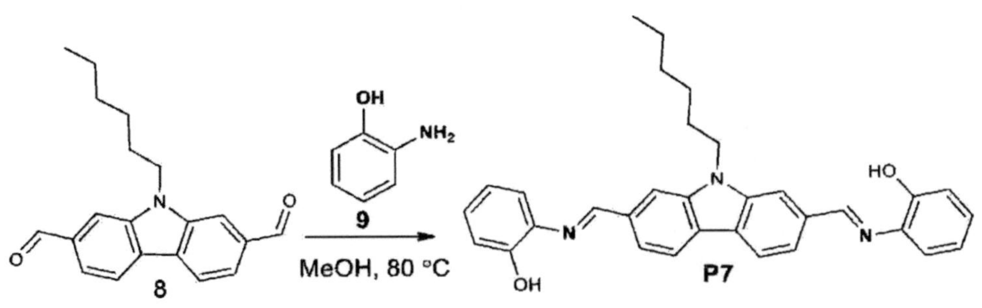

Figure 8.
*Synthetic route toward probe 7 (***P7***).*

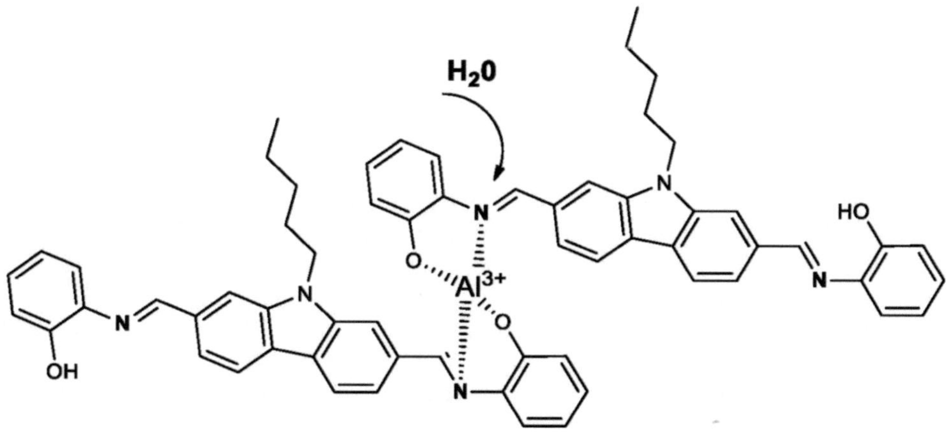

Figure 9.
Binding mode of **P7** *and* Al^{3+} *with 2:1 stoichiometry of* **P7**:Al^{3+}.

shifts and fluorescence quantum yield, which increased from 0.5% (Carb-Azo) to 13.7% (product). Furthermore, a smartphone-assisted fluorometric platform using a fiber-optic endoscope was developed for practical Al^{3+} detection in drinking water, achieving a detection limit of 0.4 μM. Recovery experiments in spiked commercial water samples gave values from 94.8 to 116.7% with RSDs <12%, validating the practical applicability of both the spectroscopic method and the portable device for real-time environmental monitoring of Al^{3+} contamination.

Inclusion of benzophenone moiety on carbazole not only can extend conjugation but also enhance the photophysical properties of probe overall, owing to its strong electron-accepting and UV-absorbing abilities. Hu and Tao group [28] recently designed an AIE-active Schiff-base probe, **P8**, tailored for selective aqueous Al^{3+} detection. The **P8** was synthesized by condensing a carbazolyl-benzophenone intermediate (**8**) with **5** in refluxing toluene. The structure was confirmed by $^{1}H/^{13}C$ NMR showing imine (–CH═N–) CH signal at ~8.7 ppm δ-value and phenolic OH at ~12.9 ppm δ-value (**Figure 10**).

P8 exhibited classic aggregation-induced emission in THF/H_2O, 1:4. As the water fraction increased from 0 to 95%, its UV–Vis π–π* band red-shifted from 380 to

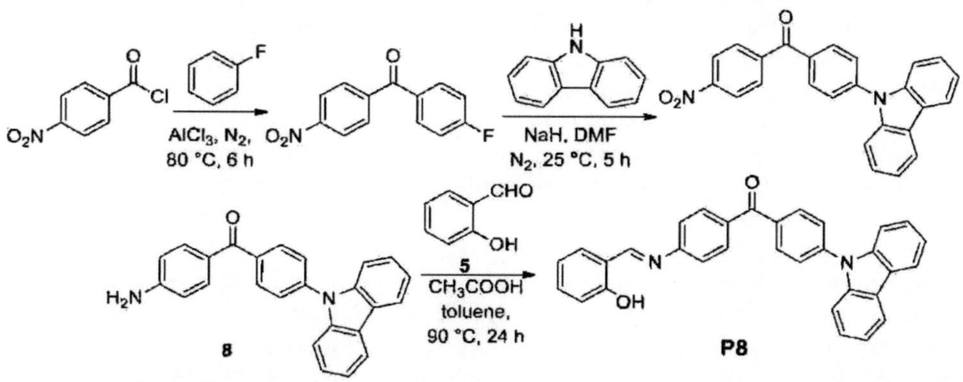

Figure 10.
Synthesis pathway toward probe **P8**.

410 nm and its emission at 545 nm intensified 6.4-fold, driven by aggregation-caused restriction of intramolecular rotations and hydrogen-bond–mediated planarization. Upon addition of Al^{3+}, a pronounced "turn-on" fluorescence at 545 nm and a new absorption at 410 nm emerged, while 18 competing metal ions (Li^{+}, Na^{+}, K^{+}, Ag^{+}, Cu^{2+}, Fe^{3+}, Zn^{2+}, Mn^{2+}, Ca^{2+}, Mg^{2+}, Cd^{2+}, Cr^{2+}, Pb^{2+}, Ba^{2+}, Ga^{2+}, La^{3+}, Ni^{2+}, In^{3+}, Ce^{3+}) elicited negligible response, owing to chelation-enhanced fluorescence (CHEF) that inhibited C═N isomerization and ESIPT quenching by coordinating Al^{3+} to the imine nitrogen and phenolic oxygen donors. FT-IR confirmed this binding by the disappearance of the 3051 cm^{-1} OH stretch and shift of the 1600 cm^{-1} C═N band to 1586 cm^{-1}, and ^{1}H-NMR titrations revealed loss of the OH singlet and downfield imine CH, verifying O/N chelation. Job's plot analysis indicated a 1:1 P8-Al^{3+} stoichiometry, and Benesi-Hildebrand fitting gave $K_a = 0.477 \times 10^3\ M^{-1}$ (**Figure 11**).

Fluorescence titrations established a linear detection range of 50–500 μM and a low detection limit of 1.486 μM, which is well below the WHO guideline of 7.41 μM. Application to real water samples showed recoveries of 100.72–102.85% with RSD <2.82%, revealing **P8**'s potential for sensitive, selective, and practical environmental monitoring of Al^{3+}.

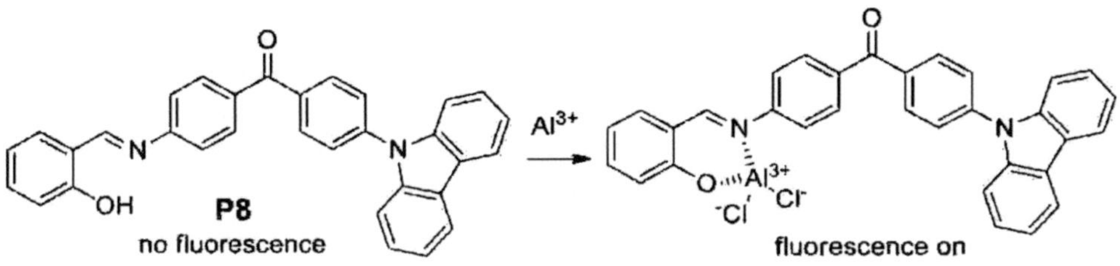

Figure 11.
Binding mechanism of **P8** *with* Al^{3+}.

2.3 Detection of Cu^{2+}

Copper is the third most abundant transition metal in the human body and is vital for the development and function of multiple organs such as the heart, central nervous system, immune system, and liver, yet its dysregulation can lead to disorders like Menkes and Wilson's diseases when deficient or overloaded [29]. Alongside its biological importance, copper has wide utilization in different industries like conducting wires, batteries, switches, transformers, and telecommunications equipment. Therefore, developing highly sensitive and selective chemosensors for trace Cu^{2+} detection is critical. Recent advances in fluorescent probes demonstrating "turn-on" mechanisms and detection limits in the micromolar to sub-micromolar range envisaged the idea of π-bridging between molecules to make their absorption and emission bands red-shifted, and then the color variation can be observed and determined via naked eyes [30, 31]. In view of this concept, Zhu and Fang *et al.* have disclosed [32] a report where they introduced the 4-aminobenzyl cyanide unit into the 3-position of carbazole to extend the π-conjugated bridge (9), the –NH_2 handle made Schiff base and water solubility boosted *via* cyano group. The N,*N*-diethyl-bearing salicylaldehyde (10) was then condensed to yield the Schiff-base probe 9 (**P9**), enabling metal coordination through the imine (**–C=N–**) nitrogen and phenolic oxygen. The whole design and synthesis are shown in **Figure 12**. The sensing ability of the final probe

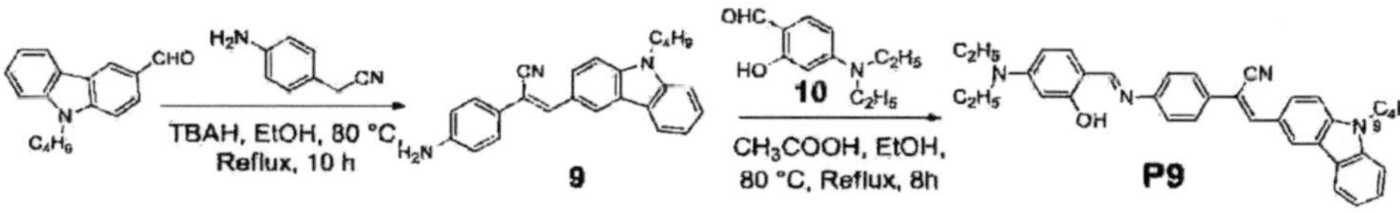

Figure 12.
Synthetic route of **P9**.

was investigated in CH_3CN-Tris (20 mM, v/v = 1:1, pH = 7.2) solution for a range of 13 metal ions (Na^+, K_+, Ca^{2+}, Mg^{2+}, Ni^{2+}, Fe^{3+}, Co^{2+}, Cu^{2+}, Zn^{2+}, Hg^{2+}, Ag^+, Cd^{2+}, Pb^{2+}, and Al^{3+}) *via* absorption spectra. The UV-Vis spectrum of **P9** displayed π–π^* and ICT absorption bands at 288 and 430 nm, respectively. On Cu^{2+} addition, the main band blue-shifted to 418 nm with markedly decreased intensity and a color change from bright yellow to near-colorless, whereas other metal ions induced no or minor spectral shifts (**Figure 13**).

In fluorescence spectroscopy, a bright-yellow fluorescence (λ_{ex} = 420 nm, λ_{em} = 520 nm) was observed, which was selectively and dramatically quenched by Cu^{2+} ion, accompanied by a visible color change from yellow-green to colorless, while other metal ions induced negligible spectral changes. UV-Vis titrations and Benesi-Hildebrand analysis revealed an association constant $K_a = 1.47 \times 10^5\ M^{-1}$ and a 1:1 P9-Cu^{2+} stoichiometry (confirmed by Job's plot), with a fluorescence-based detection limit of 1.14×10^{-5} M; the rapid (≤30 s) and reversible (*via* EDTA) response. ^1H-NMR titrations reveal downfield shifts of imine and phenolic protons, pinpointing coordination through the nitrogen of –C=N and phenolic oxygen, while ESI-MS and DFT calculations supported the formation of a non-fluorescent ground-state complex that opens a photoinduced electron transfer (PET) quenching pathway. **P9** was also deployed on test strips for naked-eye on-site Cu^{2+} detection in aqueous media and demonstrated cellular permeability *via* Cu^{2+}-responsive "turn-off" imaging in HeLa cells, highlighting its dual colorimetric and fluorometric sensing capabilities for environmental and biological applications.

Gosh *et al.* [33] envisioned another idea to find selectivity for Cu^{2+} by the use of Low Molecular Weight Gelators (LMWGs) as sensory material. For that purpose, they reacted the naphthalimide aldehyde (**11**) with **4** under reflux in ethanol to give probe 10 (**P10**) an 85% yield. The **P10** formed thermo-irreversible gels with minimum gelation concentration (5 mg mL^{-1} in DMF/H_2O, DMSO/H_2O, and 1,4-dioxane/H_2O (1:1 v/v)). Gel-sol transition temperatures (T_{9el}) were 61, 60, and 56°C, respectively. Morphology and rheology revealed fibrous networks whose packing varied with solvent.

Among the different metal ions (Hg^{2+}, Ag^+, Fe^{2+}, Cu^{2+}, Fe^{3+}, Pb^{2+}, Cd^{2+}, Al^{3+}, Ca^{2+}, Ni^{2+}, and Co^{2+}) tested, the gel underwent a gel-to-sol transition upon addition of ≥0.5 equiv. Cu^{2+} or Fe^{3+} in DMF/H_2O as solvent. Subsequent treatment with

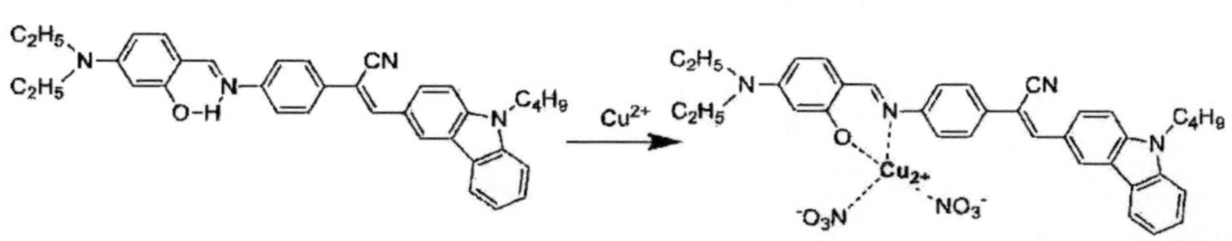

Figure 13.
Sensing mechanism and binding of **P9** *with Cu2*$^+$.

F^- distinguished Fe^{3+} (greenish-yellow) from Cu^{2+}. The solution phase sensing showed a π–π* absorption at 455 nm and green emission at 520 nm; both were selectively quenched by Cu^{2+} and Fe^{3+} (other ions inert). Job's plots indicated a 1:2 (M/P10) stoichiometry, and emission titrations yielded binding constants for Cu^{2+} and Fe^{3+} as $K_a = (4.6–6.9) \times 10^3\ M^{-1}$, $K_a = (5.3–5.8) \times 10^4\ M^{-1}$ respectively. Detection limits were 2.02×10^{-7} M (Cu^{2+}) and 2.09×10^{-7} M (Fe^{3+}). Single-crystal X-ray of the Cu^{2+}-complex revealed a square-planar coordination *via* two imines (**–C=N–**) N and two phenolate O donors, as shown in **Figure 14**. DFT/TD-DFT reproduced the disappearance of the 455 nm band upon complexation and mapped HOMO-LUMO transitions consistent with ICT-driven quenching. The extended conjugation and polar imide groups of this design enhanced the fluorescence ability and water solubility, a requirement for practical and industrial applications.

The conjugation can be extended *via* polyaromatic compounds like naphthalene and incorporating heterocyclic compounds as condensing partners to increase the chances for colorimetric recognition and improved solubility. Zhu *et al.* [34] synthesized probe 11 (**P11**) by Suzuki coupling of 3-formyl-6-iodo-N-butylcarbazole (12) with 2-naphthylboronic acid (13) to give 14, which upon condensation with 2-hydrazinylbenzothiazole (15) in ethanol yielded **P11** as shown in **Figure 15**. Structural characterization *via* ^{1}H-NMR, FT-IR, and ESI-MS confirmed the imine (-CH=N-)

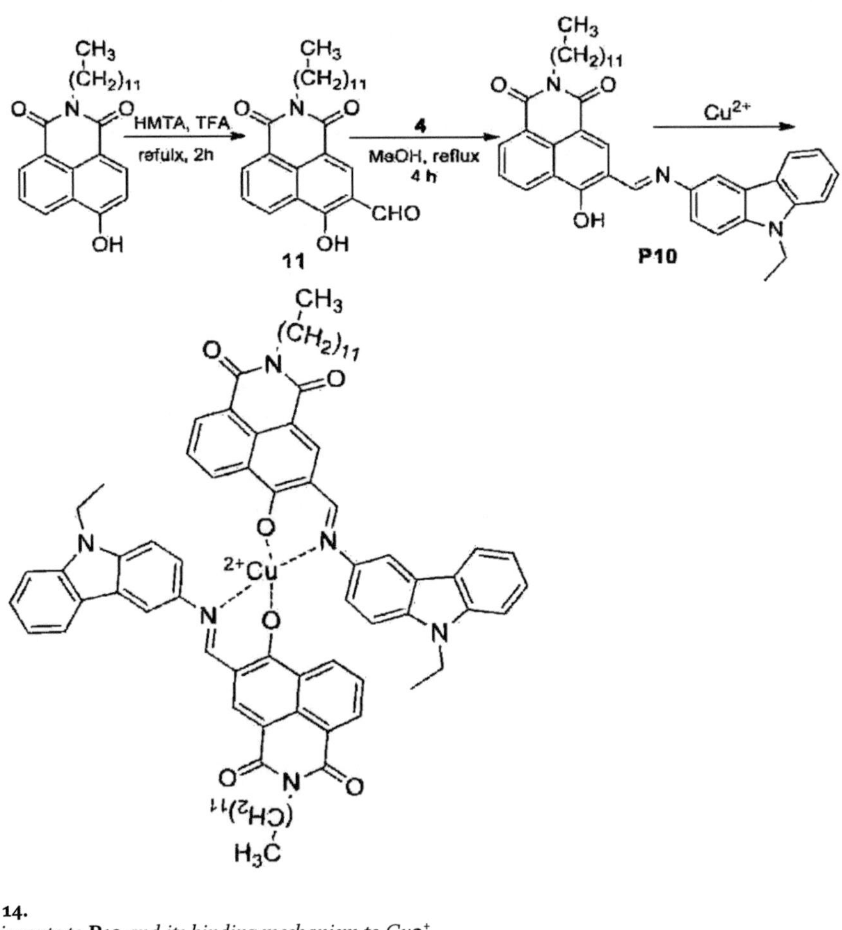

Figure 14.
Synthetic route to **P10** *and its binding mechanism to Cu2*$^+$*.*

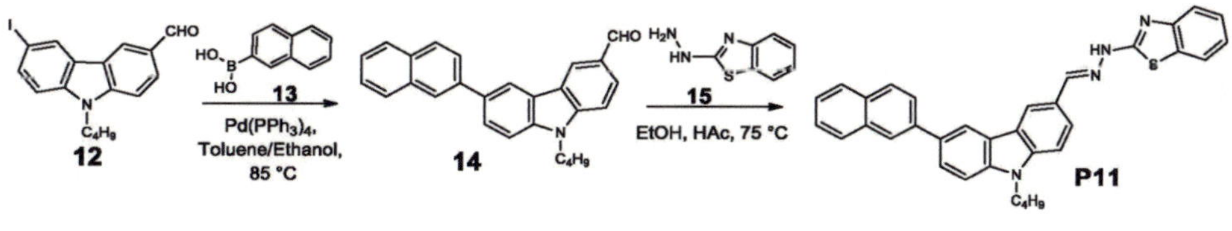

Figure 15.
Synthetic route to **P11**.

linkage and naphthalene extension, enhancing π-conjugation. They tested the sensing ability of designed probe **P11** for variety of alkali, alkaline-earth, transition, and heavy metal ions. The compound selectively detected Co^{2+} colorimetrically in DMSO/H_2O (1:1 v/v) *via* a visible color shift (*colorless → pale yellow*) with a new absorption band at 436 nm, attributed to LMCT. For Cu^{2+}, it functioned as a fluorescent "turn-on" sensor in CH_3CN, showing a 10-fold emission enhancement at 418 nm with a blue shift, driven by suppression of PET and activation of CHEF. The chemical binding ratio of CNS to Co^{2+}/Cu^{2+} was 1:1 according to Job's plot, while the LOD for Co^{2+}/Cu^{2+} was 4.31×10^{-7} and 1.98×10^{-6} $molL^{-1}$, respectively. ^1H-NMR and DFT studies identified binding sites at the imine nitrogen and thiazole ring nitrogen, with Cu^{2+} coordination stabilizing the complex via LMCT, as shown in **Figure 16**.

In another report by Mishra *et al.* [35] a turn off Cu^{2+} sensor based on bis-carbazole and bis-anthracene Schiff base was synthesized by reacting 9-hexyl-carbazole-3-carbaldehyde (16) and ethane-1,2-diamine, leading to two imino C=N– functionalities in the probe as presented in **Figure 17**. Due to relevance, only carbazole-based probe (**P12**) is discussed here. The **P12** was tested for nitrate salts of 12 different metals (Fe^{2+}, Cr^{3+}, Mn^{2+}, Cu^{2+}, Mg^{2+}, Hg^{2+}, Al^{3+}, Zn^{2+}, Cd^{2+}, Ni^{2+}, Co^{2+}, Sn^{2+}) and it demonstrated a selective "turn-off" fluorescence quenching for Cu^{2+} ions, attributing to coordination with the imine (C=N-) nitrogen as supported by DFT studies showing a reduced HOMO-LUMO gap (0.03 vs. 0.159 eV for free probe). Job's plot confirmed a 1:1 binding stoichiometry between **P12** and Cu^{2+}. The sensor

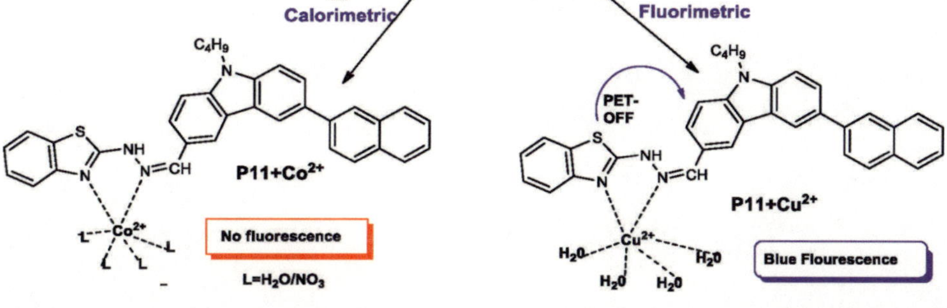

Figure 16.
Binding mechanism of **P11** *with Co2⁺ and Cu2⁺.*

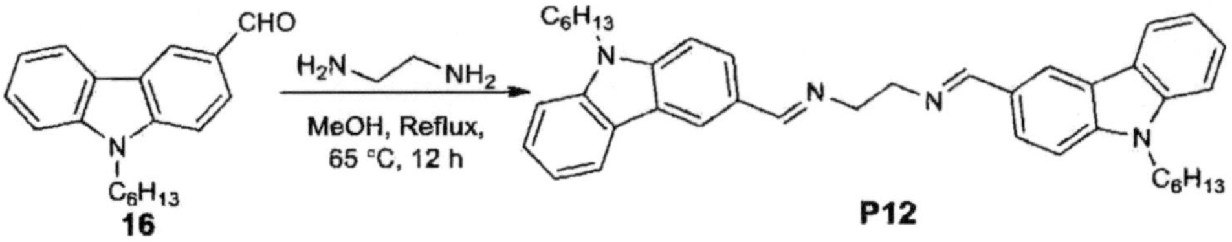

Figure 17.
Synthesis of bis-carbazole-based probe **P12.**

exhibited an LOD of 2.4 × 10^{-8} M and a high K_a value (5.1 × 10^{11} M^{-1}), revealing its sensitivity and strong binding affinity. **P12** also displayed an AIEE, with a 7.47-fold fluorescence increase in a 70% water-THF system. However, Cu^{2+} sensing by **P12** was effective only in non-aggregated states, as aggregation hindered metal interaction. The authors tested the probe for practical applications using real water samples with 94–102% recovery and reversible sensing using EDTA. TLC-based visualization under UV light enabled on-site detection.

Copper detection has been widely explored using carbazole-based Schiff bases. Another article by Hu and Tao *et al.* [36] presented the design and synthesis of an AIE-active fluorescent probe (**P13**) utilizing benzothiazole-Schiff base framework integrated with a carbazole donor group for selective Cu^{2+} detection in aqueous media. **P13** was synthesized *via* a multistep route involving a Schiff base reaction between carbazole-benzothiazole intermediate (17) and a salicylaldehyde-hydrazone (18), yielding a D-π-A structure in the **P13** (**Figure 18**). Characterization by NMR and IR confirmed the structure. The probe demonstrated high selectivity for Cu^{2+} over 20 competing metal ions, with minimal Fe^{3+} interference, and functioned effectively across a broad pH range (4–11). The probe operated *via* an ESIPT mechanism, exhibiting strong fluorescence at 553 nm in aggregated states (THF/H_2O, 1:9 v/v).

Upon Cu^{2+} binding, chelation-enhanced quenching (CHEQ) occurred due to the paramagnetic nature of Cu^{2+}, resulting in a "turn-off" response. Job's plot analysis revealed a 1:2 binding stoichiometry (**P13**:Cu^{2+}), supported by IR and ^{1}H-NMR

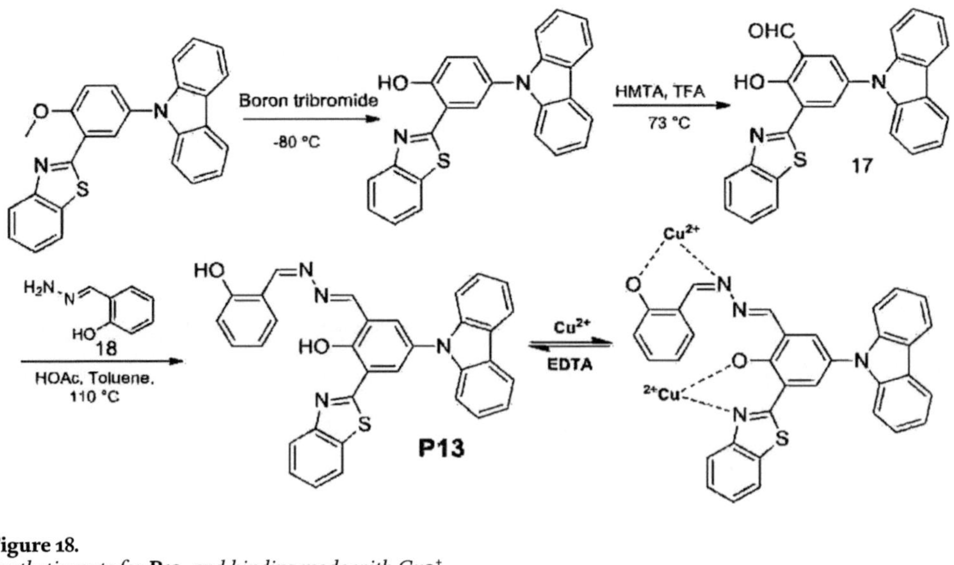

Figure 18.
Synthetic route for **P13**, *and binding mode with Cu2⁺.*

studies showing coordination of one copper ion through deprotonated hydroxyl groups and imine nitrogen of hydrazine, while the other copper bound to oxygen and nitrogen of benzothiazole moiety, as shown in **Figure 18**. Fluorescence titration established a linear detection range of 0–45 μM, a low detection limit (LOD) of 0.2 μM (3σ/k method), and a binding constant (K_a) of $1.59 \times 10^4\ M^{-1}$ via the Benesi-Hildebrand equation. Real-sample testing in environmental water (Dongting Lake, tap water) showed recovery up to 99.78–101.28%. Practical applications were demonstrated using probe-loaded filter paper strips and silica gel, enabling rapid visual Cu^{2+} detection under UV light. The probe's reversibility with EDTA and rapid response time (10 seconds) further underscore its potential for real-time environmental monitoring, meeting WHO guidelines for copper detection in drinking water.

In recent years, large biological molecules, specifically steroids, have attracted great attention in ion recognition, due to their amphiphilic and structurally rigid conformations. Carbazole and cholic acid naturally exhibit compatibility, having a rigid structure. Additionally, the hydroxyl groups present in cholic acid can facilitate subsequent derivatization, and sensors based on cholic acid could be used under physiological conditions due to biocompatibility. In such a report by Mishra *et al.* [37] an AIEE-active fluorescent probe, based on cholyl hydrazide carbazole-Schiff base (**P14**), derived from cholic acid and carbazole for selective detection of Cu^{2+} ions, was described. The probe **P14** was synthesized by refluxing cholylhydrazide (**19**) (derived from ethyl cholate and hydrazine) with **16** in ethanol, yielding 70% product (**Figure 19**). Structural confirmation was achieved *via* FT-IR, $^1H/^{13}C$-NMR, and photophysical analyses (**Figure 20**).

P14 exhibited AIEE properties in a 70% water-THF mixture, showing a 5.93-fold fluorescence enhancement compared to pure THF, attributed to restricted intramolecular rotations in aggregated states. Dynamic light scattering (DLS) and FE-SEM revealed reduced particle size (139 nm) in aggregated states, correlating with emission enhancement. The probe demonstrated selective "turn-off" fluorescence

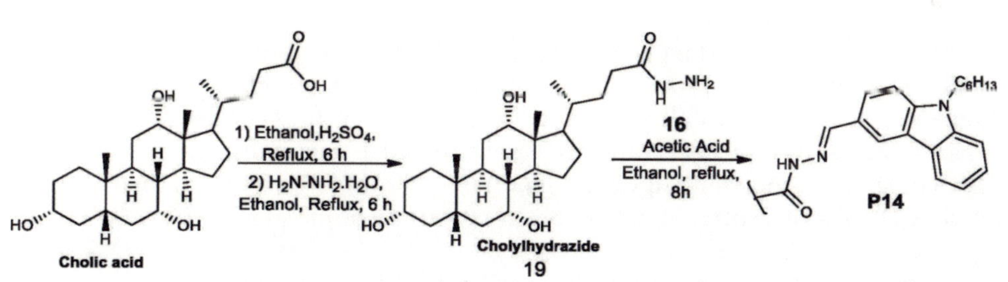

Figure 19.
Synthesis of **P14**.

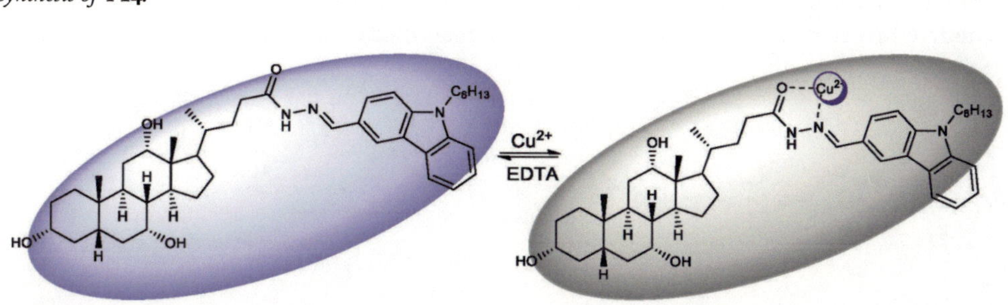

Figure 20.
Binding mode of **P14** *with Cu2+*.

quenching for Cu^{2+} *via* PET by coordinating from the imine nitrogen and carbonyl oxygen with minimal interference from other cations/anions tested. Job's plot confirmed a 1:1 binding stoichiometry, while Stern-Volmer analysis yielded a high binding constant ($K_a = 5.3 \times 10^7 M^{-1}$) and a low detection limit ($1.59 \times 10_{-7} M$). DFT studies revealed a narrowed HOMO-LUMO gap (1.42 vs. 4.13 eV free probe) upon Cu^{2+} coordination, validating the quenching mechanism. The practical applicability was demonstrated in tap and river water samples, achieving recoveries of 92.2–107.7%. Additionally, **P14** functioned as an IMP logic gate with Cu^{2+} and EDTA as inputs.

3. Detection of heavy metal ions (Hg, Pb)

Detecting heavy metal ions is crucial due to their significant impact on human health, environmental safety, and industrial processes. Many heavy metals, such as lead, mercury, and cadmium, are highly toxic even at low concentrations and can accumulate in living organisms, causing serious diseases. Industrial activities often release these ions into water, soil, and air, making their monitoring essential to prevent ecological damage. Strict regulations on heavy metals in water and food drive the need for sensitive, selective, and rapid detection methods for environmental, healthcare, and technological applications.

3.1 Detection of Hg^{2+}

Mercury is a pervasive industrial contaminant used in sectors such as petroleum refining, chemical industry, pharmaceuticals, pulp papermaking, and electronics. It bioaccumulates in brain tissues, leading to neurological damage once critical concentrations are reached. Thiourea moiety is widely used in the field of molecular recognition, especially for heavy metal ions, because of its three coordinating sites, two nitrogen atoms, and one sulfur atom (H_2NCSNH_2). The sulfur group has a special affinity for mercury, as the name mercapto for -SH group comes from the Latin words "*mercurium captans,*" meaning capturing mercury. Combining carbazole with thiourea could elevate its ability to capture Hg metal ions. Li *et al.* [38] proposed a carbazole-Schiff base (**P15**) synthesized from **20** and 4-phenylthiosemicarbazide (**21**) in ethanol, as shown in **Figure 21**. The pale-yellow compound obtained was tested by UV-Vis absorption spectroscopy for a variety of metal ions (Na^+, Ag^+, Ca2$^+$, Mg^{2+}, Zn^{2+}, Cu^{2+}, Fe^{2+}, Fe^{3+}, K^+, Cd^{2+}, Cr^{3+}, Co^{2+}, Mn^{2+}, Pb^{2+}, and Ni^{2+}), in DMF. The compound showed two absorption bands in the UV spectrum at 290 and 359 nm. **P15** specifically showed a hyperchromic shift for band at 290 nm and a large red shift or bathochromic shift from 359 to 385 nm for Hg^{2+} ion, while the color of the solution turned yellow from colorless with no obvious response for other metals, suggesting it

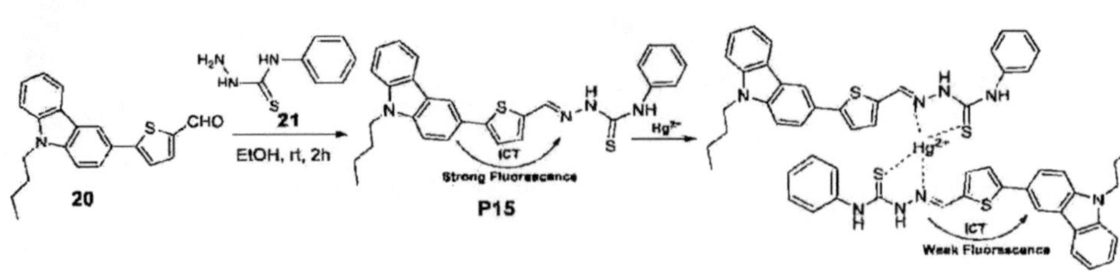

Figure 21.
Synthesis and binding mode of **P15** *toward* Hg^{2+}.

is a colorimetric chemosensor for Hg^{2+}. Further fluorescence experiments suggested that the Hg^{2+}/**P15** ratio was 1:2 during binding. The fluorescence emission from 480 to 490 nm showed a red shift only for Hg^{2+} ion, while for other metal ions, no significant change was observed. Further experiments exhibited an association constant $K_a = 2.8 \times 10^{4}$, and a detection limit of 6.5×10^{-6} mol/L as determined by Benesi-Hildebrand, Stern-Volmer, and Job's plot analyses.

Replacing amino groups with hydrazine and including heterocyclic aromatics not only increases the power of the coordination site but also the extension of π-conjugation. The article by Sekar *et al.* [39] presented the synthesis and application of a carbazole-hydrazinobenzothiazole-based fluorescent sensor (P16) for selective Hg^{2+} detection. P16 was synthesized via condensation of 9-ethyl-9H-carbazolyl-3-carbaldehyde (22) and 15 in methanol (**Figure 22**), yielding a 95% product characterized by NMR, IR, and mass spectrometry.

The sensor exhibited a turn-on fluorescence response at 433 nm upon Hg^{2+} binding, driven by an ICT mechanism from the carbazole donor to the benzothiazole acceptor. UV-Vis and fluorescence titration confirmed selectivity for Hg^{2+} over 16 competing metal ions, with a 1:1 binding stoichiometry (Job's plot) and a binding constant $K_a = 1.88 \times 105 M^{-1}$ *via* the Benesi-Hildebrand equation. The coordination involved the imine nitrogen (–C=N–) and benzothiazole sulfur, validated by NMR peak shifts (NH proton disappearance at δ 12.17 ppm upon complexation) and IR spectral changes (shifting of C=N and C–S stretches). The sensor demonstrated a low LOD of 1.41×10^{-7} M fluorometrically and 9.2×10^{-8} M electrochemically, in DMSO/H_2O (1:9) at pH 6–10. Protein binding studies with bovine serum albumin (BSA) revealed strong interaction ($K_a = 3.14 \times 10^{6} M^{-1}$), inducing conformational changes in tryptophan/tyrosine residues, as shown by 3D fluorescence. Cyclic voltammetry highlighted redox potential shifts for Hg^{2+} detection, emphasizing dual optical-electrochemical utility. The sensor's stability and applicability in biological and environmental monitoring showed its potential for Hg^{2+} sensing in complex matrices (**Figure 23**).

The thiol (-SH) group can be replaced by pyridine/quinoline nitrogen in the vicinity of methylenehydrazine to serve as a coordinating site for Hg^{2+} ions. The advantage of introducing quinoline moiety is to enhance the fluorescence capacity and quantum yield of designed probe. To this approach, Singh *et al.* [40] have reported a carbazole-quinoline tagged fluorophore **P17**, which was synthesized by condensing **22** and 2-hydrazinoquinoline (**23**) in methanol with a catalytic amount of glacial acetic acid under reflux for 1 h, followed by slow evaporation to yield yellow crystals (**Figure 24**).

Single-crystal XRD analysis revealed that **P17** crystallized in a tetragonal I41/a space group, having an almost planar molecular conformation stabilized by strong

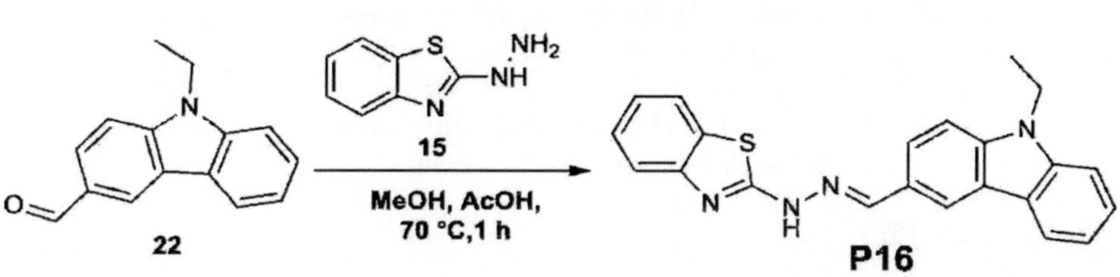

Figure 22.
Synthesis of **P16**.

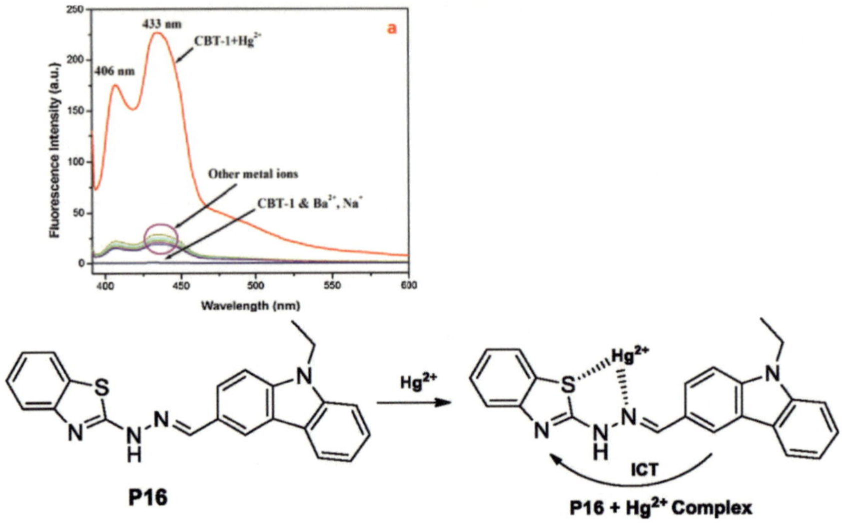

Figure 23.
Fluorescence spectrum and binding mechanism of probe **P16.**

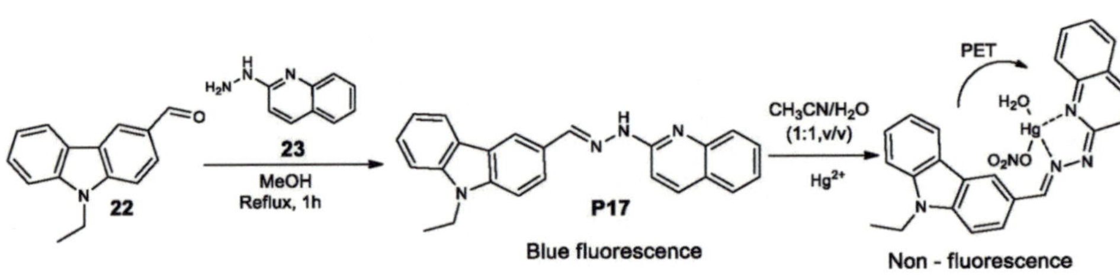

Figure 24.
Synthetic route to probe **P17** *and its binding mechanism with* Hg^{2+}.

intramolecular hydrogen bonding (2.63 Å) and intermolecular N–H···N networks. In CH_3CN/H_2O (1:1, v/v), it exhibited π–π* absorption bands at 330 and 370 nm that diminish markedly upon the addition of Hg^{2+}, accompanied by a color change from pale yellow to colorless in comparison with other 15 metals tested, including Pb and Bi. Its bright blue fluorescence emission at 485 nm underwent complete quenching, and a blue shift to 475 nm upon Hg^{2+} coordination was observed, while other competing metal ions elicit a negligible response.

Fluorescence titration and Job's plot analyses confirmed a 1:1 **P17**-Hg^{2+} stoichiometry with $K_a = 1.01 \times 10^5\ M^{-1}$ value. The LOD was 2.59×10^{-8} M. Mechanistic investigations combining FT-IR, ESI-MS, ^{1}H-NMR titration, and DFT calculations revealed bidentate coordination through azomethine-N and quinoline-N with concomitant –NH deprotonation, leading to a photo-induced electron transfer (PET) pathway (**Figure 24**). The authors tested probe **P17** for practical applicability by using P17-coated paper strips for on-site Hg^{2+} detection and successful analysis of lake and Ganga river water samples with recoveries of 94–99%.

3.2 Detection of Pb^{2+}

Recently, Schiff base formation has also become a foundation reaction for making COFs. Carbazole remains at the core of COFs materials, where the Schiff base reaction is being used to construct these 2-D or 3-D materials. The excellent crystallinity, porosity, and tunable pore size of COFs are among the ideal properties for metal ions detection and adsorption.

A carbazole-grafted covalent organic framework (COF-CB) was synthesized by Wang *et al.* [41] initially by condensation of 1,3,5-tris(4-aminophenyl)benzene and 2,5-dihydroxy-1,4-benzenedicarboxaldehyde to form COF-DhaTab, followed by post-grafting carbazole units onto the framework *via* Williamson ether reaction as shown in **Figure 25**. The resulting COF-CB exhibited an eclipsed AA-stacking structure with microporosity (3 nm pore size) and retained crystallinity after functionalization. COF-CB demonstrated a selective "turn-on" fluorescence response to Pb^{2+} ions *via* ICT upon coordination of Pb^{2+} with nitrogen (imine groups) and oxygen (phenolic hydroxyl) sites within the framework. This coordination suppressed non-radiative decay, enhancing fluorescence emission at 540 nm. The sensor showed a linear detection range of 0–80 μM Pb^{2+} with a low detection limit (LOD) of 1.48 μM (calculated via 3σ/s method) and excellent anti-interference against 15 competing metal ions (Zn^{2+}, Cr^{3+}, Fe^{2+}, Fe^{3+}, Hg^{2+}, Al^{3+}, Cu^{2+}, Ni^{2+}, Mg^{2+}, Ca^{2+}, Mn^{2+}, Ba^{2+}, Co^{2+}, Na^{+}, and $K)^{+}$. Binding stoichiometry was inferred as 1:1 (Pb^{2+}:COF-CB) based on Job's plot analysis of similar systems, though explicit stoichiometric data were not provided. Reversibility was confirmed using EDTA, which quenched fluorescence by displacing Pb^{2+}. The COF-CB's selectivity stemmed from its tailored pore structure and coordination sites, enabling Pb^{2+} detection in real water samples (tap/drinking water) with

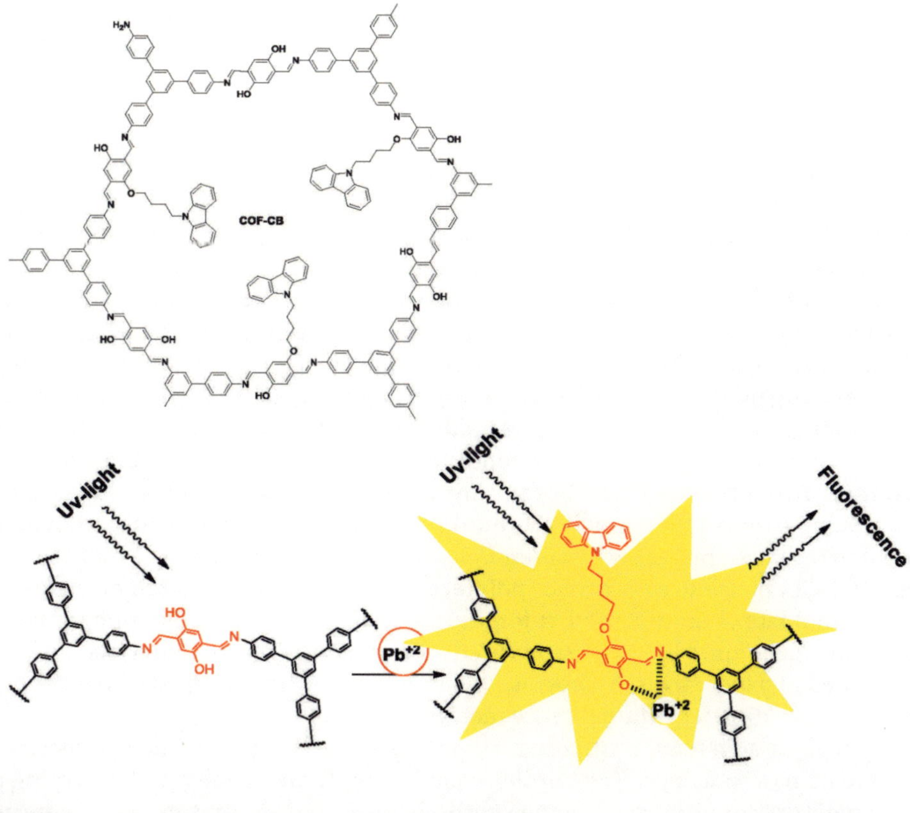

Figure 25.
Structure of carbazole-based COF and binding mechanism with Pb^{2+}.

95–98% recovery. While the association constant (K_a) was not explicitly reported, the high sensitivity and selectivity underscored strong binding affinity. This work highlights COF-CB as a robust, reusable chemosensor for environmental Pb^{2+} monitoring, leveraging its structural stability and fluorescence enhancement mechanism.

The above one is just an example of carbazole-based Schiff base COFs materials' potential toward metal ion detection; there are many exciting carbazole-based Schiff base COFs, for the selective and sensitive detection of metal ions. Since this topic deviates from the title of chapter so COFs are not discussed here.

4. Conclusions

So, the bottom line is that carbazole-based Schiff-base chemosensors unite synthetic simplicity with versatile photo-physics. By tuning the donor/acceptor balance, harnessing PET/CHEF/AIE/ESIPT mechanisms, and integrating user-friendly formats (gels, strips, smartphone read-outs), detection limits can be pushed into the nM range, to achieve rapid "turn-on" selectivity in real water samples. For common transition metals such as Fe, Cr, Cu, Al, etc., imine nitrogen (C=N) alongside other donor groups like -OH, -SH, NH2, NR2, etc. are ideal as coordinating sites while incorporating carbazole and naphthalene moieties enhance photophysical and fluorescence properties in the designed chemosensor. For heavy metals (Hg, Pb, As), the utilization of carbazole-based COFs and MOFs and nanomaterials especially nanofibers could be designed to detect such metals in the nM range. As global concerns over metal pollution and metabolic disorders escalate, carbazole-Schiff bases stand poised to play a pivotal role in advancing affordable, rapid, and reliable detection technologies.

5. Future perspectives

Although carbazole-Schiff bases could be promising candidates for developing chemosensors for a variety of sensing applications, because of their ease of formation and stapling of two moieties of tunable characteristics together, their labile hydrolysis and water sensitivity remains a challenge. In addition, the Lewis acidic behavior and catalysis ability of transition metal ions promote the Schiff bases hydrolysis in both aqueous and organic/aqueous solvents. However, a strategically designed configuration of binding sites and signaling gears holds the potential to yield a chemosensor capable of detecting the metal ions in aqueous or organic/aqueous media. So next-generation carbazole-based Schiff base sensors can utilize a push-pull architecture by combining strong donors (dialkylamino) with extended acceptors (dicyanovinyl, nitrophenyl) to red-shift absorption/emission toward the visible/NIR. To overcome water solubility problem sulfonate or poly(ethylene glycol) moieties can be introduced to eliminate organic cosolvent for sensing in full aqueous media. Integration of emerging techniques like nanostructured assemblies the sensor materials can be imbedded in paper strips, hydrogels, or nanofibers for simple dip-test, on-site colorimetry, or "dipstick" fluorescence methods.

Despite these advances, translating laboratory-scale discoveries into commercial sensors remains a challenge. Key hurdles include long-term stability under varying pH and temperature conditions, reducing synthesis costs, and integrating these materials into user-friendly devices such as paper-based strips or portable fluorimeters.

However, progress in nanotechnology and materials engineering, such as embedding carbazole-Schiff bases into polymer matrices or metal-organic frameworks (MOFs) offers promising pathways to enhance durability and scalability. Additionally, collaborations between academia and industry are critical to validate sensor performance in real-world scenarios, from monitoring industrial effluents to diagnosing metal imbalances in clinical settings.

Conflict of interest

The authors declare no conflict of interest.

Author details

Syeda Aaliya Shehzadi* and Mustaghees Ur Rehman
Sulaiman Bin Abdullah Aba Al-Khail-Centre for Interdisciplinary Research in Basic Sciences (SA-CIRBS), International Islamic University, Islamabad, Pakistan

*Address all correspondence to: aaliya.shehzadi@iiu.edu.pk

References

[1] Jia Y, Li J. Molecular assembly of Schiff base interactions: Construction and application. Chemical Reviews. 2015;**115**(3):1597-1621

[2] Faridbod F, Ganjali MR, Dinarvand R, Norouzi P, Riahi S. Schiff's bases and crown ethers as supramolecular sensing materials in the construction of potentiometric membrane sensors. Sensors. 2008;**8**(3):1645-1703

[3] Alorabi AQ, Abdelbaset M, Zabin SA. Colorimetric detection of multiple metal ions using Schiff base 1-(2-thiophenylimino)-4-(N-dimethyl) benzene. Chemosensors. 2019;**8**(1):1

[4] Reimann MJ, Salmon DR, Horton JT, Gier EC, Jefferies LR. Water-soluble sulfonate Schiff-base ligands as fluorescent detectors for metal ions in drinking water and biological systems. ACS Omega. 2019;**4**(2):2874-2882

[5] Singh A, Barman P. Recent advances in Schiff base ruthenium metal complexes: Synthesis and applications. Topics in Current Chemistry. 2021;**379**:1-71

[6] Wang L, Jiao S, Zhang W, Liu Y, Yu G. Synthesis, structure, optoelectronic properties of novel zinc Schiff-base complexes. Chinese Science Bulletin. 2013;**58**:2733-2740

[7] Sıdır İ, Sıdır YG, Berber H, Demiray F. Electronic structure and optical properties of Schiff base hydrazone derivatives by solution technique for optoelectronic devices: Synthesis, experiment and quantum chemical investigation. Journal of Molecular Structure. 2019;**1176**:31-46

[8] Amith Nayak PH, Bhojya Naik HS, Teja HB, Kirthan BR, Viswanath R. Synthesis and opto-electronic properties of green light emitting metal Schiff base complexes. Molecular Crystals and Liquid Crystals. 2021;**722**(1):67-75

[9] Sakthivel A, Jeyasubramanian K, Thangagiri B, Raja JD. Recent advances in Schiff base metal complexes derived from 4-aminoantipyrine derivatives and their potential applications. Journal of Molecular Structure. 2020;**1222**:128885

[10] Patil SK, Das D. Substituent-controlled selective and sensitive potential optical fluoride sensors based on salicylidene Schiff base derivatives. ChemistrySelect. 2017;**2**(21):6178-6186

[11] Zhang M, Gong L, Sun C, Li W, Chang Z, Qi D. A new fluorescent-colorimetric chemosensor based on a Schiff base for detecting Cr^{3+}, Cu^{2+}, Fe^{3+} and Al^{3+} ions. Spectrochimica Acta Part A: Molecular and Biomolecular Spectroscopy. 2019;**214**:7-13

[12] Duarte F, Dobrikov G, Kurutos A, Santos HM, Fernández-Lodeiro J, Capelo-Martinez JL, et al. Enhancing water sensing via aggregation-induced emission (AIE) and solvatofluorochromic studies using two new dansyl derivatives containing a disulfide bound: Pollutant metal ions detection and preparation of water-soluble fluorescent polymeric particles. Dyes and Pigments. 2023;**218**:111428

[13] RasulKhan B, Ponnaiah SK, Balasubramanian J, Periakaruppan P. Novel carbon quantum dotted reduced graphene oxide nanosheets for nano-molar range cadmium quantification. Electrocatalysis. 2022;**13**(4):435-446

[14] Musikavanhu B, Liang Y, Xue Z, Feng L, Zhao L. Strategies for improving

selectivity and sensitivity of Schiff base fluorescent chemosensors for toxic and heavy metals. Molecules. 2023;**28**(19):6960

[15] Alam MZ, Alimuddin, Khan SA. A review on Schiff base as a versatile fluorescent chemo-sensors tool for detection of Cu2+ and Fe3+ metal ion. Journal of Fluorescence. 2023;**33**(4):1241-1272

[16] Goshisht MK, Patra GK, Tripathi N. Fluorescent Schiff base sensors as a versatile tool for metal ion detection: Strategies, mechanistic insights, and applications. Materials Advances. 2022;**3**(6):2612-2669

[17] Taha A, Farooq N, Singh N, Hashmi AA. Recent developments in Schiff base centered optical and chemical sensors for metal ion recognition. Journal of Molecular Liquids. 2024;**401**:124678

[18] Afrin A, Swamy PCA. Novel Schiff base derivatives for the detection of one-to-multi metal ions and tracking the live cell imaging. Coordination Chemistry Reviews. 2023;**494**:215327

[19] Zhu W, Yang L, Fang M, Wu Z, Zhang Q, Yin F, et al. New carbazole-based Schiff base: Colorimetric chemosensor for Fe^{3+} and fluorescent turn-on chemosensor for Fe^{3+} and Cr^{3+}. Journal of Luminescence. 2015;**158**:38-43

[20] He Y, Yin J, Wang G. New selective "on-off" fluorescence chemosensor based on carbazole Schiff base for Fe^{3+} detection. Chemistry of Heterocyclic Compounds. 2018;**54**:146-152

[21] David CI, Prabakaran G, Karuppasamy A, Veetil JC, Kumar RS, Almansour AI, et al. A single carbazole based chemosensor for multiple targets: Sensing of Fe^{3+} and arginine by fluorimetry and its applications. Journal of Photochemistry and Photobiology A: Chemistry. 2022;**425**:113693

[22] Inan-Eroglu E, Ayaz A. Is aluminum exposure a risk factor for neurological disorders? Journal of Research in Medical Sciences. 2018;**23**(1):51

[23] Bryliński Ł, Kostelecka K, Woliński F, Duda P, Góra J, Granat M, et al. Aluminium in the human brain: Routes of penetration, toxicity, and resulting complications. International Journal of Molecular Sciences. 2023;**24**(8):7228

[24] Klotz K, Weistenhöfer W, Neff F, Hartwig A, van Thriel C, Drexler H. The health effects of aluminum exposure. Deutsches Ärzteblatt International. 2017;**114**(39):653-659

[25] Kolcu F, Kaya I. Carbazole-based Schiff base: A sensitive fluorescent 'turn-on'chemosensor for recognition of Al (III) ions in aqueous-alcohol media. Arabian Journal of Chemistry. 2022;**15**(7):103935

[26] Erdener D, Kaya İ. Synthesis and characterization of a carbazole based Schiff base capable of detection of Al^{3+} in organic/aqueous media. Journal of Fluorescence. 2022;**32**(6):2097-2106

[27] Bhiri F, Chemli M, Sakly N, Bahrouni Y, Wazzan N, Majdoub M. New chromogenic and fluorogenic symmetric carbazole Schiff base-derivative: Synthesis, sensing performances and smartphone-assisted fluorometric assay for selective Al^{3+} detection in drinking water. Journal of Molecular Structure. 2025;**1322**:140548

[28] Lv C, Hu B, Tao Y. A novel AIE-active salicylaldehyde-Schiff base probe with carbazole group for Al^{3+} detection in aqueous solution. Journal of Fluorescence. 2024:1-10

[29] Fieten H, Gill Y, Martin AJ, Concilli M, Dirksen K, van Steenbeek FG, et al. The Menkes and Wilson disease genes counteract in copper toxicosis in Labrador retrievers: A new canine model for copper-metabolism disorders. Disease Models & Mechanisms. 2016;**9**(1):25-38

[30] Yang L, Zhu W, Fang M, Zhang Q, Li C. A new carbazole-based Schiff-base as fluorescent chemosensor for selective detection of Fe^{3+} and Cu^{2+}. Spectrochimica Acta Part A: Molecular and Biomolecular Spectroscopy. 2013;**109**:186-192

[31] Zhu WJ, Wu ZY, Song JM, Li C. Preparation and properties of two new soluble carbazole-containing functional polyacetylenes. Chinese Chemical Letters. 2011;**22**(9):1017-1020

[32] Liu X, Xu P, Zhao X, Ge J, Huang C, Zhu W, et al. A novel dual-function chemosensor for visual detection of Cu^{2+} in aqueous solution based on carbazole and its application. Inorganica Chimica Acta. 2019;**495**:118975

[33] Ghosh S, Dege N, Kumar A, Misra N, Ghosh K. Naphthalimide-linked carbazole derivative: Synthesis, gelation and naked eye detection of Cu^{2+} and Fe^{3+}-ions under different conditions. Inorganica Chimica Acta. 2024;**572**:122280

[34] Liu L, Liu X, Guo C, Fang M, Li C, Zhu W. Carbazole-based dual-functional chemosensor: Colorimetric sensor for Co^{2+} and fluorescent sensor for Cu^{2+} and its application. Journal of the Chinese Chemical Society. 2021;**68**(12):2368-2377

[35] Vyas S, Barot YB, Mishra R. Novel anthracene and carbazole based aggregation induced enhanced emission active Schiff base as a selective sensor for Cu^{2+} ions. Journal of Fluorescence. 2025;**35**:2903-2915

[36] Lv C, Hu B, Tao Y. A novel AIE-active benzothiazole-Schiff base probe with carbazole group for Cu^{2+} detection in aqueous solution. Journal of Molecular Structure. 2025;**1328**:141259

[37] Bariya D, Vyas S, Mishra R, Mishra S. AIEE active Cholyl Hydrazide Carbazole-Schiff base for selective detection of copper (II) ion. Journal of Photochemistry and Photobiology A: Chemistry. 2024;**451**:115519

[38] Yin FF, Zhu WJ, Fang M, Wu ZY, Xu Y, Wang HL, et al. A new fluorescence chemosensor for Hg^{2+} based on carbazole and thiourea. Heterocycles. 2017;**94**(9):1719-1727

[39] Leslee DBC, Karuppannan S, Kothottil MM. Carbazole-hydrazinobenzothiazole a selective turn-on fluorescent sensor for Hg^{2+} ions–its protein binding and electrochemical application studies. Journal of Photochemistry and Photobiology A: Chemistry. 2021;**415**:113303

[40] Singh AK, Singh A, Yadav P, Singh AK, Singh VP. Carbazole-quinoline based ultrasensitive fluorometric sensor for detection of Hg^{2+} in aqueous medium: Crystal structure, DFT and real sample application. Journal of Molecular Structure. 2025;**1337**:142197

[41] Zhang N, Wei B, Ma T, Tian Y, Wang G. A carbazole-grafted covalent organic framework as turn-on fluorescence chemosensor for recognition and detection of Pb^{2+} ions with high selectivity and sensitivity. Journal of Materials Science. 2021;**56**:11789-11800

Section 5

Corrosion Inhibition and Advanced Materials

Chapter 10

Utilizing Schiff Bases for Surface Coating and Corrosion Prevention

Ghalia A. Gaber

Abstract

Schiff bases are more common than other chemical compounds as anticorrosion inhibitors. Important metal corrosion protection has been developed by mixing an inhibitor with the corrosive solution. It is an eco-friendly technology, inexpensive, easy to use, and offers a great deal of security. The lone pair electrons in the organic compounds corrosion inhibitors can overcome rigorous chemisorption by forming a dative covalent bond with the empty orbital of the transition metal. Adsorbing organic molecules to metal surfaces is the function of these heteroatoms and their conjugated structure. Physical adsorption occurs when the metal surface and inhibitor molecules make electrostatic contact, while chemical adsorption occurs when a coordinating bond is formed between the inhibitor and the metal's unoccupied d-orbital. A protective layer is created when chemical inhibitors are adsorbed at the metallic contact.

Keywords: corrosion inhibition, metal and alloys, corrosive media, Schiff base, adsorption

1. Introduction

Corrosion, a natural consequence of metals reacting with their surroundings, is a problem that the industrial sector regularly monitors [1, 2]. Preventive measures and techniques are vital given the significant difficulties that corrosion poses in traditional production processes [3]. Metals lose a lot of money due to electrochemical corrosion. Adsorption of organic molecules onto metal surfaces is a process that inhibits corrosion. Organic compounds with sulfur, nitrogen, and oxygen have been thoroughly researched as metal corrosion inhibitors. The reduction of metal waste in engineering materials depends on these inhibitors [4].

Schiff bases, a family of organic compounds with a functional group created when a primary amine and a carbonyl molecule condense, have demonstrated significant promise in the prevention of corrosion [5]. Schiff bases have been widely used in recent research as corrosion inhibitors because of their remarkable adsorption and film-forming capabilities. Because they interact with metal surfaces to generate protective layers that stop further degradation, they have important applications in corrosion protection, especially on metallic surfaces. Schiff base-derived metal complexes have been used by numerous researchers as efficient corrosion inhibitors for metals in a variety of settings [6–8].

Using mild steel in a solution of hydrochloric acid, Taha et al. [9] examined the corrosion-inhibiting capabilities of two Schiff base complexes, Cd(II) and UO2(II), which contain the ligand N-carbamimidoyl-4-((4-chlorobenzylidene)-amino) benzenesulfonamide. At 500 ppm and 25°C, weight loss assessments showed that the Cd(II) and UO2(II) complexes had protection efficiencies of 57.56 and 98.48%, respectively. The corrosion inhibition characteristics of two Schiff base ligands generated from 2-aminophenol and 2-hydroxy-1-naphthaldehyde, along with their corresponding Co(II) and Mn(II) complexes, were investigated by Haruna et al. [10] on copper in 0.1 M HCl solution. Because of their contact with the copper surface, which resulted in adsorption, these chemicals' inhibitory efficacy rose as concentration increased.

Using electrochemical techniques, Tahani et al. [11] examined the anticorrosion capabilities of the H2L Schiff base and its metal complexes, $[LCu(H_2O)_2]_2H_2O$, $[(HL)Co(H_2O)Cl]5H_2O$, and $[LNi(H_2O)_2]_2H_2O$, on carbon steel in 1.0 mol/L H_2SO_4 solution. The Schiff base H2L and its metal complexes, L-Cu, L-Co, and L-Ni, showed a considerable improvement in corrosion prevention efficiency, reaching 72.92, 63.88, and 60.06%, respectively.

In 1 M H_2SO_4, Gaber et al. [8] investigated 2-cyano-N-(4-morpholinobenzyldine) acetohydrazide as a corrosion inhibitor for 304 stainless steel (304 SS) and galvanized steel (GS). The results obtained were utilized to compute the adsorption properties, corrosion rate (CR), and inhibitory efficacy. The inhibitors stick to the Langmuir adsorption isotherm on the various surfaces under investigation. To help address the dearth of knowledge in the literature regarding the inhibitory effects of complexes and their corresponding ligands on corrosion. This research ultimate overview objective of utilizing Schiff bases for surface coating and corrosion prevention to mitigate metals/alloys corrosion and rusting in various industrial environment.

2. Schiff bases

Since Schiff bases were discovered, a great deal of research has been done on designing different Schiff bases and analyzing their varied structures and wide range of uses. Schiff bases have been used as good ligands to coordinate with late transition metals and transition metals. These Schiff base substances may be ligands that are mono-, di-, or polydentate [12]. Schiff base compounds result from the condensation of carbonyl compounds (aldehyde or ketone) with primary amines after water molecules are removed. Schiff base production is typically favored by the presence of a dehydrating agent.

Different kinds of Schiff base complexes with Schiff base preferred ligands include transition metal complexes with ligand atoms of oxygen, sulfur, or nitrogen. Because these Schiff bases can bond with different metal centers involving different coordination sites and enable the effective creation of metal complexes of varying geometries, they have gained prominence. According to a review of the literature, these complexes are substances with biological activity. Imine or azomethine groups of Schiff bases are crucial for exhibiting superior biological activity [13, 14].

When the amino group ($-NH_2$) of a primary amine reacts with the carbonyl group (C=O) of a carbonyl compound, like an aldehyde or a ketone, as seen in **Figure 1**, a carbon-nitrogen double bond (C=N) is formed, which is the general structure of a Schiff base [15, 16]. R and R' stand for organic groups that are joined to the nitrogen and carbon atoms, respectively, in this reaction. This C=N link gives Schiff bases

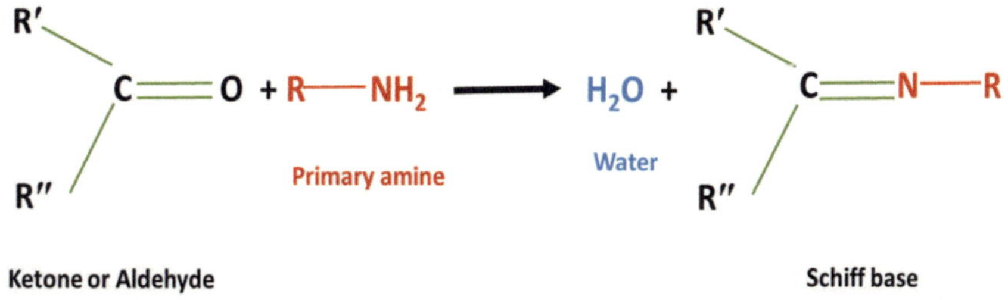

Figure 1.
Synthesis of Schiff bases by condensation method.

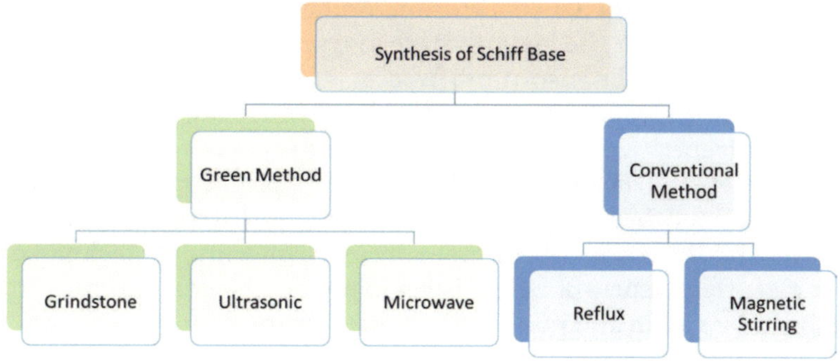

Figure 2.
Alternative techniques for Schiff base synthesis.

special chemical characteristics that make them adaptable ligands in coordination complexes with transition metals. Since these compounds frequently have intriguing and useful characteristics, Schiff bases are significant in a number of domains, including corrosion inhibition, as discussed in the review study [17, 18]

Conventional techniques, such as magnetic stirring and reflux, can be used to create Schiff bases [19]. The ease with which this family of chemicals can be made by condensation of amines with carbonyl compounds has kept several research groups interested. Although traditional synthesis has been the most popular method, green synthetic approaches such as sonication, microwave-assisted, and natural acid-catalyzed, as well as the use of ionic liquid solvents or deep eutectic solvents, have also been investigated for this reaction. The various techniques for synthesizing Schiff bases are displayed in **Figure 2**.

3. Schiff base corrosion inhibition mechanism

Schiff bases mainly prevent corrosion by covering the metal surface with a protective coating. The oxidation reactions that cause corrosion are slowed down by this layer, which prevents corrosive substances from reaching the metal. Reactive centers, like the (-C=N-) group in Schiff bases, are essential for adsorbing onto the metal surface and creating a solid, impermeable protective layer. Schiff bases can continually interact with the metal surface by dynamic and reversible adsorption process, which provides long-lasting corrosion protection [20–22].

The adsorption-related interactions between Schiff bases and metal surfaces. Schiff bases contain functional groups, including the nitrogen atoms from the azomethine group, which easily coordinate with metal ions on the surface to produce a coordinated protective layer. In addition to preventing corrosive species from penetrating, this interaction stabilizes the metal surface, lowering the possibility of additional oxidation [23, 24].

Schiff bases differ from traditional corrosion inhibitors in a number of distinctive ways. Their adaptability in structural alteration is one noteworthy feature. Schiff bases' corrosion prevention qualities can be fine-tuned by adjusting their chemical structure, which maximizes their efficacy for particular metal kinds and environmental circumstances. Furthermore, Schiff bases frequently have antioxidant qualities due to the presence of several functional groups, offering a dual method of defense against chemical and electrochemical erosion. Additionally, Schiff bases can establish a dynamic equilibrium with the metal surface in response to modifications in the corrosion environment since the adsorption process is reversible. This flexibility is a clear benefit, particularly under changing circumstances where other inhibitors may show only patchy effectiveness [25–27]. In conclusion, the process by which Schiff bases prevent corrosion entails the dynamic adsorption of a protective layer onto metal surfaces. Their distinct chemical makeup and interaction patterns set them apart as potent corrosion inhibitors, providing benefits over conventional techniques. The adaptability and effectiveness of Schiff bases in preventing corrosion are attributed to their reversible adsorption and possibilities for structural change.

The method by which Schiff bases control corrosion is complex and makes use of their special chemical characteristics. Comprehending this mechanism is essential to understanding how well Schiff bases work to stop metal deterioration. The mechanism by which Schiff bases reduce corrosion is explained in **Figure 3**. It highlights the creation of a protective layer that prevents corrosive processes and shows how Schiff bases interact with metal surfaces. Comparing the absolute values of adsorption enthalpy in an exothermic process allows one to differentiate between physisorption and chemisorption. Chemisorption's absolute value frequently approaches or surpasses 100 kJ/mol, whereas physisorption is usually less than 40 kJ/mol [28].

The measured ΔH°_{ads} values of 37–87 kJ/mol indicate the possibility of physical adsorption. Negative ΔS°_{ads} values accompany the adsorption of inhibitor molecules, presumably as a result of the metal surface forming an ordered layer. In addition to demonstrating the strong binding of the inhibitor molecules onto the metal surface, the negative ΔG°_{ads} values also reveal that the adsorption process of these compounds is spontaneous [5]. Electrostatic interactions (physisorption) between charged molecules and the charged metal surface are frequently credited with preventing metal corrosion.

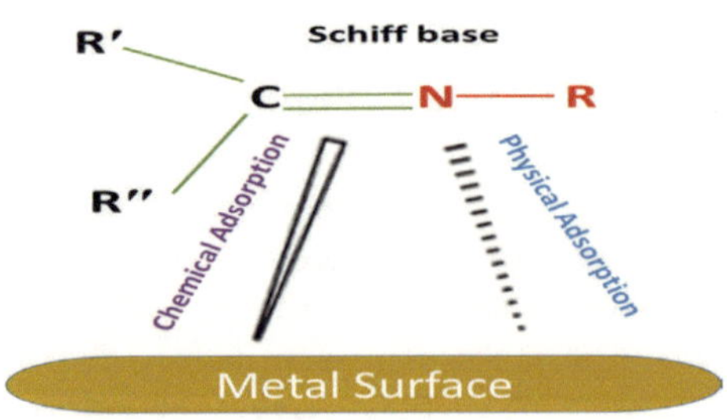

Figure 3.
Schiff base corrosion inhibition mechanism.

Schiff bases as corrosion inhibitor	Metal	Media	IE %	Ref.
Cd(II) – N-carbamimidoyl-4-((4-chlorobenzylidene)-amino)benzenesulfona mide	Mild Steel	1 M HCl	57.56	[9]
UO_2(II) – N-carbamimidoyl-4-((4-chlorobenzylidene)-amino)benzenesulfonamide	Mild Steel	1 M HCl	98.48	[9]
2-cyano-N-(4 morpholinobenzyldine) acetohydrazide	304 SS	1 M H_2SO_4	59.40	[8]
2-Cyano-N-(4 morpholinobenzyldine) acetohydrazide	Galvanized Steel	1 M H_2SO_4	62.60	[8]
2-Cyano-N-(4-morpholino benzyl dine) acetohydrazide (CMBAH)	1.5% Cr-Nb alloy	1 M H_2SO_4	54.57	[30]
4-(Morpholin-4-yl) benzylidenyl]thiosemicarbazide (MBT)	304 SS	1 M HCl	84.98	[5]
1,13-Bis-[(2-hydroxynaphtaldehyde) 4,7,10-trioxatridecane diimine] (HNTTD)	Copper	1 M HCl	87.36	[31]
(N1E)-N1,N2-bis(4-(dimethylamino)benzylidene)-ethane-1,2-diamine, DMAB	Carbon Steel	1 M HCl	97.70	[32]
2,2′-((1E,1′E) (cyclohexane-1,2 diylbis (azanylylidene)) bis (methanylylidene)) diphenol	Magnesium	0.01 M HCl	95.10	[33]

Table 1.
Compares the inhibitory effectiveness of previously reported Schiff bases on different metals.

4. Impacts of Schiff bases on different metals

In a variety of settings, including neutral and acidic media, Schiff bases have been demonstrated to successfully prevent iron and steel from corroding. Numerous industries, including chemical processing, mining, pipeline building, petroleum production and refinement, nuclear and fossil fuel power plants, marine applications, and metal processing equipment, heavily rely on carbon steel. Its high conductivity, low cost, ease of manufacture, and exceptional mechanical strength are the reasons for its appeal [29]. A Schiff base and its metal complexes Mn(II) and Cu(II) were synthesized by Emrayed et al. [4] These chemicals were tested for corrosion inhibition efficiency (% IE) on carbon steel (CS) in 1 M HCl. By lowering corrosion brought on by chloride ions and other harsh conditions, Schiff base derivatives can serve as protective coatings for aluminum and its alloys. By reducing oxidation and generating stable, protective complexes, Schiff bases can prevent corrosion in copper. **Table 1** compiles the comparison of the inhibitory efficacy of previously published Schiff bases on various metals.

5. Synergistic impact of Schiff bases with other inhibitors

When coupled compounds function better than when used alone, this is known as the synergistic inhibition effect. Increasing inhibitive efficacy, lowering the number of inhibitors in the protective system, and varying the use of inhibitors in acidic situations can all be accomplished through synergism. In earlier research, the combination of organic inhibitors with metal ions and inorganic materials, as well as the synergy of organic inhibitors and halide ions on metal corrosion in an acidic solution, were noted [5]. Effective corrosion prevention requires selecting the right corrosion inhibitor.

Synergistic Schiff bases as corrosion inhibitor	Metal	Media	IE %	Ref.
4-(morpholin-4-yl) benzylidenyl]thiosemicarbazide (MBT) + KI	304 SS	1 M HCl	95.48	[5]
1-dodecanethiol upon inhibition of Schiff bases *N,N′*-ethylen-bis and *N, N′-ortho*-phenylen-bis	Carbon Steel	0.5 M H_2SO_4		[36]
Pyrazole Schiff bases CPMNA	Mild Steel	1 M HCl	94.60	[34]
Pyrazole Schiff bases CMPMA	Mild Steel	1 M HCl	91.99	[34]
Schiff bases with additive Na_2SO_4	Aluminum	2 N H_2SO_4	95.02	[37]
Chi-SP/TiO_2 nanocomposite	Cu-Ni alloy	1 M HCl	79.65	[28]
N-benzylidene-5-phenyl-1,3,4-thiadiazol-2-amine (BPTA) + titanium dioxide nanoparticles (TiO NPs)	Mild Steel	1 M HCl	> 96	[38]

Table 2.
Compares the inhibitory effectiveness of previously reported synergistic Schiff bases on different metals.

According to published data, a good corrosion inhibitor should possess a number of desired characteristics, such as a benzene ring, numerous heteroatoms, a pi-electron structure, and a planar dimension. According to reports, heterocyclic rings and organic molecules with N, S, O, and P atoms can effectively suppress corrosion in a range of corrosive environments for a variety of metals [34].

Through chemical or physical adsorption, organic inhibitors attach to metal and transfer electrons, preventing corrosion. Additionally, the inhibitors electron-rich cores attach to the metal surface efficiently, preventing corrosion. Pyrazole analogues are useful ingredients in medications with a variety of biological uses, including antibacterial, antifungal, anti-tubercular, anti-inflammatory, anticancer, and antiviral qualities. They also have potential as metal corrosion inhibitors. Unlike dangerous chemicals, these molecules have biological activity and control corrosion without being harmful [35]. In their investigation of the inhibitory effects of pyrazole Schiff bases, which are significant, affordable, possess sufficient N, O, and aromatic rings, and have a strong propensity to disintegrate in acidic media, Khanna et al. [34] also looked at the corrosion-controlling capabilities of synthetic Schiff bases. Synergistic Schiff bases compatibility with different metal substrates and environmental circumstances is evaluated in **Table 2**.

6. Schiff base benefits in coating and corrosion prevention

Schiff bases offer protection for a variety of industries and applications and present a viable, environmentally benign alternative to corrosion inhibition. Their significance in corrosion prevention technologies will only grow with future study aimed at improving their performance and conquering obstacles in challenging situations. Compared to many synthetic corrosion inhibitors, which may contain heavy metals or other hazardous materials, Schiff bases are comparatively green. Their non-toxicity and biodegradability make them a more environmentally friendly choice for industrial coatings. Schiff base coatings are a safer substitute for conventional coatings that contain dangerous materials like lead or chromium, which are bad for the environment and human health. Schiff bases have been demonstrated to have a high effectiveness in inhibiting corrosion, particularly in neutral and acidic environments.

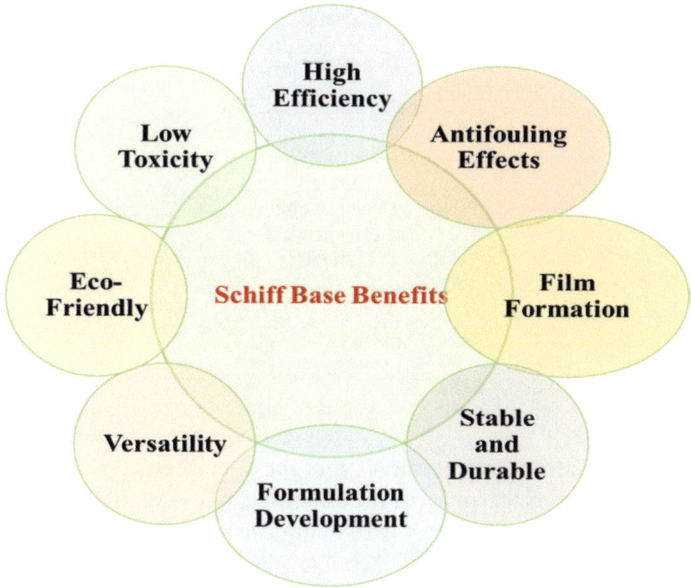

Figure 4.
Schiff base benefits in corrosion prevention.

Under typical operating conditions, the protective layers created by Schiff bases are stable and long-lasting, offering sustained corrosion prevention. Schiff bases can be made to fit particular metal surfaces and corrosion conditions by changing the amine or carbonyl components. Further study into the long-term performance of Schiff base-based formulations, including coatings and additives for real-world applications, as well as the possibility of combining them with additional additives like polymers or nanoparticles for improved qualities, is necessary. Working along with other additives or inhibitors can improve corrosion protection overall. Performance enhancement conducted to increase the mechanical qualities, anticorrosion capabilities, and wear resistance of Schiff base-based coatings by adding nanoparticles, polymers, or other additives. By preventing marine creature attachment, Schiff bases used in antifouling coatings lower the chance of localized corrosion and damage associated with biofouling. **Figure 4** illustrates the benefits of Schiff base in preventing corrosion.

7. Schiff base complex characterization

Understanding how Schiff bases interact with metal surfaces is crucial to determining how efficient they are at preventing corrosion. The adsorption process, film development, and overall effect of Schiff bases on metal corrosion behavior are investigated using a variety of analytical, chemical, and electrochemical approaches (**Figure 5**). Spectroscopic techniques, such as UV-visible spectroscopy and infrared spectroscopy (FT-IR), are essential for describing the interaction between Schiff bases and metal surfaces [28, 39]. When assessing the electrochemical behavior of metal surfaces in the presence of Schiff bases, chemical and electrochemical techniques are crucial [8, 40].

Weighed samples were submerged in corrosive media both with and without the studied Schiff bases using the weight loss method. Following temperature

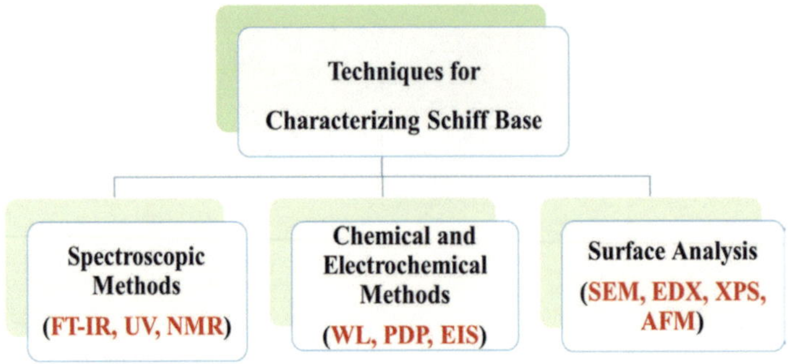

Figure 5.
Techniques for characterization of Schiff bases.

adjustments and the required immersion time, the specimens were taken out of the test solution, dried in a desiccator without moisture, and weighed again. The weight reduction was determined in each instance using the weight difference. The corrosion rates were calculated using Eq. (1) [39–42]:

$$CR = (\Delta W \times 534) / Atd \tag{1}$$

where A stands for surface area (cm^2), t for exposure time (h), d for metal surface density (g/cm^3), CR for corrosion rate (mmpy), and $\Delta W(g)$ for specimen weight before and after submersion in the experimental fluid. Using Eq. (2), the degree of surface covering (θ) was calculated:

$$\theta = \frac{CR_{\circ} - CR_i}{CR_{\circ}} \tag{2}$$

where CR_i and CR_o represent the corrosion rates of alloys in inhibited and uninhibited solutions, respectively. Eq. (3) was used to calculate the inhibitory efficiency [43]:

$$IE\% = \theta \times 100 \tag{3}$$

Using the potentiodynamic polarization (PDP) approach, the corrosion rate of metals is measured by monitoring the current-voltage relationship. Schiff bases can affect the kinetics of corrosion reactions by changing the polarization curves. Using Faraday's law, the corrosion rate, CR, can be calculated [44]. The observations of the corrosion rate were used to calculate the inhibition efficiency [45].

A metal interfaces reaction to a tiny alternating current is measured by electrochemical impedance spectroscopy (EIS). The corrosion inhibition process and the protective effectiveness of Schiff bases are better understood when impedance parameters are changed. Using Eq. (4) [46]. The inhibition efficiency (IE%) and surface coverage (θ) can be calculated based on the impedance readings:

$$IE\% = \theta \times 100 = 1 - \frac{R_{ct}}{R_{ct(inh)}} \times 100 \tag{4}$$

where Rct and $Rct_{(inh)}$ are the charge transfer resistance in the absence and presence of an inhibitor, respectively.

Techniques for surface investigation offer direct insights into the content and morphology of the protective layer that Schiff bases form [43, 46]. Metal surface morphology may be seen both before and after exposure to the Schiff base, thanks to scanning electron microscopy (SEM). It facilitates the evaluation of the uniformity, coverage, and any changes brought about by the corrosion inhibition process. X-ray Photoelectron Spectroscopy (XPS) is used to identify the chemical states and elemental makeup of the adsorbed species and the metal surface. It gives thorough details about the makeup of the shield that Schiff bases create. AFM provides high-resolution imaging of the metal surface, making it possible to see changes and characteristics at the nanoscale brought about by the adsorption of Schiff bases. In conclusion, the interaction between Schiff bases and metal surfaces may be thoroughly understood by combining spectroscopic, electrochemical, and surface analysis methods. By using these techniques, the corrosion inhibition mechanism is clarified and the protective effectiveness of Schiff bases in various corrosive situations is evaluated.

8. Applications of Schiff bases as corrosion inhibitors and coating

Schiff bases have demonstrated immense potential as corrosion inhibitors, offering a versatile and effective solution across various industries. Their unique chemical properties and reactivity make them applicable to diverse metal substrates and corrosive environments [47, 48]. On iron, Schiff bases show a high inhibitory efficiency; however, on copper and aluminum, their effectiveness differs. Insights for customized applications are influenced by structural alterations and the particular corrosive environment [8, 30]. **Figure 6** displays a variety of Schiff bases' uses as coatings and corrosion inhibitors.

Schiff bases are essential for preventing corrosion in mild steel components used in the metal sector. Protecting mild steel, an alloy that is frequently used in many different applications, is essential for preserving structural integrity and halting material deterioration [9]. In pipeline coatings, Schiff bases are used as corrosion inhibitors. By creating a barrier on the pipeline's interior, these inhibitors stop the corrosive medium from coming into direct contact with the mild steel. This prolongs the pipelines' lifespan and lowers the rate of corrosion [49].

The corrosion resistance of metals like copper, aluminum, steel, and alloys is known to be enhanced by Schiff base coatings. In harsh environments like seawater, acidic solutions, and industrial chemicals, these coatings work very well. A few Schiff base coatings have the ability to mend themselves. Over time, the corrosion protection is maintained by the coating's ability to self-repair tiny scratches and damages. Schiff bases strengthen the link between the polymer coating and the metal substrate underneath, improving the coating's performance under real-world circumstances. The ability of Schiff base-functionalized polymers to provide coatings that tolerate higher temperatures makes them appropriate for industrial, automotive, and aerospace applications where heat resistance is essential. In order to reduce the rate of industrial

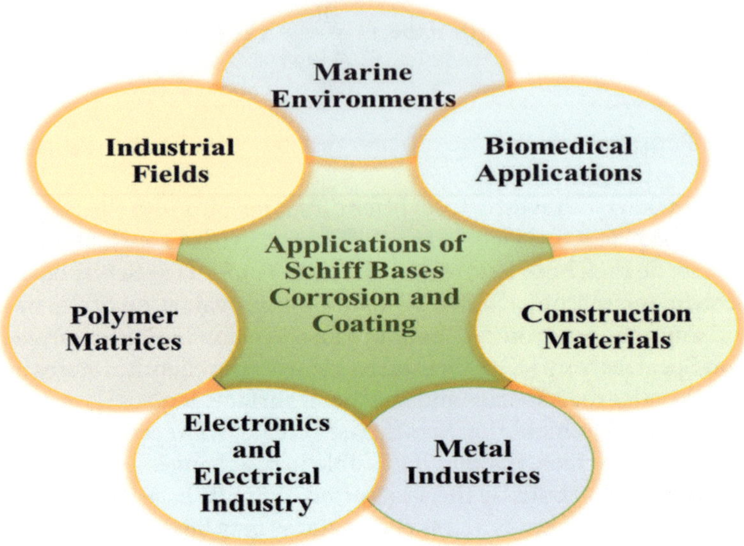

Figure 6.
Applications of Schiff bases as corrosion inhibitors and coating.

corrosion of Cu-Ni alloys, Dardeer et al. [28] developed a novel chitosan-Schiff base and its derivatives, which is a unique and attractive synthetic pathway for green corrosion inhibitors. Consequently, a new chitosan-sodium pyruvate (Chi-SP) polymer was effectively created by the condensation of sodium pyruvate and chitosan.

The role of Schiff base in corrosion prevention materials face special difficulties in marine environments because seawater is caustic. Schiff bases, which are useful for preserving different metal components utilized in maritime structures, are recognized for their adaptability and inhibitory qualities. The preservation of mild steel parts in ships, offshore platforms, and undersea constructions is one well-known example [50]. The antifouling coatings used on ship hulls contain Schiff bases. These coatings have two functions: they slow down corrosion and stop marine life, including algae and barnacles, from adhering, which speeds up deterioration. The structural elements of offshore platforms are coated with Schiff bases, which act as corrosion inhibitors. These coatings offer a barrier of defense, stopping corrosion and guaranteeing the platform's dependability under challenging marine conditions. Coatings for underwater constructions use Schiff bases. By creating a barrier that prevents seawater's corrosive effects, these coatings maintain the structural integrity of the submerged parts.

Schiff bases have drawn interest because of their possible uses in biomedicine, especially in protecting biocompatible materials against corrosion. In this context, the application of Schiff bases goes beyond conventional corrosion prevention because they also help to guarantee the durability and safety of materials meant for biological applications [51]. To guarantee long-term biocompatibility within the human body, materials used in biomedical implants, such as cardiovascular stents and orthopedic implants, must have strong corrosion resistance [52]. Because Schiff bases and their complexes are biocompatible, they can be used in biological fluids and tissues. By lowering the biotoxicity linked to certain metals used in implants, Schiff bases can improve the general safety of biomedical devices. The application of a multifunctional titanium dioxide (TiO_2) coating on the metal surface to give biomedical materials anticorrosive, antibacterial, and bioactive qualities has been established. The TiO_2

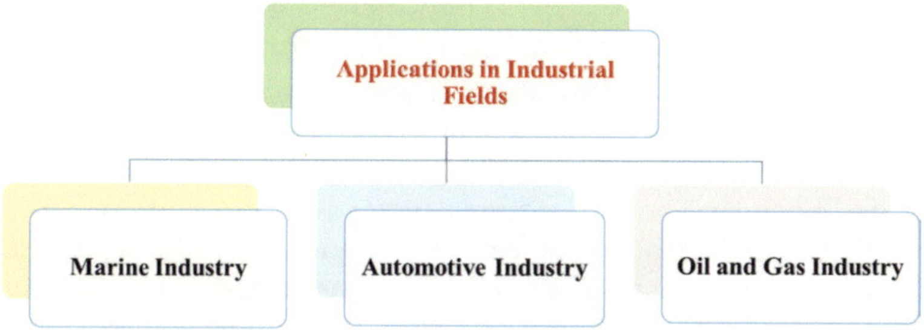

Figure 7.
Applications of Schiff bases in industrial fields.

compound's stable bonding structure is responsible for this [53]. The antibacterial properties of Schiff base derivatives make them advantageous for use in medical coatings. By inhibiting bacterial growth on surfaces including as implants, catheters, and surgical tools, these coatings lower the risk of infection.

Schiff bases are used to protect mild steel components from corrosion in the electronics and electrical industries, where component performance and dependability are crucial. In electrical and electronic devices, the usage of Schiff bases helps to preserve the longevity and integrity of vital parts [54]. Schiff base coatings are used on electronic components and gadgets to prevent corrosion brought on by external elements including humidity and salt exposure. These coatings are very helpful for safeguarding circuit boards, connections, and other delicate electronic components. Schiff base coatings not only resist corrosion but also offer electrical insulation, which keeps delicate electronic applications safe from short circuits and other electrical failures.

Schiff base coatings are used to shield steel reinforcing in concrete constructions against corrosion brought on by moisture and harsh chemicals like chlorides. Bridges, structures, and tunnels all have a longer lifespan because to this. Galvanized steel and other building materials are also coated with Schiff base-based coatings, which stop deterioration and increase the materials' overall longevity [55, 56].

Figure 7 depicts Schiff bases in industries such as the oil and gas, automotive, and marine sectors. Schiff bases are used to prevent corrosion on the metal surfaces of ships and offshore structures caused by salty conditions and seawater. These aids in safeguarding automobile parts, particularly those that are subjected to severe conditions, such as engines and exhaust systems. In order to stop pipelines and other metallic structures exposed to harsh chemicals and high pressure from corroding, the oil and gas sector uses Schiff bases. Because pipelines are exposed to severe and corrosive conditions, protecting them from corrosion is an important consideration. In the oil and gas industry, pipelines are subjected to corrosive substances such moisture, carbon dioxide, and hydrogen sulfide. Materials used in pipelines may deteriorate as a result of these corrosive conditions [57]. The oil and gas firms save money on maintenance because of the longer pipeline lifespan brought about by Schiff base protection. Because fewer repairs and replacements are required, this has a positive economic impact.

9. Conclusion

From preventing metal corrosion to enhancing surface protection, Schiff base coatings provide a very adaptable and efficient solution. They are a promising material for many sectors due to their adaptable qualities, environmental friendliness, and improved performance. The variety of synthetic techniques, including both traditional and green ones, shows how adaptable Schiff bases are, which is consistent with the increasing focus on ecologically friendly and sustainable corrosion inhibition techniques. Examining transition metal complexes with Schiff bases broadens the range of uses for these compounds. Schiff bases extensive use in sectors such metal manufacturing, oil and gas, marine settings, biomedical applications, constructional infrastructure and the electronics and electrical industry highlights their importance in corrosion inhibition. From the demanding requirements of biomedical implants to the severe environments of offshore platforms, these applications demonstrate the adaptability of Schiff bases in tackling corrosion issues in a variety of contexts.

Acknowledgements

In-kind support received from the Department of Chemistry, Faculty of Science (Girls), Al-Azhar University is gratefully acknowledged. The author acknowledge the use of Quillbot AI tool for language polishing of the manuscript.

Conflict of interest

The author declare that they have no conflict of interest.

Appendixes and nomenclature

304 SS	304 stainless steel
GS	galvanized steel
CS	carbon steel
CR	corrosion rate
UV	UV-visible spectroscopy
FT-IR	infrared spectroscopy
SEM	scanning electron microscopy
XPS	X-ray photoelectron spectroscopy
AFM	atomic force microscopy
EDX	energy dispersive X-ray spectroscopy
NMR	nuclear magnetic resonance
PDP	potentiodynamic polarization
EIS	electrochemical impedance spectroscopy
WL	weight loss
IE	inhibition efficiency
θ	surface coverage

Author details

Ghalia A. Gaber
Department of Chemistry, Faculty of Science (Girls), Al-Azhar University,
Nasr City, Cairo, Egypt

*Address all correspondence to: ghaliaasaid@azhar.edu.eg

References

[1] Gaber GA, Mohamed LZ, Abdelfatah A. Study the corrosion issues on galvanized steel induced in water tanks. Chemical Papers. 2023;**77**:7539-7549. DOI: 10.1007/s11696-023-03043-4

[2] Abdelfatah A, Mohamed LZ, Gaber GA. Corrosion inhibition of hybrid Ti/Ce salt for zinc in HCl solution. Egyptian Journal of Chemistry. 2023;**66**(13):1069-1079. DOI: 10.21608/ejchem.2023.202393.7785

[3] Abdelfatah A, Mohamed LZ, El-Hadad S, Moussa ME, Gaber GA. Comprehensive investigation of Si additions and nanocomposite inhibitors on microstructure/corrosion performance of cast AX53 alloy in 3.5% NaCl solution. Surface Review and Letters. 2024;**31**:2550050. DOI: 10.1142/S0218625X25500507

[4] Emrayed HF, Amraga EA, Ibrahim DM, Fouda AE-AS. Corrosion inhibition effect of Schiff base and its metal complexes with [Mn (II) and Cu (II)] on carbon steel in hydrochloric acid solution: Experimental and surface studies. International Journal of Electrochemical Science. 2025;**20**(1):100912. DOI: 10.1016/j.ijoes.2024.100912

[5] Hosny S, Abdelfatah A, Gaber GA. Synthesis, characterization, synergistic inhibition, and biological evaluation of novel Schiff base on 304 stainless steel in acid solution. Scientific Reports. 2024;**14**:470. DOI: 10.1038/s41598-023-51044-w

[6] Massoud AA, Hefnawy A, Langer V, Khatab MA, Öhrstrom L, Abu-Youssef MAM. Synthesis, X-ray structure and anti-corrosion activity of two silver(I) pyrazino complexes. Polyhedron. 2009;**28**(13):2794-2802. DOI: 10.1016/j.poly.2009.05.064

[7] Gupta SR, Mourya P, Singh MM, Singh VP. Structural, theoretical and corrosion inhibition studies on some transition metal complexes derived from heterocyclic system. Journal of Molecular Structure. 2017;**1137**:240-252. DOI: 10.1016/j.molstruc.2017.02.047

[8] Gaber GA, Hosny S, Mohamed LZ. Experimental and theoretical studies of 2-cyano-N-(4-morpholinobenzyldine) acetohydrazide as corrosion inhibitor for galvanized steel and 304 stainless steel in 1M H_2SO_4 solution. International Journal of Electrochemical Science. 2021;**16**(12):211214. DOI: 10.20964/2021.12.14

[9] Taha RH, Gaber GA, Mohamed LZ, Ghanem WA. Corrosion inhibition of two Schiff base complexes on the mild steel in 1M HCl solution. Egyptian Journal of Chemistry. 2019;**62**(Special Issue (Part 1) Innovation in Chemistry):367-381. DOI: 10.21608/ejchem.2019.11232.1722

[10] Haruna A, Rumah MM, Sani U, Ibrahim AK. Synthesis, characterization and corrosion inhibition studies on Mn (II) and Co (II) complexes derived from 1-{(Z)-[(2-hydroxyphenyl) imino]methyl}naphthalen-2-ol in 1M HCl solution. International Journal of Biological Physical and Chemical Studies. 2021;**3**(1):09-18. DOI: 10.32996/ijbpcs.2021.3.1.2

[11] Kashar TI, Emran KM, Mo'ala A. Novel metal complexes as antimicrobial and anticorrosion in acid media. Arabian Journal of Chemical and Environmental Research. 2020;7(1):29-49

[12] Abd-El-Aziz A et al. Advances in coordination chemistry of Schiff base complexes: A journey from nanoarchitectonic design to biomedical applications. Topics in Current Chemistry. 2025;**383**(1):8. DOI: 10.1007/s41061-025-00489-w

[13] Kumar D, Chadda S, Sharma J, Surain P. Syntheses, spectral characterization, and antimicrobial studies on the coordination compounds of metal ions with Schiff base containing both aliphatic and aromatic hydrazide moieties. Bioinorganic Chemistry and Applications. 2013;**2013**:981764. DOI: 10.1155/2013/981764

[14] Nagar S, Raizada S, Tripathee N. A review on various green methods for synthesis of Schiff base ligands and their metal complexes. Results in Chemistry. 2023;**6**:101153. DOI: 10.1016/j.rechem.2023.101153

[15] Mahdi BS et al. Corrosion inhibition of mild steel in hydrochloric acid environment using terephthaldehyde based on Schiff base: Gravimetric, thermodynamic, and computational studies. Molecules. 2022;**27**(15):4857. DOI: 10.3390/molecules27154857

[16] Saha SK, Murmu M, Murmu NC, Banerjee P. Synthesis, characterization and theoretical exploration of pyrene based Schiff base molecules as corrosion inhibitor. Journal of Molecular Structure. 2021;**1245**:131098. DOI: 10.1016/j.molstruc.2021.131098

[17] Stanly Jacob K, Parameswaran G. Corrosion inhibition of mild steel in hydrochloric acid solution by Schiff base furoin thiosemicarbazone. Corrosion Science. 2010;**52**(1):224-228. DOI: 10.1016/j.corsci.2009.09.007

[18] Kaur M, Kaur K, Kaur H. Quest of Schiff bases as corrosion inhibitors: A first principle approach. Journal of Physical Organic Chemistry. 2021;**34**(11):e4260. DOI: 10.1002/poc.4260

[19] Gopi D, Govindaraju KM, Kavitha L. Investigation of triazole derived Schiff bases as corrosion inhibitors for mild steel in hydrochloric acid medium. Journal of Applied Electrochemistry. 2010;**40**(7):1349-1356. DOI: 10.1007/s10800-010-0092-z

[20] Jasim AN et al. Schiff's base performance in preventing corrosion on mild steel in acidic conditions. In: Progress in Color Colorants Coating. Vol. 16. 2023. pp. 319-329

[21] Alamiery A, Shaker LM, Allami T, Kadhum AH, Takriff MS. A study of acidic corrosion behavior of Furan-derived Schiff base for mild steel in hydrochloric acid environment: Experimental, and surface investigation. Materials Today: Proceedings. 2021;**44**(Part 1):2337-2341. DOI: 10.1016/j.matpr.2020.12.431

[22] Şafak S, Duran B, Yurt A, Türkoĝlu G. Schiff bases as corrosion inhibitor for aluminium in HCl solution. Corrosion Science. 2012;**54**:251-259. DOI: 10.1016/j.corsci.2011.09.026

[23] Andreeva NP, Kuznetsov YI, Shikhaliev KS. The use of ellipsometry for studying adsorption of organic corrosion inhibitors from aqueous solutions on metals. Review. Part 1. Methods for obtaining adsorption isotherms. International Journal of Corrosion and Scale Inhibition. 2022;**11**(4):1716-1733. DOI: 10.17675/2305-6894-2022-11-4-20

[24] Kuznetsov YI. Organic corrosion inhibitors: Where are we now? A review. Part II. Passivation and the role of chemical structure of carboxylates.

International Journal of Corrosion and Scale Inhibition. 2016;**5**(4):282-318. DOI: 10.17675/2305-6894-2016-5-4-1

[25] Kuznetsov YI, Andreeva NP, Agafonkina MO. On co-adsorption on passive iron from aqueous 1,2,3-benzotriazole and sodium phenylundecanoate. Russian Journal of Electrochemistry. 2010;**46**:560-564. DOI: 10.1134/S1023193510050101

[26] Salman AZ, Jawad QA, Ridah KS, Shaker LM, Al-Amiery AA. Selected BIS-Thiadiazole: Synthesis and corrosion inhibition studies on mild steel in HCL environment. Surface Review and Letters. 2020;**27**(12):2050014. DOI: 10.1142/S0218625X20500146

[27] Abdulazeez MS et al. Corrosion inhibition of low carbon steel in HCl medium using a thiadiazole derivative: Weight loss, DFT studies and antibacterial studies. International Journal of Corrosion and Scale Inhibition. 2021;**10**(4):1812-1828. DOI: 10.17675/2305-6894-2021-10-4-27

[28] Dardeer HM, Abbas SA, Ghobashy MM, Gaber GA, Aly MF. Synthesis and characterization of novel chitosan-sodium pyruvate polymer and its derivatives for corrosion feature evaluation of Cu-Ni alloy. Inorganic Chemistry Communications. 2023;**157**:111308. DOI: 10.1016/j.inoche.2023.111308

[29] Boulechfar C et al. Corrosion inhibition of Schiff base and their metal complexes with [Mn (II), Co (II) and Zn (II)]: Experimental and quantum chemical studies. Journal of Molecular Liquids. 2023;**378**:121637. DOI: 10.1016/j.molliq.2023.121637

[30] Gaber GA, Mohamed LZ, Aly HA, Hosny S. Corrosion potential and theoretical studies of fabricated Schiff base for carbidic austempered ductile iron in 1M H_2SO_4 solution. BMC Chemistry. 2024;**18**(1):170. DOI: 10.1186/s13065-024-01278-0

[31] Benabid S, Douadi T, Issaadi S, Penverne C, Chafaa S. Electrochemical and DFT studies of a new synthesized Schiff base as corrosion inhibitor in 1MHCl. Measurement. 2017;**99**:53-63. DOI: 10.1016/j.measurement.2016.12.022

[32] Saleh MGA et al. Retardation of the C-steel destruction in hydrochloric acid media utilizing an effective Schiff base inhibitor: Experimental and theoretical computations. ACS Omega. 2024;**9**(27):29666-29681. DOI: 10.1021/acsomega.4c03135

[33] Seifzadeh D, Basharnavaz H, Bezaatpour A. A Schiff base compound as effective corrosion inhibitor for magnesium in acidic media. Materials Chemistry and Physics. 2013;**138**(2):794-802. DOI: 10.1016/j.matchemphys.2012.12.063

[34] Khanna R et al. Synergistic experimental and computational approaches for evaluating pyrazole Schiff bases as corrosion inhibitor for mild steel in acidic medium. Journal of Molecular Structure. 2024;**1297**:136845. DOI: 10.1016/j.molstruc.2023.136845

[35] Gaber GA, Hassan AY, Kadh MS, Saleh NM, Abou-Amra ES, Hyba AM. A computational and experimental investigation of novel synthesis fused pyrazolopyrimidine as zinc corrosion inhibitor in 1 M HNO3. Chemical Papers. 2024;**78**(5):3189-3203. DOI: 10.1007/s11696-024-03303-x

[36] Ehteshamzadeh M, Shahrabi T, Hosseini M. Synergistic effect of 1-dodecanethiol upon inhibition of Schiff bases on carbon steel corrosion in sulphuric acid media. Anti-Corrosion

Methods and Materials. 2006;**53**(3):147-152. DOI: 10.1108/00035590610665545

[37] Sethi T, Chaturvedi A, Upadhyaya RK, Mathur SP. Synergistic inhibition between Schiff's bases and sulfate ion on corrosion of aluminium in sulfuric acid. Protection of Metals and Physical Chemistry of Surfaces. 2009;**45**(4):466-471. DOI: 10.1134/S2070205109040169

[38] Al-Taweel SS, Al-Janabi KWS, Luaibi HM, Al-Amiery AA, Gaaz TS. Evaluation and characterization of the symbiotic effect of benzylidene derivative with titanium dioxide nanoparticles on the inhibition of the chemical corrosion of mild steel. International Journal of Corrosion and Scale Inhibition. 2019;**8**(4):1149-1169. DOI: 10.17675/2305-6894-2019-8-4-21

[39] El Refay HM, Hyba AM, Gaber GA. Fabrication, characterization and corrosion feature evaluation of mild steel in 1 M HCl by nanoparticle-modified activated carbon. Chemical Papers. 2022;**76**(2):813-825. DOI: 10.1007/s11696-021-01895-2

[40] Abdel-karim AM, Shahen S, Gaber G. 4-Aminobenzenesulfonic acid as effective corrosion inhibitor for carbon steel in hydrochloric acid. Egyptian Journal of Chemistry. 2021;**64**(2):825-834. DOI: 10.21608/ejchem.2020.44335.2897

[41] Hyba AM, El Refay HM, Shahen S, Gaber GA. Comparison fabrication, identification and avoidance of corrosion potential of M-CuO NPs/S-CuO NPs to suppress corrosion on steel in an acidic solution. Chemical Papers. 2023;**77**:5395-5407. DOI: 10.1007/s11696-023-02871-8

[42] Gaber GA, Soliman MM, Nasr ZA, Hyba AM. Comprehensive investigation of sustainable corrosion inhibitors on Cu–Zn alloy in simulated cooling water: Electrochemical explorations, SEM/EDX analysis, and DFT/molecular simulations utilizing expired Bepotastine-B as a green inhibitor. Sustainable Chemistry and Pharmacy. 2024;**37**:101340. DOI: 10.1016/j.scp.2023.101340

[43] Fouda AEAS, Nageeb M, Gaber GA, Ahmed AS, El-Hossiany AA, Atia MF. Carob fruit extract as naturally products corrosion inhibitor for copper-nickel alloys in brine solutions. Scientific Reports. 2024;**14**(1):29290. DOI: 10.1038/s41598-024-80589-7

[44] Gaber GA et al. Study of the corrosion-inhibiting activity of the green materials of the Posidonia oceanica leaves' ethanolic extract based on PVP in corrosive media (1 M of HCl). Green Processing and Synthesis. 2021;**10**(1):555-568. DOI: 10.1515/gps-2021-0055

[45] Gaber GA, Maamoun MA, Ghanem WA. Evaluation of the inhibition efficiency of a green inhibitor on corrosion of Cu-Ni alloys in the marine application. Key Engineering Materials. 2018;**786**:174-194. DOI: 10.4028/www.scientific.net/KEM.786.174

[46] Nageeb MM, Gaber GA, Ahmed ASI, Fouda AEAS. Application of lentil seed extract as an inhibitor to assess the corrosion properties of copper–nickel alloys in a NaCl environment. RSC Advances. 2024;**14**(38):28044-28057. DOI: 10.1039/D4RA04758C

[47] Noor EA, Al-Moubaraki AH. Thermodynamic study of metal corrosion and inhibitor adsorption processes in mild steel/1-methyl-4[4′(-X)-styryl pyridinium iodides/hydrochloric acid systems]. Materials Chemistry and Physics.

2008;**110**(1):145-154. DOI: 10.1016/j.matchemphys.2008.01.028

[48] Taha Mohamed M et al. Revolutionizing corrosion Defense: Unlocking the power of expired BCAA. Progress in Color Colorants Coating. 2024;**17**(2):97-111. DOI: 10.30509/pccc.2023.167156.1228

[49] Shwetha KM, Praveen BM, Devendra BK. A review on corrosion inhibitors: Types, mechanisms, electrochemical analysis, corrosion rate and efficiency of corrosion inhibitors on mild steel in an acidic environment. Results in Surfaces and Interfaces. 2024;**16**:100258. DOI: 10.1016/j.rsurfi.2024.100258

[50] Salman TA et al. New environmental friendly corrosion inhibitor of mild steel in hydrochloric acid solution: Adsorption and thermal studies. Cogent Engineering. 2020;7:1826077. DOI: 10.1080/23311916.2020.1826077

[51] Abdel Hameed RS et al. Green recycling of poly (ethylene terphthalate) waste as corrosion inhibitor for steel in marine environment. Egyptian Journal of Chemistry. 2021;**64**(5):2685-2695. DOI: 10.21608/EJCHEM.2021.54262.3145

[52] Mohamed LZ, Abd Elmomen SS, Ibrahim KM, Gaber GA. Corrosion characteristics of high-temperature oxidized TC21 alloy in simulated body fluid containing proline coated with Nb/Ag. Journal of Materials Engineering and Performance. 2024;**1**:1-13. DOI: 10.1007/s11665-024-10297-w

[53] Gaber GA, Mohamed LZ, Järvenpää A, Hamada A. Enhancement of corrosion protection of AISI 201 austenitic stainless steel in acidic chloride solutions by Ce-doped TiO_2 coating. Surface and Coatings Technology. 2021;**423**:127618. DOI: 10.1016/j.surfcoat.2021.127618

[54] Montemor MF. Functional and smart coatings for corrosion protection: A review of recent advances. Surface and Coatings Technology. 2014;**258**:17-37. DOI: 10.1016/j.surfcoat.2014.06.031

[55] Abo Hashem IA, Gaber GA, Ahmed ASI, Abdel Ghany NA. Assessment of blended cement containing waste basalt powder: physicomechanical and electrochemical impedance spectroscopy investigations. Discover Applied Sciences. 2024;**6**(8):380. DOI: 10.1007/s42452-024-06075-x

[56] Gaber GA. MOFs in construction and infrastructure. In: Advancing Corrosion Control with Metal-Organic Frameworks: Beyond Rust, ACS Symposium Series. Vol. 1503. American Chemical Society; 2025. pp. 245-261. DOI: 10.1021/bk-2025-1503.ch012

[57] Tamalmani K, Husin H. Review on corrosion inhibitors for oil and gas corrosion issues. Applied Sciences. 2020;**10**(10):3389. DOI: 10.3390/app10103389

Chapter 11

Polymer-Based Schiff Bases and Their Applications

Mai Saud Abdullah Alsubaie

Abstract

This research explores the intricate relationship between polymers and Schiff bases. Schiff bases, characterized by the presence of an imine group (-C=NH), have emerged as key components in modern polymer chemistry due to their structural flexibility and dynamic covalent bonding properties. Schiff base polymers are thermally stable, mechanically strong, and chemically reactive, responding to various environmental parameters, including temperature, pH, and electromagnetic fields. This chapter provides further insight into the applications of Schiff base polymer materials as drug carriers in drug delivery systems (DDS), smart polymers, biomedical devices, and conductive systems. The chapter also provides a comprehensive analysis of the synthetic methods, chemical structures, and spectroscopic techniques employed in the study of Schiff base polymers.

Keywords: Schiff bases, polymers, applications, synthesis, characterization

1. Introduction

Polymers-Schiff base composites are among the most significant pillars in organic and material chemistry. Macromolecules are composed of repeating structural monomer units that exhibit a broad range of physicochemical properties. These composite materials are widely used in various industries, including packaging, electronics, biomedicine, and construction [1]. The developed innovative materials, which combine the synergistic functionality of both the (C=NH) imine group and polymer, exhibit chelating ability and tunable reactivity due to the advanced combined properties of the Schiff base and the polymer. Examples of Schiff base polymers are polyvinyl alcohol-salicylaldehyde-ethylenediamine) and chitosan-vanillin/2-hydroxy-acetophenone. Schiff base linkages are embedded in soft polymer matrices, such as polyurethane (PU) or polyethylene glycol (PEG). Combining Schiff bases with other supramolecular interactions enhanced material performance. All the primary amines used in the design strategy incorporated aromatic aldehydes, and the primary amines form stable imine bonds [2]. Phenylenediamine reacts with benzaldehyde and gives polyimine in good yield. A simple, efficient synthesis of functional Schiff base polymers is represented in the following reaction.

$$C_6H_4(NH_2)_2 + 2C_6H_5CHO \rightarrow \left[C_6H_4(N=CHC_6H_5)_2\right]_n \quad (1)$$

1.1 Preparation of Schiff bases and corresponding polymer

The Schiff base compound was discovered by Hugo Schiff in 1864, formed by a condensation reaction.

Primary amines with carbonyl (C=O) compounds, aldehydes, or ketones contain either the CHO or C=O functional group, respectively. For instance (**Figure 1**) represents the condensation reactions between 2-amino-dibenzothiophene with various aromatic compounds, yielding various Schiff bases.

Thiophene Schiff bases are readily prepared from the available precursors, as shown in **Figure 2** by adding the primary 2-amino-dibenzothiophene to benzaldehyde, salicylaldehyde, and 2,4-dihydroxy-benzaldehyde. All reagents have an analytical-grade purity.

Piperidine (0.85 g, 10 mmol) was slowly added to cyclohexanone solution (2.94 g, 30 mmol), malononitrile (1.32 g, 20 mmol), and elemental sulfur (0.64 g, 20 mmol) in 20 mL ethanol at a constant temperature of 50°C. The reaction mixture was heated under reflux for 2 h and then left to cool overnight. The precipitated (compound A) product was filtered and washed twice with 10 mL of cold ethanol. To the compound A (0.89 g, five mmol) solution in ethanol, an equimolar amount of the corresponding aldehyde (R is either R1 = phenol or R2 = furan) was added dropwise. The reaction mixture was stirred at room temperature until the product precipitated from the solvent. The reaction mixture was stirred for an additional time until the TLC showed complete consumption of these reactants. The product was filtered off, washed with cold ethanol (2 × 5 mL), and recrystallized from hot ethanol. Two different colored Schiff bases were formed.

Reaction: Salicylaldehyde-polyethylene glycol diamine (PEG-diamine) → R-CH=N-R'-N=CH-R as a cross-linked network Schiff base polymer.

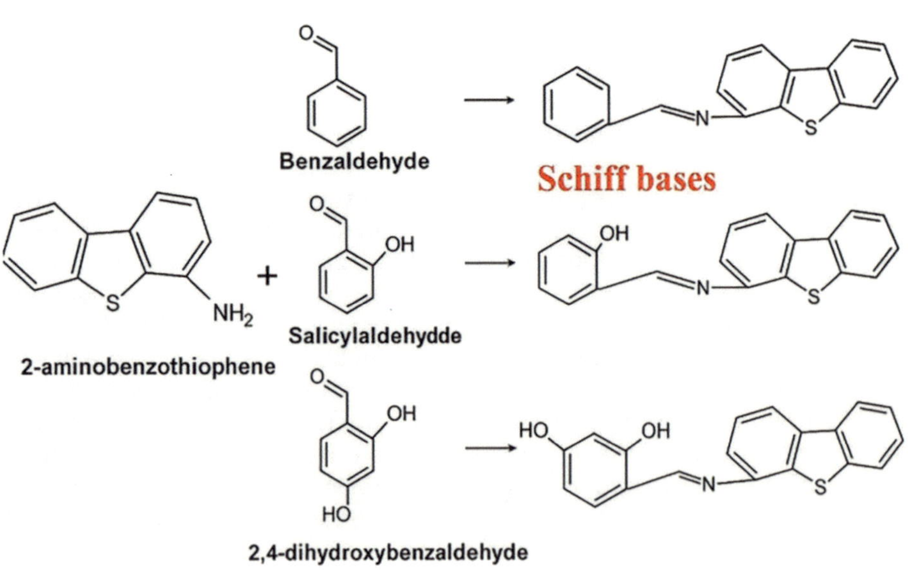

Figure 1.
Preparation of Schiff bases following a condensation reaction.

DOI: http://dx.doi.org/10.5772/intechopen.1012665

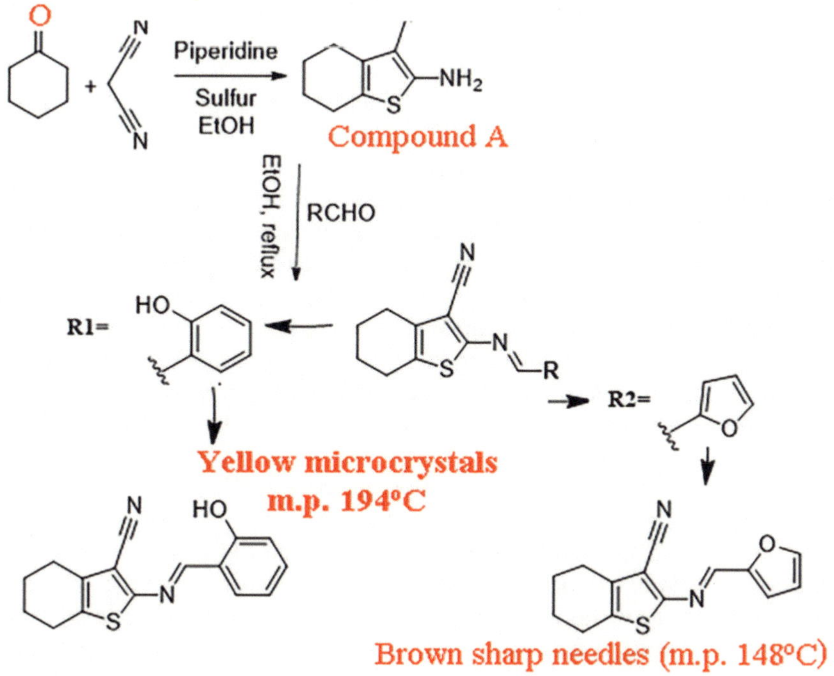

Figure 2.
Preparation of Schiff bases: 2-((2-hydroxy-benzylidene)-amino)-4, 5, 6, 7 tetra-H-benzo [b]thiophene-3-carbonitrile, 2-((Furan-2-ylmethylene) amino)-4,5,6,7-tetra-H-benzo [b]thiophene-3-carbonitrile [4].

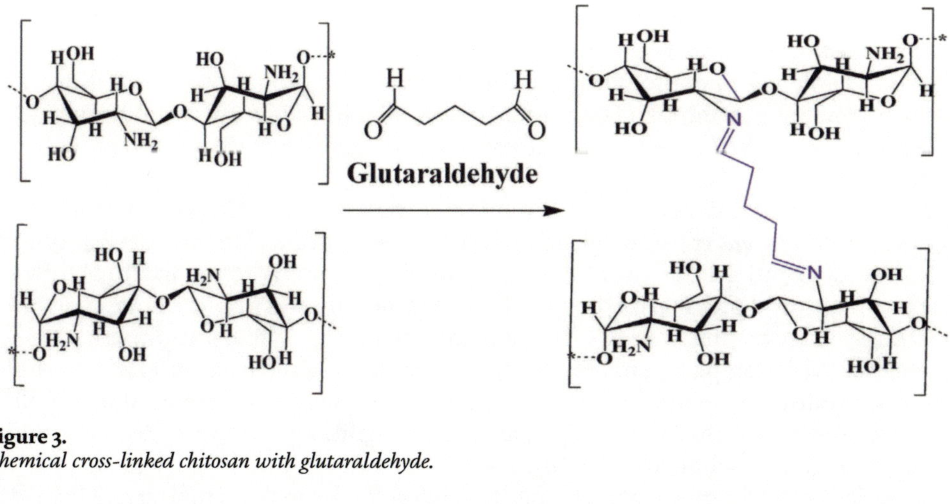

Figure 3.
Chemical cross-linked chitosan with glutaraldehyde.

Another example (**Figure 3**) is the chemical cross-linking of chitosan (CT) by glutaraldehyde (the common cross-linking agent). Glutaraldehyde linked neighboring polymeric chains into a 3D network structure, reducing segmental mobility, enhancing mechanical strength, chemical resistance, and improving the adsorption capacity of CT. All the physicochemical properties of the CT adsorbent improved. The optimized cross-linking degree enhances the adsorption capacity and efficiency [3].

Glutaraldehyde has two reactive CHO functional groups, which act as bridges by forming covalent bonds with polymeric chains of CT. Schiff base bridges restricted the motion of CT chains by the new interconnected chains, which then formed a 3D network structure. Composite features of CT-Schiff base depend on the cross-linking, the ratio of moles of repeating units of CT polymer to the moles of the cross-linker.

Chemically cross-linked CT is an efficient adsorbent that has good mechanical strength due to irreversible chemical links and is used in water treatment for the removal of reactive organic dyes. CT is a natural amino polymer containing a high content of hydroxyl and amino groups that give high adsorption capacity toward various pollutants, such as dyes and heavy metals (HMs), especially in neutral solutions. pH of the solution controls binding capacity. The CT polymer Schiff base has high potential in pollutant removal by adsorption. It is non-toxic, abundant, biocompatible, hydrophilic, antibacterial, environmentally friendly, versatile, and biodegradable bio-adsorbent. Schiff base binding overcame the dissolution problem of CT at low pH.

The CT-benzaldehyde Schiff base composite is an efficient adsorbent for removing heavy metals from wastewater. The imine group acts as a chelating site, improving the removal efficiency [4]. CT-Schiff base composites are used in wastewater treatment. CT modification with benzaldehyde enhanced the adsorption capacity of heavy metals. The imine group serves as a chelating site, thereby improving removal efficiency. Schiff bases are incorporated as cross-linkers, functional pendant groups, or integral parts of the polymer backbone, providing tailored reactivity, enhancing stimulus-response, and improving mechanical performance [5].

Innovative hybrid Schiff base polymer materials are utilized in energy storage and membrane technologies for renewable energy systems, including polyethylene glycol (PEG)-Schiff base composites formed by reacting PEG-diamine with salicylaldehyde which are incorporated into polymer matrices. These are approached through various strategies, each offering control over the structure and functionality of the final Schiff base polymer material. The post-functionalization process involved the formation of a polymer bearing CHO or amino groups, which react with complementary reagents to yield Schiff base linkages, thereby allowing for surface modification or cross-linking [6].

In situ polymerization: Schiff base monomers polymerized directly into the polymer backbone *via* free radical, condensation, or coordination polymerization. In the one-pot synthesis, as an effective and simple processing route, in situ Schiff base formation and polymerization yield higher amounts and provide better control over the polymer architecture [7]. The template-assisted polymerization, utilizing a metal ion (M^{+n}) template, enables the organization of Schiff base units within the polymer, resulting in an ordered, functional network that yields advanced materials with dynamic bonds, thermal stability, and functionalities such as responsive stimuli, metal binding, or biocompatibility [8].

Functionalized amines react with electrophilic centers to form innovative hybrid polymer materials that respond to sunlight, making them suitable for energy storage and membrane technologies in renewable solar cell energy systems. **Figures 4** and **5** demonstrate the synthesis of s-triazine-centered dendrimer Schiff base mono- and di-hydroxy-polymers, prepared without employing protection and deprotection processes [9].

Schiff base monomers prepared *via* reactions: 4-amino-benzoic acid and substituted hydroxyl-benzaldehydes in refluxed ethanolic solutions (**Figure 6**).

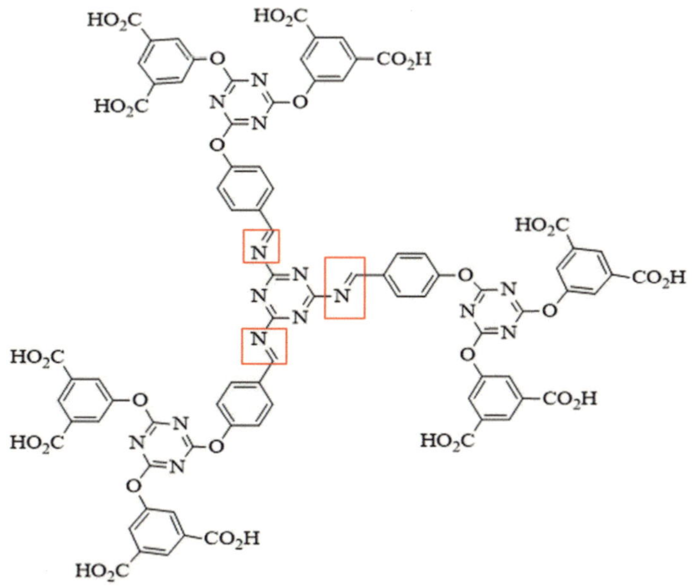

Figure 4.
s-Triazine-centered *dendrimer Schiff base polymer: dicarboxylic groups at the meta-position.*

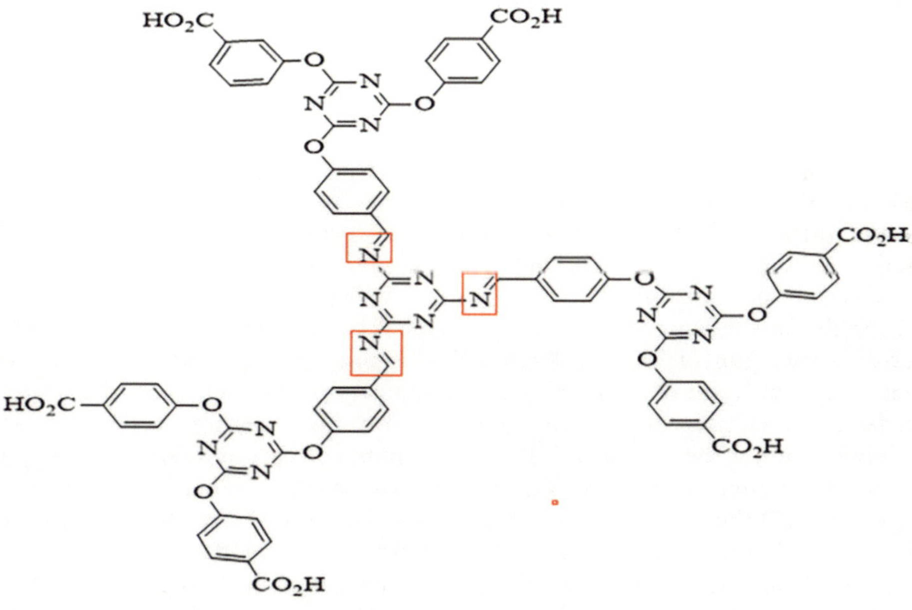

Figure 5.
s-Triazine-centered *dendrimer Schiff base-polymer: monocarboxylic groups at the meta-position.*

Classical polymerization *via* esterification was carried out using p-chloride [11]. Polymerization under atmospheric pressure air involved DBD plasma [12]. Monomers are exposed to plasma-reactive species, which induce polymerization and form a thin film polymer. This approach used reactive species, charged particles, and UV photons

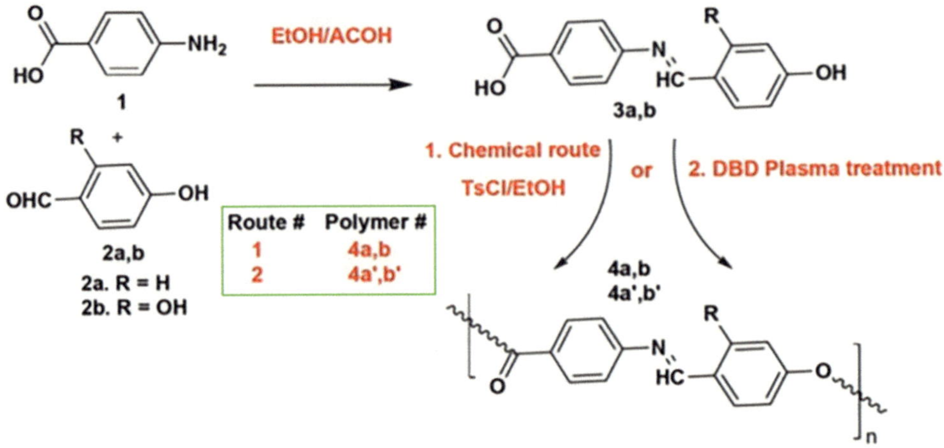

Figure 6.
Synthesis route of Schiff base polymer macromolecules via *innovative plasma and classical approaches [10].*

to activate monomers. Classical polymerization yielded polymers with higher molecular weights (Mw), while plasma polymerization produced polymers with somewhat lower Mw but still successful polymer chains. Thermal behavior revealed multi-stage degradation related to the polymer structure, with plasma radiation (activation pathways involving protonic and radical ion species from plasma-treated polymers, partially cross-linked or rearranged [12].

1.2 Characterization of Schiff base polymers

Spectroscopic and physical characterizations are used in elucidating structural, thermal, and electronic properties. Chemical structure, confirmed by carbon, hydrogen, and nitrogen (CHN) elemental analysis The expected theoretical calculated percentages and the experimental findings confirmed the suggested molecular structural formulas. Fourier Transform Infrared Spectroscopy (FTIR) confirms the formation of the Schiff base linkage –C=N– by detecting its characteristic stretching vibration at a frequency region of 1600–1650 cm^{-1}. UV-Vis. Spectroscopy was used to explore $\pi\pi^*$ and $n\pi^*$ transitions of the imine group, conjugation, and metal-ligand charge transfer. NMR spectroscopy (^1H and ^{13}C) confirmed the structure, particularly for the imine proton at the chemical shift (δ) ~8–9 ppm and carbon signal at ~160 ppm [3]. Thermogravimetric analysis (TGA) assesses thermal stability and decomposition temperatures. Differential scanning calorimetry (DSC) determines glass transition temperature (Tg), crystallinity, and melting behavior. X-ray Diffraction (XRD) enabled evaluation of the crystalline or amorphous nature. Scanning and transmission electron microscopy are used to explore the 3D surface morphology and nanostructural features.

1.3 Chemical reactivity and stability

Aromatic Schiff bases from aromatic aldehydes, such as salicylaldehyde or benzaldehyde, are more chemically and thermally stable than aliphatic Schiff bases (from aliphatic aldehydes or ketones), due to resonance stabilization. Heterocyclic Schiff bases contain N, O, and F heteroatoms in the ring structure and improve the biological and catalytic activity of the blended polymer. Chemically, Schiff bases exhibit nucleophilic and electrophilicity enhancing further reactions of functionalized polymers: coordinate bond formation with M^{+n} forming stable complexes such as polyethyleneimine-metal complexes with various electronic and magnetic properties applied in efficient catalysis, sensors, and biological activity. Schiff bases are used as cross-linkers in polymers as functional pendant groups, or an integral part of the polymer backbone, granting tailored reactivity, stimuli-response, and increased mechanical strength [3, 5]. Biological activity depends on functional groups. Many substituted thiophene derivatives are antibacterial ((against antibiotic-resistant bacteria, *E.Coli*), and antifungal. It is applied in medicinal chemistry.

Multidentate Schiff bases possessing multiple imine groups (chelating sites for metal ion (M^{+n}) give coordinating metal-polymers complexes): magnets, catalysts, and electronic and advanced functional composites. Moreover, semiconductors, sensors, batteries, and conductive coatings. Imine -C=N-group provides reactive sites coordinating metals and participating in various chemical reactions. Schiff base polymers are chemically stable, soluble in organic solvents, and undergo reversible bonding and self-healing in the protective anticorrosive smart coating. The optical and electronic properties belonged to some Schiff base polymers exhibit fluorescence, conductivity, and redox activity. This polymeric composite is used in optoelectronic applications as a sensor and smart material [13].

2. Discussion

Distinct thermal, mechanical, and chemical properties are enhanced depending on the nature of the Schiff base and polymer matrix. The most relevant properties include thermal stability, which is attributed to the presence of aromatic rings and conjugation systems in Schiff base polymers. Thermogravimetric analysis (TGA) and differential scanning calorimetry (DSC) demonstrate higher decomposition temperatures, glass temperature (Tg), and melting temperature. High tensile mechanical strength and elasticity are achieved mainly when Schiff bases act as cross-linking agents or participate in supramolecular interactions. Rigid imine structures reinforce the polymer framework [9].

2.1 Application of Schiff-base polymers

2.1.1 Renewable energy applications

Schiff base polymer used as a bright coating, liquid crystals materials (fluidity and molecular order between liquids and solids), enhances light absorbance and the orientation in the flexible organic solar cells (innovative energy systems), energy storage systems (batteries, and supercapacitors due to light weight and mechanical properties [14]. This innovative polymer material is used in the conversion of temperature difference into electricity in thermal to electrical converters. Renewable energy systems with polymers, such as Schiff base polymers, are an interdisciplinary research area that integrates chemistry, physics, and materials science for sustainable innovation, and is utilized in displays, sensors, and solar energy systems [2]. Schiff bases yielded mesogenic molecules, as the rigid C=N linkage promotes order. Example: 4-methoxybenzaldehyde-4-amino-biphenyl Schiff base polymer liquid crystal that is used in bright optics and display technologies [15].

Organic light-emitting diodes (OLEDs) using fluorescent polymer-Schiff base dye as emitting materials have a general multilayered structure ordered as cathode, electron-injection layer, electron-transporting layer, emissive layer, hole-transporting layer, hole-injection layer, and anode (**Figure 7**). Phosphorescent organic light-emitting diodes have also received a great deal of attention because of their high internal efficiency [16].

Schiff base polymer 2,4,6-tris(di-aryl-amino)-1,3,5-triazine derivatives, as shown in **Figure 8**, are used as the host electron source layer. The transport material gave a high external 10% quantum efficiency. The glass-forming performance and unique optical/electrochemical properties make these molecules potentially suitable hosts in OLEDs, exhibiting a high triplet bandgap of 2.91 eV and a high glass transition temperature of 148°C. Good charge transport properties in single-carrier devices have proven that the host potential acts as a blue phosphorescent emitter at long wavelengths [17]. White electroluminescent devices fabricated using a red-orange emitter unit and 8-hydroxyl-quinolinolato Li (blue-green emitter). An additional agent adjusts charge carrier mobility introduced between layers, improving the stability of the white emission color [18].

N-phenyl carbazole substituted 2,4,6-trisphenyl-*s*-triazine host material used for solution-processed green phosphorescent organic light-emitting devices exhibits a high Tg of 165°C and triplet energy 2.63 eV which is suitable for the fabrication of efficient single-layer green OLEDs [19, 20].

Star-shaped Schiff-base polymer electron transport (ET)-type host materials for green phosphorescent organic light-emitting devices. The meta–meta linkage between *s*-triazine core and peripheral aryl moieties limited the effective extension of pi-conjugation, giving high triplet energies 2.80 and 2.69 eV individually and good electron mobilities [21].

Organic fluorescent heterocyclic chromophores in Schiff-base polymers are utilized as molecular probes, fluorescent markers, organic light-emitting diodes (OLEDs), and photovoltaic cells, as well as in both traditional textile and polymer applications. Electron donors, such as triphenylamine, diphenylamine, 1,3,5-triazines, and carbazoles, which exhibit high electron mobility, thermal, and photochemical stability, are commonly used as hole-transporting materials

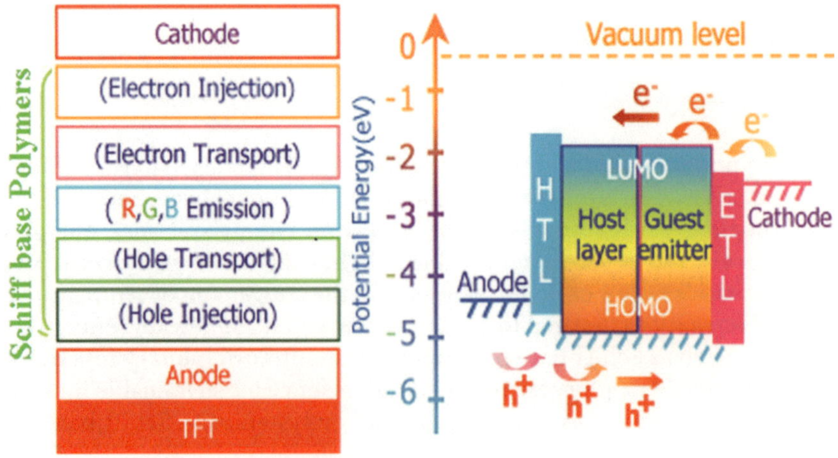

Figure 7.
The structure arrangement of organic light-emitting diodes.

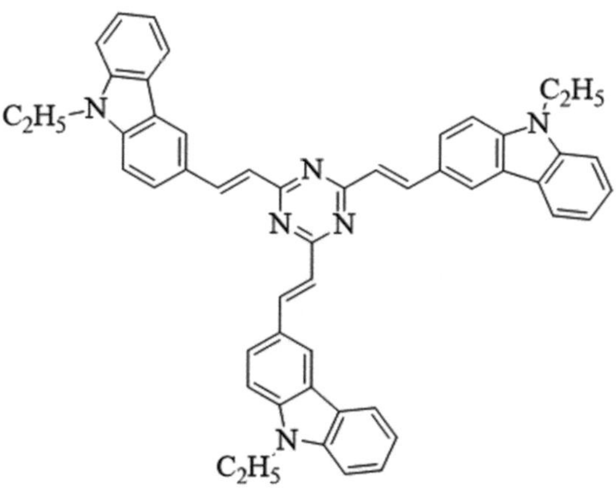

Figure 8.
Red-orange emitting Schiff-base polymer material of branched molecular structure.

or light-emitting materials. Star-shaped compounds contain an electron-deficient *s*-triazine-core scaffold, and electron-rich chromophores extend pi-conjugation of the heterocyclic triazine-core to the aromatic donor-substituents. Fluorenyl-and carbazolyl-substituted triazine derivatives are highly fluorescent in solutions and rigid polymer matrices [19–21]. Schiff-base polymer hybrid materials are used in ion-conducting membranes, enhancing charge transport and thermal/chemical stability. Conjugation is electrically conductive with facile electron transport.

Polymer solar cells are one type of organic solar cell [22], offering advantages such as flexibility, low cost, color, lightweight design, and the ability to be manufactured in any shape or size, making them an eco-friendly option. Schiff base overcame the downside of this type (low efficiency due to limited absorbance in the solar spectrum, low carrier mobility for both donor and acceptor, and short diffusion length, low stability, affected by oxygen, moisture, and UV). The conjugated Schiff base polymer acts as an electron donor.

Polymer solar cells yielding 5% power conversion efficiencies. Example: poly(2-tri-hydroxy-5-(3′,7′-dimethyloctyloxy)-1,4-phenylenevinylene) showed strong $\pi\pi^*$ optical transitions (absorption coefficients above $\sim 10^5$ cm^{-1}) in the visible region. Molecular design enables the synthesis that controls their optical $\pi\pi^*$ transition on absorption of photon energy. The electron donor part excites and creates an exciton with a high binding energy of 0.1–1.4 eV in organic semiconductors because the electron (e) and the hole (h) wave functions are localized. Dielectric constants are low ($\varepsilon \sim$ 3–4 Debye), making a strong coulomb e-h interaction. Inorganic materials have an exciton binding energy of just a few milli eV. Excitons migrate to an interface with an acceptor where fast electron transfer occurs, causing charge separation. Then, separated es in the acceptor LUMO level and separated holes in the donor HOMO level are transported through the device to the appropriate contact by the internal field created by electrodes of different work functions (**Figure 9**) [22].

Researchers have created new device architectures to increase the efficiency of polymeric solar cell bilayer heterojunction (**Figure 10**). Low efficiency, 1% limited by the exciton diffusion length, which is the distance an exciton must travel before undergoing recombination. In most organic semiconductors, the thickness is ~3–10 nm, which is shorter than the optical absorption length (50–100 nm). In a bilayer donor-acceptor heterojunction structure, donor-acceptor layers must be thin

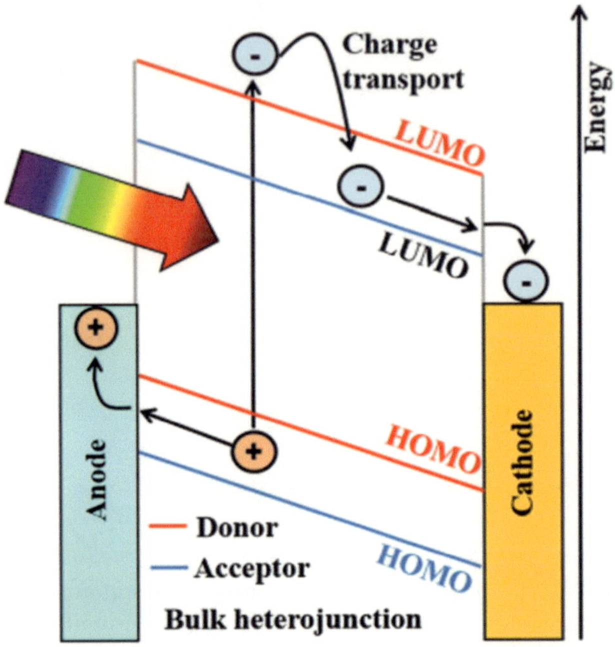

Figure 9.
Generation photo current mechanism in bulk heterojunction.

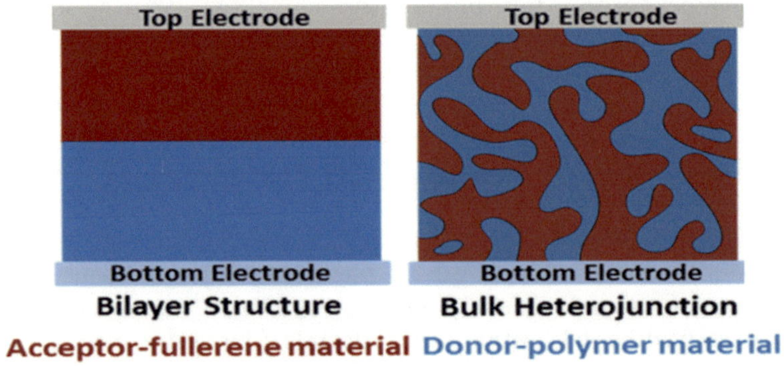

Figure 10.
Different architecture of the active layer in polymer solar cells [23].

to allow photo-generated excitons to reach the interface with high probabilities before undergoing recombination. So, this limits the light conversion efficiency. This limitation was overcome *by* increasing the interface area between donor and acceptor. With the discovery of fast electron transfer from conjugated polymer poly[2-methoxy-5-(2-ethylhexyloxy)]-1,4-phenylenevinylene (MEH-PPV) to fullerene (C_{60}), a promising bulk heterojunction (BHJ) has been developed. In polymer-based bulk heterojunction (BHJ) solar cells, the donor and the acceptor materials are mixed in organic solvents, by applying different deposition techniques such as thermal annealing and solvent annealing, a self-assembly. Hence, an interpenetrating network formed. By optimizing the morphology of active layers, the power conversion efficiency (PCE) reaches 12%.

To make bulk heterojunction (BHJ) more efficient, two materials must be mixed at a scale less than the exciton diffusion length to increase the probability of charge transfer at the interface.

2.1.2 Application of polymer-Schiff base in biomedicine

2.1.2.1 Antibacterial biomaterials

Biocompatible biodegradable Schiff base-polymers applied in tissue engineering and wound healing. In DDS, the designed Schiff base polymer drug carrier responds to pH or temperature, allowing for controlled and targeted release of the drug. Reversible bonding capabilities are key to this function. Multifunctionality of Schiff base-containing polymers, enabling innovation in innovative materials and biocompatible systems [24]. Schiff's bases of *s*-triazin (**Figure 11**), are biologically active against various bacterial and fungal strain cultures. Results showed that triazine-based Schiff's base derivatives are promising candidates as anti-microbial drugs [25].

A series of the tri-substituted *s*-triazine hydrazones (**Figure 12**), is an antimicrobial and antimycobacterial agent that has moderate inhibition against strains used [26].

2.1.2.2 Schiff base polymers in drug delivery systems

In the DDS, Schiff base polymers sense and respond to pH and temperature. This character enables the controlled and targeted release of chemotherapeutic agents. Reversible bonding capabilities are key to this function. These applications highlight the multifunctionality of Schiff base polymers, allowing the innovation of smart, biocompatible drug carriers [27]. As biomaterials: biocompatibility, antibacterial, and biodegradable Schiff base applied in tissue engineering and wound healing.

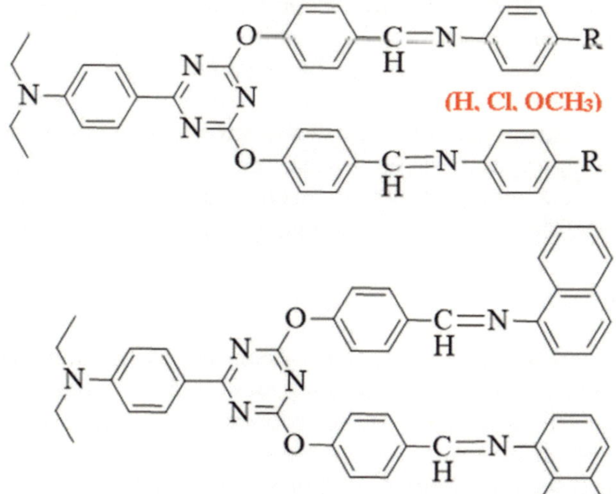

Figure 11.
Moderate to good antimicrobial Schiff's bases of s-triazine polymer.

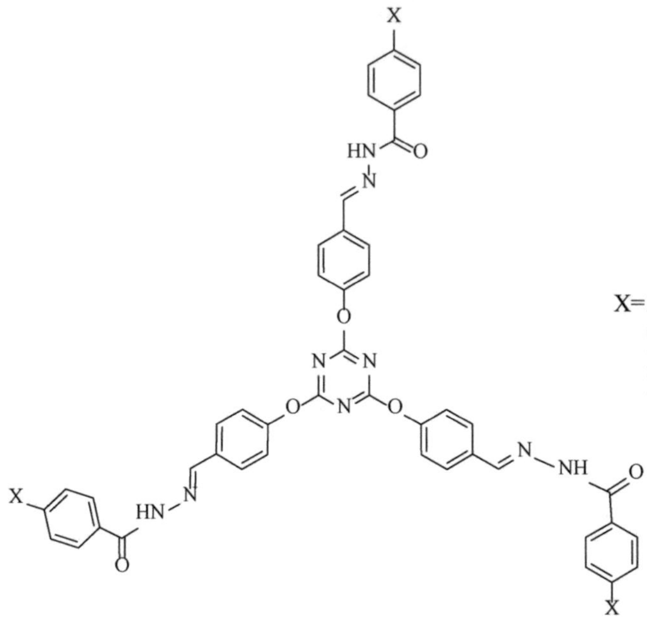

Figure 12.
Antimicrobial tri-substituted s*-triazine hydrazones.*

In DDS-controlled drug release with long-term stability and degradation under environmental conditions. Another 3D-printable formulated Schiff base polymer used in biomedical and sensor applications [27].

2.2 Schiff base polymers: Sustainability and recycling

Due to the dynamic and reversible imine bond, Schiff base polymers' eco-friendly performance. Green synthesis under mild, solvent-free, or aqueous conditions, reducing the environmental footprint of production. Biodegradability and biocompatibility: many Schiff base polymers, especially those derived from natural sources such as chitosan or starch, demonstrate biodegradability and are non-toxic, making them ideal for biomedical and packaging applications, as well as electrolysis membranes. The C=N bond functionalizes polymers, coordinating with metals to provide active catalytic sites and electronic properties. These materials are used in catalysis and magnetic materials [28].

2.2.1 Recyclability

The reversible C=N imine bond enables depolymerization under specific conditions, allowing materials to be recycled or reprocessed without requiring extensive energy input such as photocatalysts in the production of hydrogen from water.

Incorporating Schiff base functionality facilitates the design of closed-loop life cycles for polymer products, aligning with principles of a circular economy. Eco-friendly materials are utilized in water purification, controlled-release systems, and environmental sensors—examples include CT-Schiff base hydrogels, which are used for the adsorption of various pollutants. Polyurethane-Schiff base materials are used in green coatings. Bio-based Schiff base resins are utilized in the production of compostable packaging. Schiff base polymers are tunable through the selection of aldehyde/amine components. Dynamic imine bonds impart self-healing and reprocessing.

2.2.2 Smart and self-healing

Polymer-Schiff base composites responding to external environmental stimuli and autonomously repairing damage due to the dynamic reversible –C=N– imine bond. Reforming after mechanical damage by heat or moisture, broken bonds reconnect, restoring structural integrity. Respond to pH, temperature, or light. Imine linkage forms or breaks reversibly, enabling smart behaviors such as shape memory or targeted drug release. Widely used in drug delivery, sensors, green packaging, and catalysis.

Schiff base polymeric systems are functional materials with tailored properties, exhibiting unique thermal stability, mechanical strength, and chemical tunability. Additionally, there are excellent candidates for various industrial and biomedical applications. For example, polymers incorporating salicylaldehyde and diamines form stable imine-linked networks, which are applicable in coatings and innovative packaging. Phenylenediamine-aromatic aldehydes enhanced thermal resistance in the membrane technologies.

Bio-based aldehydes and amines enhance sustainability. The properties of the resulting polymers (thermal, mechanical, and chemical) were improved. Polymers-Schiff base functionalities exhibit distinct and enhanced properties depending on the nature of the Schiff base and the polymer matrix. The most relevant properties include the following:

Thermal Properties: Schiff base-containing polymers are thermally stable due to the presence of aromatic rings and conjugated systems. TGA and DSC often demonstrate higher decomposition temperatures and Tg. Mechanical properties: High tensile strength and elasticity when Schiff bases are cross-linking agents or participate in supramolecular interactions. Rigid imine structures reinforce the polymer framework. Chemically, imine group (-C=N-) provides reactive sites for coordination with metals and participation in various chemical reactions. Schiff base polymers often show enhanced chemical resistance and solubility in organic solvents. Reversible bonding improved self-healing behavior. Optical and electronic properties: Some Schiff base polymers exhibit fluorescence, conductivity, and redox activity, making them suitable for use in optoelectronic applications, sensors, and innovative materials.

Integrated Schiff base units into polymers impart multifunctional capabilities, enabling applications requiring thermal durability, mechanical robustness, and chemical versatility. The industrial and biomedical applications of Schiff base polymers are primarily attributed to their structural versatility and functional properties.

Electrical conductivity: Conjugation facilitated electron transport. When Schiff base polymers complexed with M^{+n} gave rise to semiconductors used in sensors, batteries, and conductive coatings, Poly (vinyl alcohol)-salicylaldehyde-ethylenediamine, CT-Schiff base derivatives, such as vanillin or 2-hydroxyacetophenone, and polyacrylamide-terephthalaldehyde-4-aminopyridine-aminoantipyrine are promising advanced materials. Physicochemical behavior facilitates optimizing targeted applications.

Table 1 compares some applications among the reported references in this book chapter, covering the main features, synthesis approaches, characterizations, and applications.

No.	Ref.	Sample	Synthesis	Character	Applications	Notable outcomes
1	Gao and Chen [1]	SWCNT/poly-Schiff base composite	One-pot condensing glyoxal and p-phenylenediamine- SWCNT	Electrical conductivity, Seebeck, power factor	Thermoelectric flexible films	Metal complexing; max power factor 77.7$\mu Wm^{-1} K^{-2}$
2	Mighani [2]	Various Schiff base polymers	Simple acid/base/ heat-catalyzed condensing aldehydes/ketones + amines	FTIR, NMR, TGA,	Antimicrobial, anticancer; catalysts, dyes, UV stabilizers	Pharma/industry
3	Nematidil et al. [3]	Chitosan-g-glutaraldehyde/ NaMMTNPs/APTES Schiff base	Grafting CT with nanoparticles	Adsorbent (Pb^{2+}, Hg^{2+}),	Water treatment	Efficient removal Pb^{2+}, Hg^{2+}
4	Jawad et al. [4]	CT-glutaraldehyde-activated charcoal-Schiff base composite	Cross-linking	Adsorption kinetic optimizing	Cationic dye removal Senor	Optimized by response surface methodology
5	Agha et al. [5]	Algae-modified Schiff base–chitosan benzaldehyde composite	Natural algae with CT-benzaldehyde, Schiff base	Adsorption kinetics, thermodynamics	Remove cationic dye (methyl violet 2B)	Food-grade, green synthesis, statistical optimization of adsorption
6	Molina et al. [6]	Amphiphilic polypyrrole-poly(Schiff base) copolymers	Copolymerization	Electrochemistry, bacteriostatic, FTIR, NMR	Biosensors, biomedicine	Catalyzes serotonin oxidation, biocompatible PEG chains
7	Xu et al. [7]	Covalent organic polymer in situ via Schiff reaction	In situ reaction in a living cell	—	Regulating biological function	Hollow covalent organic polymers bio-control
8	Tang et al. [8]	Metallo-supramolecular Schiff base hydrogels.	Incorporated amphiphilic polymers	Responsive testing	Self-healing, hydrogels	Rapid self-healing, multi-stimuli response
9	Al-Rasheed et al. [9]	s-Triazine-Schiff base polymers	Condensation polymerization	Thermal stability, degradation kinetics	High thermal stability polymers	Detailed kinetics studies of degradation
10	Tsao et al. [11]	Reactions of Schiff bases, benzaldehyde, and other aromatics with supercritical water	Supercritical water chemistry	Kinetics, reaction intermediates	Green chemistry, organo-transformations	Mechanistic reactivity under supercritical co

Table 1.
Various Schiff base and poly-Schiff base materials, composites, and their associated properties and applications.

3. Conclusion

Schiff base polymers are multifunctional materials widely applied in industry and biomedicine due to their antibacterial activity, thermal stability, durability, mechanical strength, chemical and structural versatility, and functional properties. Polyethyleneimine-Schiff base metal complexes are efficient catalysts. Schiff base improved the physicochemical behavior of polymers and targeting applications. Schiff bases enhanced the sustainability of designed polymers, encouraging more environmentally responsible practices in materials science. Synthesis, properties, and applications, illustrated with representative chemical reactions and corresponding analysis. The imine N=CH bonds increased mechanical strength and thermal stability, allowing for self-healing via hydrolysis of the imine exchange under moist or thermal conditions into the corresponding aldehyde and amine. Reversible equilibrium core enabled self-healing. Such simple reactions underline the efficient and straightforward synthesis functionality of Schiff base polymers.

The integration of Schiff base moieties into polymeric systems opened broad avenues for designing functional materials with tailored properties. Schiff base polymers exhibit unique thermal stability, mechanical strength, and chemical tunability, making them excellent candidates for various industrial and biomedicine applications. Polymers incorporating salicylaldehyde-diamines form stable imine-linked networks that are useful in coatings and smart packaging. Other systems using phenylenediamine and aromatic aldehydes showed enhanced thermal resistance and are innovative for membrane technologies.

4. Recommendations

Exploring bio-based aldehydes and amines to enhance sustainability. Study Schiff base-containing polymers in controlled drug release systems. Investigate the long-term stability and degradation of materials under various environmental conditions. More 3D-printable formulations are required for biomedical and sensor applications. Improving liquid crystals as materials exhibiting properties between liquids and solids, characterized by fluidity and molecular order, to be used in displays, sensors, and solar cells. More optimized Schiff bases are needed in the synthesis of mesogenic molecules due to their rigid C=N linkage, which promotes order. The substituent is improved in methoxybenzaldehyde-4-amino-biphenyl Schiff base liquid crystal for bright optics and display technologies to enhance light absorption and orientation in organic solar cells.

Author details

Mai Saud Abdullah Alsubaie
Research Department, Health Sciences Research Center, Princess Nourah bint Abdulrahman University, Kingdom of Saudi Arabia

*Address all correspondence to: maisaudalsubaie@gmail.com

References

[1] Gao C, Chen G. A new strategy to construct thermoelectric composites of SWCNTs and poly-Schiff bases with 1, 4-diazabuta-1, 3-diene structures acting as bidentate-chelating units. Journal of Materials Chemistry A. 2016;**4**(29):11299-11306

[2] Mighani H. Schiff Base polymers: Synthesis and characterization. Journal of Polymer Research. 2020;**27**(6):168

[3] Nematidil N, Sadeghi M, Nezami S, Sadeghi H. Synthesis and characterization of Schiff-base based chitosan-g-glutaraldehyde/NaMMTNPs-APTES for removal Pb2+ and Hg2+ ions. Carbohydrate Polymers. 2019;**222**:114971

[4] Jawad AH, Abdulhameed AS, Wilson LD, Hanafiah MAKM, Nawawi WI, ALOthman ZA, et al. Fabrication of Schiff's base chitosan-glutaraldehyde/activated charcoal composite for cationic dye removal: Optimization using response surface methodology. Journal of Polymers and the Environment. 2021;**29**(9):2855-2868

[5] Agha HM, Abdulhameed AS, Jawad AH, Sidik NJ, Aazmi S, Wilson LD, et al. Food-grade algae modified Schiff base-chitosan benzaldehyde composite for cationic methyl violet 2B dye removal: RSM statistical parametric optimization. International Journal of Phytoremediation. 2024;**26**(4):459-471

[6] Molina BG, Cianga L, Bendrea AD, Cianga I, del Valle LJ, Estrany F, et al. Amphiphilic polypyrrole-poly (Schiff base) copolymers with poly (ethylene glycol) side chains: Synthesis, properties and applications. Polymer Chemistry. 2018;**9**(31):4218-4232

[7] Xu HB, Chen HY, Lv J, Chen BB, Zhou ZR, Chang S, et al. Schiff base reaction in a living cell: in situ synthesis of a hollow covalent organic polymer to regulate biological functions. Angewandte Chemie International Edition. 2023;**62**(44):e202311002

[8] Tang L, Chen X, Wang L, Qu J. Metallo-supramolecular hydrogels based on amphiphilic polymers bearing a hydrophobic Schiff base ligand with rapid self-healing and multi-stimuli responsive properties. Polymer Chemistry. 2017;**8**(32):4680-4687

[9] Al-Rasheed HH, Mohammady SZ, Dahlous K, Siddiqui MR, El-Faham A. Synthesis, characterization, thermal stability and kinetics of thermal degradation of novel polymers based-s-triazine Schiff base. Journal of Polymer Research. 2020;**27**(1):10

[10] Okasha RM, Mohamed AAH, Elhenawy AA, Alsehli MH, Alsaedi WH, Alblewi FF, et al. A novel avenue in the successful synthesis of Schiff base macromolecules via innovative plasma and classical approaches. Scientific Reports. 2025;**15**(1):10840

[11] Tsao CC, Zhou Y, Liu X, Houser TJ. Reactions of supercritical water with benzaldehyde, benzylidenebenzylamine, benzyl alcohol, and benzoic acid. The Journal of Supercritical Fluids. 1992;**5**(2):107-113

[12] Hosseini H. An overview of chemical reactions activated by plasma. Industrial & Engineering Chemistry Research. 2024;**63**(45):19418-19434

[13] Dong YB, Smith MD, zur Loye HC. Metal-containing ligands for mixed-metal polymers: Novel Cu (II)– Ag (I) mixed-metal coordination polymers

generated from [Cu (2-methylpyrazine-5-carboxylate) 2 (H2O)]⊙ 3H2O and silver (I) salts. Inorganic Chemistry. 2000;**39**(9):1943-1949

[14] Liu H, Fu ZE, Xu K, Cai HL, Liu X, Chen MC. Preparation and characterization of high-performance Schiff-base liquid crystal diepoxide polymer. Materials Chemistry and Physics. 2012;**132**(2-3):950-956

[15] Alamro FS, Gomha SM, Shaban M, Altowyan AS, Abolibda TZ, Ahmed HA. Optical investigations and photoactive solar energy applications of new synthesized Schiff base liquid crystal derivatives. Scientific Reports. 2021;**11**(1):15046

[16] Singh S, Singh A, Kaur N. Imine-linked receptors decorated ZnO-based dye-sensitized solar cells. Bulletin of Materials Science. 2016;**39**(6):1371-1379

[17] Nosova EV, Lipunova GN, Zyryanov GV, Charushin VN, Chupakhin ON. Functionalized 1, 3, 5-triazine derivatives as components for photo-and electroluminescent materials. Organic Chemistry Frontiers. 2022;**9**(23):6646-6683

[18] Inomata H, Goushi K, Masuko T, Konno T, Imai T, Sasabe H, et al. High-efficiency organic electrophosphorescent diodes using 1, 3, 5-triazine electron transport materials. Chemistry of Materials. 2004;**16**(7):1285-1291

[19] Huang B, Jiang W, Tang J, Ban X, Zhu R, Xu H, et al. A novel, bipolar host based on triazine for efficient solution-processed single-layer green phosphorescent organic light-emitting diodes. Dyes and Pigments. 2014;**101**:9-14

[20] Poriel C, Rault-Berthelot J. Designing host materials for the emissive layer of single-layer phosphorescent organic light-emitting diodes: Toward simplified organic devices. Advanced Functional Materials. 2021;**31**(24):2010547

[21] Chongyang Z, Ping H, Biqin W, Wenyan F, Keqing Z, Bertrand D, et al. Star-shaped triphenylene-triazine multi-stimuli responsive discotic liquid crystals: Synthesis, properties and applications. Acta Chimica Sinica. 2023;**81**(5):469

[22] Li G, Zhu R, Yang Y. Polymer solar cells. Nature Photonics. 2012;**6**(3):153-161

[23] Lu L, Zheng T, Wu Q, Schneider AM, Zhao D, Yu L. Recent advances in bulk heterojunction polymer solar cells. Chemical Reviews. 2015;**115**(23):12666-12731

[24] Zhang Z, Song Q, Jin Y, Feng Y, Li J, Zhang K. Advances in Schiff base and its coating on metal biomaterials—A review. Meta. 2023;**13**(2):386

[25] Hao YM, Wang MY, Liao HY, Liu ML, Zhang DZ, Xie F. The Applications of s-Triazine Based Compounds as Potential Antifungal Agents: A Mini-Review. Basel, Switzerland: Multidisciplinary Digital Publishing Institute (MDPI); 2024

[26] Al Rasheed HH, Malebari AM, Dahlous KA, Fayne D, El-Faham A. Synthesis, anti-proliferative activity, and molecular docking study of new series of 1, 3-5-triazine Schiff base derivatives. Molecules. 2020;**25**(18):4065

[27] Jalalvandi E, Hanton LR, Moratti SC. Schiff-base based hydrogels as degradable platforms for hydrophobic drug delivery. European Polymer Journal. 2017;**90**:13-24

[28] Subramaniyan S, Najjarzadeh N, Vanga SR, Liguori A, Syrén PO, Hakkarainen M. Designed for circularity: Chemically recyclable and enzymatically degradable biorenewable Schiff base polyester-imines. ACS Sustainable Chemistry & Engineering. 2023;**11**(8):3451-3465